Book **3**

3rd edition

STP *Caribbean* MATHEMATICS

CE Layne
F W Ali
L Bostock
S Chandler
A Shepherd
E Smith

OXFORD
UNIVERSITY PRESS

OXFORD
UNIVERSITY PRESS

Great Clarendon Street, Oxford, OX2 6DP, United Kingdom

Oxford University Press is a department of the University of Oxford.
It furthers the University's objective of excellence in research, scholarship,
and education by publishing worldwide. Oxford is a registered trade mark of
Oxford University Press in the UK and in certain other countries

First published by Nelson Thornes Ltd in 2005
This edition published by Oxford University Press in 2014

British Library Cataloguing in Publication Data
Data available

978-0-7487-9089-0

10 9 8 7 6 5 4 3 2

Printed in India by Multivista Global Pvt. Ltd

Acknowledgements

Illustrations: Peters and Zabransky, Rupert Besley, Steve Ballinger, A & R
Nelson, Guyana, Linda Jeffrey
Page make-up: Tech-Set Ltd

Although we have made every effort to trace and contact all
copyright holders before publication this has not been possible in all
cases. If notified, the publisher will rectify any errors or omissions at
the earliest opportunity.

CONTENTS

Contents

INTRODUCTION

This book has been designed for those of you who are hoping to attempt the highest CXC General Proficiency papers in Mathematics. The book contains a good deal of revision of earlier work. As is the case with earlier books there are plenty of straightforward questions and the exercises are divided into three types of question:

The first type, identified by plain numbers, e.g. **12**, helps you to see if you understand the work. These questions are considered necessary for every chapter you attempt.

The second type, identified by a single underline, e.g. **12**, are extra, but not harder, questions for quicker workers, for extra practice or for later revision.

The third type, identified by a double underline, e.g. **12**, are for those of you who manage Type 1 questions fairly easily and therefore need to attempt questions that are a little harder.

Most chapters end with 'mixed exercises'. These will help you revise what you have done, either when you have finished the chapter or at a later date.

Finally a word of advice: when you are doing numerical work, whether with the help of a calculator or not, remember always to ask yourself 'Is my answer a reasonable one for the question that was asked?'

To the teacher

The general aims of the series are:

(1) to help students to
- attain solid mathematical skills
- connect mathematics to their everyday lives and understand its role in the development of our contemporary society
- see the importance of thinking skills in everyday problems
- discover the fun of doing mathematics and reinforce their positive attitudes to it.

(2) to encourage teachers to include historical information about mathematics in their programme.

In writing this four book series the authors attempted to present topics in such a way that students will understand the connections in Mathematics, be encouraged to see and use mathematics as a means to help make sense in the real world.

Topics from the history of mathematics have been incorporated to ensure that mathematics is not dissociated from its past. This should lead to an increase in the levels of enthusiasm, interest and fascination for mathematics, and should also enrich the teaching of it.

Careful grading of exercises makes the books approachable.

Some suggestions:

(1) Before each lesson give a brief outline of the topic to be covered in the lesson. As examples are given refer back to the outline to show how the example fits into it.

(2) List terms on the chalkboard that you consider new to the students.
Solicit additional words from the class and encourage students to read from the text and make their own vocabulary.
Remember that mathematics is a foreign language. The ability to communicate mathematically must involve the careful use of the correct terminology.

(3) When possible have students construct alternative ways to phrase questions. This ties in with seeing mathematics as a language. Students tend to concentrate on the numerical or 'maths' part of the question and pay little attention to the instructions which give information which is required to solve the problem.

(4) When solving problems have students identify their own problem solving strategies and listen to others. This practice should create an atmosphere of discussion in the class centred around different approaches to the same problem.
As the students try to solve problems on their own they will make mistakes. This is healthy, as this was the experience of the inventors of mathematics: they tried, guessed, made many mistakes and worked for hours, days and sometimes years before reaching a solution.
There are enough problems in the exercises to allow the students to try and try again. The excitement, disappointment and struggle with a problem until a solution is found provide a healthy classroom atmosphere.

To the student

These books are written for you. As you study:

Try to break up the material in a chapter into manageable bits.

Always have paper and pencil when you study mathematics.
When you meet a new word write it down together with its meaning.
Read your questions carefully and rephrase them in your own words.
The information which you need to solve your problem is given in the wording of the problem, not the number part only.
Your success in mathematics may be achieved through practice.
You are therefore advised to try to solve as many problems as you can.
Always try more problems than those set by your teacher for homework.
Remember that the greatest cricketer or netball player became great by practising for many hours.

We have provided enough problems in the books to allow you to practice. Above all do not be afraid to make mistakes as you are learning. The greatest mathematicians all made many mistakes as they tried to solve problems.

You are now on your way to success in mathematics – GOOD LUCK!

Picture Credit

Fibonacci, page 63, Science Photo Library

1 BASIC ARITHMETIC

Did you know that the ancient Greeks did not think of unity as a number? To them 3 was the first odd number.

Prime numbers

A number greater than 1 that can be divided exactly only by itself and 1 is called a prime number; for example, 7 is a prime number (as 7 can only be 7×1) but 6 is not a prime number (because $6 = 6 \times 1$ or 3×2).

The smallest prime number is 2.

The Greek mathematician Euclid (born about 300 BC) showed that the number of primes is infinite.

Exercise 1a

1 Write down the first four prime numbers, starting with 2.

2 Write down the prime numbers that are less than 10.

3 Write down the prime numbers that are between 10 and 20.

4 In the set {2, 3, 4, 5, 6, 7, 8}, which members are prime numbers?

5 In the set {5, 7, 9, 11, 13, 15}, which members are prime numbers?

6 In the set {12, 13, 14, 15, 16, 17}, which members are prime numbers?

7 In the set {20, 21, 22, 23, 24, 25}, which members are prime numbers?

8 Apart from 2, are there any even numbers that are prime numbers?

Factors

A factor of a number divides exactly into that number, for example, 2 is a factor of 10 because 2 divides into 10 exactly 5 times.

On the other hand 3 is not a factor of 10 because 3 does not divide exactly into 10.

Exercise 1b

1 Is 2 a factor of
   a 4   b 9   c 12   d 21   e 84?

2 How can you tell whether 2 is a factor of a number?

3 Is 5 a factor of
   a 8   b 10   c 15   d 25   e 32   f 85?

4 How can you tell whether 5 is a factor of a number?

5 Write down those members of the set {8, 16, 40, 35, 41, 206, 515} for which
   a 2 is a factor   b 5 is a factor.

6 Is 3 a factor of
   a 6   b 9   c 10   d 15   e 16?

7 The number 3 is a factor of 51. Add the digits of 51.

Is 3 a factor of your answer?

8 The numbers 75 and 102 are each divisible by 3.

    a Add the digits of 75. Is your answer divisible by 3?

    b Add the digits of 102. Is your answer divisible by 3?

From the last exercise we see that

- any even number is divisible by 2

- any number ending in 0 or 5 is divisible by 5

- when the sum of the digits of a number is divisible by 3, the number itself is divisible by 3.

Exercise 1c

Write down all the factors of 12.

1, 2, 3, 4, 6, 12 are all factors of 12.

(Note that factors of a number are not always prime.)

Write down all the factors of:

1 6	**3** 10	**5** 9	**7** 15	**9** 21	**11** 51
2 8	**4** 4	**6** 18	**8** 16	**10** 26	**12** 19

Prime factors

Sometimes we need to know just the prime factors of a number and to be able to express the number as the product of its prime factors.

For example, the prime factors of 6 are 2 and 3

and $\qquad 6 = 2 \times 3$

For larger numbers we need a more systematic approach.

Consider the number 84.

The lowest prime factor of 84 is 2,

so we divide 2 into 84 $\qquad 84 \div 2 = 42$

We then repeat the process with 42 $\qquad 42 \div 2 = 21$

We repeat the process with 21
but 3 is the lowest prime factor of 21 $\qquad 21 \div 3 = 7$

Repeating the process again $\qquad 7 \div 7 = 1$

Therefore the prime factors of 84 are 2, 2, 3 and 7

i.e. $\qquad 84 = 2 \times 2 \times 3 \times 7$

Did you know that $2^{25964951} - 1$ is the largest prime number discovered so far (Feb 18, 2005)? It is predicted that by the year 2025, a prime number consisting of one billion digits will be discovered.

It is easier to keep track of the prime factors if the division is set out like this:

$$2\overline{)84}$$
$$2\overline{)42}$$
$$3\overline{)21}$$
$$7\underline{)\ 7}$$
$$1$$

Now we can see that $84 = 2 \times 2 \times 3 \times 7$

Exercise 1d

Express the following numbers as products of their prime factors.

1 10   **3** 35   **5** 8   **7** 60   **9** 36   **11** 126

2 21   **4** 12   **6** 28   **8** 50   **10** 66   **12** 108

Multiples

If a number divides exactly into a second number, the second number is a multiple of the first.

For example, the first six multiples of 3 are 3, 6, 9, 12, 15, 18.

Exercise 1e

1 From the set

{2, 3, 5, 6, 8, 10, 12, 14, 15, 16, 18, 20}

write down the members that are

    a multiples of 2   **b** multiples of 3   **c** multiples of 4

    d multiples of 5   **e** multiples of 6   **f** multiples of 8

2 Write down the first four multiples of

    a 7   **b** 5   **c** 8   **d** 10   **e** 12   **f** 15

3 Write down the multiples of 9 between 50 and 100.

4 Write down the multiples of 7 between 10 and 50.

5 Write down the multiples of 11 between 10 and 100.

Investigation

Sieve of Erastothenes

This is a way of finding prime numbers:

Start with an array of whole numbers:

```
 1   2   3   4   5   6   7   8   9  10
11  12  13  14  15  16  17  18  19  20
21  22  23  24  25  26  27  28  29  30
31  32  33  34  35  36  37  38  39  40
41  42  43  44  45  46  47  48  49  50
51  52  53  54  55  56  57  58  59  60
61  62  63  64  65  66  67  68  69  70
71  72  73  74  75  76  77  78  79  80
81  82  83  84  85  86  87  88  89  90
91  92  93  94  95  96  97  98  99 100
```

1 is not a prime number – cross it out.

Cross out all numbers, apart from 2, that are divisible by 2 (some have been done).

Then cross out all numbers, apart from 3, that are divisible by 3 (some have been done).

4 has been crossed out. The next number is 5, cross out all numbers apart from 5 that are divisible by 5 and so on for 7, 11, etc.

The numbers left are the prime numbers less than 100.

Extend this idea to find the prime numbers between 100 and 200.

Squares

A square number, or a perfect square, can be written as the product of two equal numbers.

For example, 16 is a perfect square because $16 = 4 \times 4$.

Exercise 1f

1 Write down the perfect squares between 2 and 10.

2 Write down the perfect squares between 10 and 101.

3 In the set {2, 4, 6, 8, 10, 12, 16, 18, 20}, which members are perfect squares?

4 In the set {2, 4, 8, 16, 32, 64, 128}, which members are perfect squares?

5 In the set {3, 9, 27, 81}, which members are perfect squares?

6 A perfect cube is the product of three equal numbers, e.g. $27 = 3 \times 3 \times 3$.
Find the prime factors of each number in the set {6, 8, 25, 81, 125}. Hence write down those members of the set that are perfect cubes.

Multiplying a string of numbers

Consider $2 \times 3 \times 6$; this means multiply 2 by 3 and then multiply the result by 6,

therefore
$$2 \times 3 \times 6 = 6 \times 6$$
$$= 36$$

Similarly $\quad 2 \times 3 \times 4 \times 5 = 6 \times 4 \times 5 = 24 \times 5 = 120$

Exercise 1g

Calculate:

1 $2 \times 4 \times 3$	4 $7 \times 2 \times 2$	7 $6 \times 2 \times 3$	10 $4 \times 2 \times 4 \times 3$
2 $3 \times 2 \times 5$	5 $3 \times 5 \times 2$	8 $3 \times 7 \times 2$	11 $3 \times 7 \times 1 \times 5$
3 $5 \times 4 \times 2$	6 $4 \times 5 \times 3$	9 $2 \times 5 \times 2 \times 3$	12 $5 \times 2 \times 3 \times 4$

Mixed operations

When a calculation involves a mixture of addition, subtraction, multiplication and division, we do the multiplication and division first.

For example, $\quad 3 + 4 \times 2 = 3 + 8 = 11$

It helps if brackets are put round the parts that are to be done first.

In this case we would write $3 + (4 \times 2) = 3 + 8$

Exercise 1h

Calculate:

1 $8 - 2 \times 2$	10 $5 \times 3 - 5$	
2 $3 \times 4 - 6$	11 $7 + 12 \div 2$	**Tip** Put brackets round the multiplications and the divisions.
3 $12 \div 4 + 3$	12 $24 \div 8 - 2$	
4 $5 + 2 \times 3$	13 $3 + 4 - 6 \div 3$	19 $3 + 1 \times 3 - 2$
5 $3 \times 7 - 8$	14 $2 \times 7 + 8 \div 4$	20 $12 \div 4 + 3 \times 2$
6 $10 \div 2 + 2$	15 $3 + 2 \times 4 - 5$	21 $6 + 3 \times 2 - 4$
7 $3 - 4 \div 2$	16 $7 - 10 \div 2 + 3$	22 $10 - 15 \div 3 + 2$
8 $8 + 2 \times 2$	17 $5 \times 2 - 3 + 1$	23 $7 \times 3 - 6 + 2$
9 $16 - 3 \times 2$	18 $18 \div 3 + 6 \div 2$	24 $14 - 2 \times 4 + 7$

Mixed exercises

Exercise 1i

1 From the set

{1, 2, 5, 9, 12, 18, 21, 36, 39, 41}

write down the numbers which are

  a prime   b odd   c even
  d multiples of 2   e multiples of 3   f factors of 36

2 Find the highest number which is a factor of

  a both 6 and 4   b both 15 and 20   c both 8 and 20

3 Find the lowest number that is a multiple of

  a both 2 and 3   b both 4 and 5   c both 3 and 4

4 Calculate:

  a $2 \times 4 \times 3$ b $2 + 4 \times 3$ c $2 \times 4 + 3$

5 Calculate:

  a $12 \div 6 + 4$ b $3 + 4 \times 2$ c $7 - 4 \div 2$

Exercise 1j

1 From the set

{4, 5, 6, 8, 12, 15, 20, 21, 23, 27, 29}

write down the numbers which are

  a odd   b multiples of 5   c factors of 40
  d prime   e even   f multiples of 4

2 Calculate:

  a $6 \times 2 \times 4$ b $6 \times 2 - 4$ c $16 - 2 \times 4$

3 Calculate:

  a $12 - 5 \times 2$ b $4 \div 2 + 3$ c $3 \times 2 \times 4 \times 3$

Exercise 1k

In this exercise you are given several alternative answers. Write down the letter that corresponds to the correct answer.

1 The prime numbers in the set {1, 2, 4, 7, 11, 15} are

  A 1, 2, 7, 11   B 2, 4   C 2, 7, 11   D 2, 7, 11, 15

2 The value of $2 + 3 \times 2$ is

    A 7   **B** 10   **C** 6   **D** 8

3 The square numbers (perfect squares) in the set {2, 4, 8, 9, 10, 16, 20} are

    A 4, 9, 16   **B** 4, 8, 10, 16, 20   **C** 2, 4, 8, 9, 16

4 The highest number which is a factor of both 4 and 6 is

    A 2   **B** 4   **C** 1   **D** 12

5 The value of $3000 - 3$ is

    A 2007   **B** 2997   **C** 2907   **D** 2700

Odd numbers – About 550 BC the Pythagoreans considered the odd numbers, except 13, to be masculine, heavenly and symbols for good luck. The luckiest odd number at that time was 3. Odd numbers were said to be 'strong numbers'.

Even numbers were considered to be feminine and earthly. The even numbers represented unlucky situations and were associated with poverty and sorrow. These were said to be 'weak numbers'.

Prime numbers were considered to be mysterious and divine. Two, the only even prime number, represented dissension – it takes two to argue.

IN THIS CHAPTER...

you have seen that:

- a prime number can be divided exactly only by 1 and itself

- an even number can be divided exactly by 2

- any number that ends in 5 or 0 can be divided exactly by 5

- any number can be divided exactly by 3 if the sum of its digits can be divided by 3

- a multiple of a number can be divided exactly by that number

- a perfect square is the product of two equal numbers

- when a calculation involves a mixture of operations, do the multiplication and division first.

2 MAKING SURE OF ARITHMETIC

AT THE END OF THIS CHAPTER...

you should be able to:

1   Distinguish between different types of number.

2   Write the reciprocal of a given number.

3   Change decimals to fractions and fractions to decimals.

4   Use the dot notation to write recurring decimals.

5   Multiply and divide decimals.

6   Write numbers using zero and negative indices.

6   Use standard form notation to write a given number.

8   Approximate a number to a given number of decimal places or significant figures.

Did you know that the numbers we use today 1, 2, 3, 4, ... are called Arabic numerals?

The Islamic people of the Middle East did not invent them – they came from India – but they did bring them to Europe.

BEFORE YOU START

you need to know:
- ✓ how to use set notation and the meaning of a subset
- ✓ the meaning of mixed fractions and improper fractions
- ✓ how to add, subtract and multiply decimals

KEY WORDS

common fraction, decimal fraction, denominator, equivalent fraction, improper fraction, index (plural indices), integer, least common multiple (LCM), mixed number, natural number, numerator, product, rational number, real number, reciprocal, recurring decimal, significant figures, standard form, the symbols < and >

Names for numbers

As young children we start learning about numbers by counting objects: 1, 2, 3, ... 10, ...

The numbers that we use for counting are called the whole numbers or *natural numbers* and we use the symbol $\mathbb{N}$ for the set of natural numbers.

The natural numbers, together with 0, i.e. 0, 1, 2, ... are called whole numbers and are denoted by $\mathbb{W}$.

Later we learn about negative numbers. The numbers, ..., $-3, -2, -1, 0, 1, 2, 3, \ldots$ are called *integers*. The set of integers is denoted by the symbol $\mathbb{Z}$, so $\mathbb{Z} = \{ \ldots, -2, -1, 0, 1, 2, \ldots\}$. Notice that $\mathbb{N}$ is a subset of $\mathbb{Z}$.

We next learn that whole objects can be divided into parts, or fractions, for example, half an orange, one and a half bars of chocolate.

We use *common fractions*, e.g. $\frac{1}{2}, \frac{3}{4}, \frac{3}{2}, \ldots$ to describe the size of the part. The *denominator* (bottom number) describes the number of equal parts of the object and the *numerator* (top number) tells us how many parts we have. For example, $\frac{3}{4}$ means that we have 3 of the 4 equal parts of the whole object.

A fraction such as $\frac{3}{2}$, where the numerator is bigger than the denominator, is called an *improper fraction*. The improper fraction $\frac{3}{2}$ can be written as $1\frac{1}{2}$ and in this form it is called a *mixed number*. Note that $1\frac{1}{2}$ means $1 + \frac{1}{2}$.

Any number that can be written as $\frac{a}{b}$ where a and b are integers is called a *rational number*, and we use the symbol $\mathbb{Q}$ for the set of rational numbers.

Any integer can be written as a common fraction, e.g. $4 = \frac{4}{1}$ and $-3 = \frac{-3}{1}$, so the set of rational numbers includes the integers, i.e. $\mathbb{Z} \subset \mathbb{Q}$.

Much later we learn that there are other numbers, such as π and $\sqrt{2}$, that cannot be written exactly as common fractions. These numbers are called irrational numbers and when they are added to the set of rational numbers, we have all the possible types of number that can be shown as points on a number line. The set of all possible numbers that can be shown on a number line is called the set of real numbers and is denoted by $\mathbb{R}$.

It follows from this that $\mathbb{N}$, $\mathbb{W}$, $\mathbb{Z}$ and $\mathbb{Q}$ are all subsets of $\mathbb{R}$.

Exercise 2a

1 3 and 2 are both members of $\mathbb{N}$. Which of the following are members of $\mathbb{N}$? Explain your answers.

   a $3 + 2$ **b** 3×2 **c** $3 \div 2$ **d** $3 - 2$ **e** $2 - 3$

2 2, -5, $\frac{3}{4}$, π

   Choose one of this list of numbers to complete the following.

   Use each number once only.

   a $\subset \mathbb{Q}$ **b** $\subset \mathbb{Z}$ **c** $\subset \mathbb{N}$ **d** $\subset \mathbb{R}$

3 Describe the set given by $\mathbb{Z} \cap \mathbb{N}$.

4 $A = \{2, 4, 6, 8, 10, \ldots\}$ and $B = \{1, 3, 5, 7, 9, \ldots\}$ Describe the set $A \cup B$.

5 Describe the set $\mathbb{R} \cap \mathbb{Q}$.

Addition and subtraction of fractions

We cannot add apples to oranges unless we reclassify them both as, say, fruit. In much the same way, we cannot add tenths to quarters unless we change them both into the same kind of fraction, i.e. change them so that both have the same, or common, denominator. To do this we use the following fact.

> The value of a fraction is unaltered if both numerator and denominator are multiplied by the same number.

To find a common denominator for, say, $\frac{1}{4}$ and $\frac{3}{10}$, we look for the lowest number that both 4 and 10 divide into exactly. This number is called the *lowest common multiple* (LCM) of 4 and 10 and in this case it is 20.

Then, as $\qquad 20 = 4 \times 5 \quad$ we have $\quad \frac{1}{4} = \frac{1 \times 5}{4 \times 5} = \frac{5}{20}$

and as $\qquad 20 = 10 \times 2 \quad$ we have $\quad \frac{3}{10} = \frac{3 \times 2}{10 \times 2} = \frac{6}{20}$

Therefore $\qquad \frac{1}{4} + \frac{3}{10} = \frac{5}{20} + \frac{6}{20} = \frac{11}{20}$

Exercise 2b

Find the LCM of the following sets of numbers:

1 3, 7	**3** 2, 8, 10	**5** 2, 6, 3	**7** 4, 6	**9** 2, 3, 7	**11** 24, 8, 6
2 2, 9	**4** 3, 4, 6	**6** 5, 4	**8** 3, 5, 12	**10** 18, 6, 9	**12** 12, 36, 8

> Find $\frac{5}{11} + \frac{2}{5} + \frac{7}{10}$
>
> You must write each fraction as an equivalent fraction with the LCM of 11, 5 and 10 as the denominator then
>
> $$\frac{5}{11} = \frac{5 \times 10}{11 \times 10} = \frac{50}{110}, \quad \frac{2}{5} = \frac{2 \times 22}{5 \times 22} = \frac{44}{110} \text{ and } \frac{7}{10} = \frac{7 \times 11}{10 \times 11} = \frac{77}{110}$$
>
> so $\qquad \frac{5}{11} + \frac{2}{5} + \frac{7}{10} = \frac{50 + 44 + 77}{110}$
>
> $$= \frac{171}{110}$$
>
> $$= \frac{110 + 61}{110} = 1\frac{61}{110}$$

Find:

13 $\frac{2}{3} + \frac{7}{8}$	**16** $\frac{2}{7} + \frac{1}{2} + \frac{3}{14}$	**19** $\frac{1}{3} + \frac{5}{9}$	**22** $\frac{2}{7} + \frac{1}{9} + \frac{1}{6}$
14 $\frac{3}{5} + \frac{3}{10}$	**17** $\frac{3}{16} + \frac{3}{4} + \frac{5}{12}$	**20** $\frac{5}{6} + \frac{2}{3} + \frac{1}{4}$	**23** $\frac{5}{12} + \frac{3}{8} + \frac{3}{4}$
15 $\frac{3}{5} + \frac{5}{8} + \frac{1}{2}$	**18** $\frac{1}{4} + \frac{2}{3}$	**21** $\frac{3}{4} + \frac{13}{20} + \frac{4}{5}$	**24** $\frac{8}{21} + \frac{1}{2} + \frac{2}{3}$

Find $\frac{7}{15} - \frac{5}{12}$

60 is the LCM of 15 and 12 so express each fraction as an equivalent fraction with a denominator of 60.

$$\frac{7}{15} = \frac{7 \times 4}{15 \times 4} = \frac{28}{60} \text{ and } \frac{5}{12} = \frac{5 \times 5}{12 \times 5} = \frac{25}{60}$$

so

$$\frac{7}{15} - \frac{5}{12} = \frac{28}{60} - \frac{25}{60} = \frac{28 - 25}{60}$$

$$= \frac{3}{60} = \frac{1}{20}$$

25 $\frac{7}{9} - \frac{5}{12}$ **26** $\frac{1}{4} - \frac{2}{9}$ **27** $\frac{3}{10} - \frac{1}{15}$ **28** $\frac{1}{10} - \frac{1}{20}$ **29** $\frac{2}{5} - \frac{3}{8}$ **30** $\frac{5}{6} - \frac{5}{9}$

Find   **a** $1\frac{2}{3} + \frac{1}{6} + 2\frac{1}{2}$ **b** $2\frac{1}{4} - \frac{3}{5}$

a $1\frac{2}{3} + \frac{1}{6} + 2\frac{1}{2} = 3 + \frac{2}{3} + \frac{1}{6} + \frac{1}{2}$ (adding whole numbers)

$$= 3 + \frac{4 + 1 + 3}{6} \quad \text{(expressing the fractions as equivalent fractions}$$
$$= 3 + \frac{8}{6} \quad \text{with a denominator of 6.)}$$

$$= 3 + 1\frac{1}{3} = 4\frac{1}{3}$$

b $2\frac{1}{4} - \frac{3}{5} = 2 + \frac{5 - 12}{20}$

$$= 1 + \frac{20 + 5 - 12}{20} \quad \left(\text{changing 1 unit into } \frac{20}{20}\right)$$

$$= 1 + \frac{13}{20} = 1\frac{13}{20}$$

Find:

31 $2\frac{3}{5} + 1\frac{1}{8}$ **34** $2\frac{3}{4} + 1\frac{1}{2} - \frac{1}{3}$ **37** $2\frac{3}{5} + 1\frac{8}{15}$

32 $4\frac{2}{9} - 3\frac{5}{6}$ **35** $3\frac{1}{8} - 2\frac{3}{4} + 4\frac{1}{2}$ **38** $1\frac{3}{8} + 1\frac{1}{4} - 2\frac{1}{2}$

33 $2\frac{1}{5} - 4\frac{1}{8} + 1\frac{7}{10}$ **36** $1\frac{3}{10} - \frac{9}{20}$ **39** $2\frac{1}{4} - 1\frac{5}{6} + \frac{2}{3}$

Multiplication of fractions

Remember that $\frac{2}{3} \times \frac{1}{5}$ means $\frac{2}{3}$ of $\frac{1}{5}$.

Now $\frac{1}{3}$ of $\frac{1}{5}$ is $\frac{1}{15}$, so $\frac{2}{3}$ of $\frac{1}{5}$ is $\frac{2}{15}$,

i.e. $\frac{2}{3} \times \frac{1}{5} = \frac{2 \times 1}{3 \times 5} = \frac{2}{15}$

> To multiply fractions, multiply the numerators together and multiply the denominators together.

Any mixed numbers must be changed into improper fractions, and factors that are common to the numerator and denominator should be cancelled before multiplication.

Find $2\frac{1}{2} \times \frac{3}{15}$

$$2\frac{1}{2} \times \frac{3}{15} = \frac{5}{2} \times \frac{3}{15} \quad \text{(Write } 2\frac{1}{2} \text{ as an improper fraction)}$$

$$= \frac{\cancel{5}^1 \times \cancel{3}^1}{2 \times \cancel{15}_1}$$

$$= \frac{1}{2}$$

Find:

1 $\frac{2}{3} \times \frac{5}{6}$

2 $1\frac{1}{2} \times \frac{8}{9}$

3 $4 \times \frac{3}{8}$

4 $\frac{2}{3} \times \frac{1}{4} \times \frac{3}{5}$

5 $1\frac{1}{3} \times \frac{1}{2} \times \frac{5}{7}$

6 $\frac{3}{4} \times \frac{2}{5}$

7 $\frac{2}{5} \times 1\frac{3}{7}$

8 $\frac{2}{5} \times \frac{7}{8} \times \frac{10}{11}$

Tip Write $1\frac{1}{2}$ as an improper fraction first.

Tip Write 4 as $\frac{4}{1}$.

In questions **9** to **14** find the missing numbers:

9 $\frac{3}{4} \times \frac{3}{6} = \frac{1}{4}$

10 $\frac{1}{3} \times \frac{}{2} = \frac{1}{2}$

11 $\frac{2}{} \times \frac{1}{4} = \frac{1}{6}$

12 $\frac{}{} \times \frac{2}{3} = 1$

13 $\frac{3}{4} \times \frac{}{} = 1$

14 $\frac{}{} \times \frac{7}{8} = 1$

Tip The bottom is 6 times too big. What is the missing number for the top to be 6 times too big?

Reciprocals

If the product of two numbers is 1 then each number is called the reciprocal of the other.

We know that $\frac{1}{3} \times 3 = 1$ so

$\frac{1}{3}$ is the reciprocal of 3   and   3 is the reciprocal of $\frac{1}{3}$.

To find the reciprocal of $\frac{3}{4}$ we require the number which, when multiplied by $\frac{3}{4}$, gives 1

Now $\frac{4}{3} \times \frac{3}{4} = 1$

so $\frac{4}{3}$ is the reciprocal of $\frac{3}{4}$

In all cases the reciprocal of a fraction is obtained by turning the fraction upside down.

Any number can be written as a fraction, e.g. $3 = \frac{3}{1}$, $2.5 = \frac{2.5}{1}$, ...

so   the reciprocal of $\frac{3}{1}$ is $\frac{1}{3}$ or $1 \div 3$,

the reciprocal of $\frac{2.5}{1}$ is $\frac{1}{2.5}$ or $1 \div 2.5$

The reciprocal of a number is 1 divided by that number.

Division by a fraction

Consider $\frac{2}{5} \div \frac{3}{7}$

This can be interpreted as $\frac{2}{5} \times 1 \div \frac{3}{7} = \frac{2}{5} \times \left(1 \div \frac{3}{7}\right)$

Now $1 \div \frac{3}{7}$ is the reciprocal of $\frac{3}{7}$, i.e. $\frac{7}{3}$.

Therefore $\qquad \frac{2}{5} \div \frac{3}{7} = \frac{2}{5} \times \frac{7}{3} = \frac{14}{15}$

i.e. to divide by a fraction we multiply by its reciprocal.

Exercise 2d

Write down the reciprocals of the following numbers.

 1 4

 2 $\frac{1}{2}$

3 $\frac{2}{5}$

4 10

5 $\frac{1}{8}$

6 $\frac{3}{11}$

7 100

8 $\frac{2}{9}$

9 $\frac{15}{4}$

> **Tip** The reciprocal of a number is 1 divided by that number.

> **Tip** The reciprocal of a fraction is the fraction turned upside down.

Find $3\frac{1}{2} \div \frac{7}{8}$

(Before multiplying or dividing, mixed numbers must be changed into improper fractions.)

$$3\tfrac{1}{2} \div \tfrac{7}{8} = \tfrac{7}{2} \div \tfrac{7}{8}$$

$$= \frac{\cancel{7}^1}{\cancel{2}_1} \times \frac{\cancel{8}^4}{\cancel{7}_1} = 4$$

Find:

 10 $\frac{2}{3} \div \frac{1}{2}$

11 $1\frac{2}{3} \div \frac{5}{6}$

12 $2\frac{1}{2} \div 4$

13 $5 \div \frac{4}{5}$

14 $\frac{2}{9} \div 1\frac{2}{7}$

15 $\frac{1}{2} \div \frac{3}{4}$

16 $\frac{3}{7} \div 1\frac{3}{4}$

17 $\frac{5}{9} \div 10$

18 $3 \div \frac{2}{3}$

> **Tip** Multiply $\frac{2}{3}$ by the reciprocal of $\frac{1}{2}$.

> **Tip** Change $1\frac{2}{3}$ to an improper fraction first.

Find $2\frac{1}{2} + \frac{3}{5} \div 1\frac{1}{2} - \frac{1}{2}\left(\frac{3}{5} + \frac{1}{3}\right)$

(Remember that brackets are worked out first, then multiplication and division and lastly addition and subtraction.)

$$2\tfrac{1}{2}+\tfrac{3}{5}\div 1\tfrac{1}{2}-\tfrac{1}{2}\left(\tfrac{3}{5}+\tfrac{1}{3}\right)=2\tfrac{1}{2}+\tfrac{3}{5}\div 1\tfrac{1}{2}-\tfrac{1}{2}\left(\tfrac{9+5}{15}\right)$$

$$=2\tfrac{1}{2}+\tfrac{3}{5}\div \tfrac{3}{2}-\tfrac{1}{2}\times\tfrac{14}{15}$$

$$=2\tfrac{1}{2}+\tfrac{3}{5}\times\tfrac{2}{3}-\tfrac{1}{2}\times\tfrac{14}{15}$$

$$=2\tfrac{1}{2}+\tfrac{2}{5}-\tfrac{7}{15}$$

$$=2+\tfrac{15+12-14}{30} \quad\text{(the LCM of 2, 5 and 15 is 30)}$$

$$=2+\tfrac{13}{30}=2\tfrac{13}{30}$$

19 $1\tfrac{2}{3}\times\tfrac{1}{2}-\tfrac{2}{5}$

20 $\tfrac{3}{7}+\tfrac{1}{4}\div 1\tfrac{1}{3}$

21 $\tfrac{2}{5}\div\left(\tfrac{1}{2}+\tfrac{3}{4}\right)$

22 $5\tfrac{1}{2}\div 3+\tfrac{2}{9}$

23 $\tfrac{4}{5}\div\tfrac{1}{6}+\tfrac{1}{3}\times 1\tfrac{1}{2}$

24 $\tfrac{9}{11}-\tfrac{2}{5}\times\tfrac{3}{4}$

25 $2\tfrac{1}{2}\div\tfrac{7}{9}+1\tfrac{1}{3}$

26 $\tfrac{3}{5}\left(1\tfrac{1}{4}-\tfrac{2}{3}\right)$

27 $3\tfrac{1}{2}-\tfrac{2}{3}\times 6$

28 $2\tfrac{1}{3}+\tfrac{1}{2}\left(2\div\tfrac{4}{5}\right)$

29 $\tfrac{3}{4}\left(5\tfrac{1}{3}-2\tfrac{1}{5}\right)\div\tfrac{7}{9}$

30 $\left(\tfrac{2}{3}-\tfrac{1}{2}\right)\div\left(\tfrac{3}{4}-\tfrac{1}{3}\right)$

31 $\tfrac{7}{9}-\tfrac{1}{3}$ of $1\tfrac{2}{7}$

32 $\tfrac{1}{2}+\left(\tfrac{3}{4}\div\tfrac{1}{6}\right)$ of 3

33 $\dfrac{\tfrac{1}{3}+\tfrac{1}{4}}{\tfrac{5}{6}-\tfrac{3}{4}}$

34 $\dfrac{1\tfrac{1}{5}-\tfrac{3}{4}}{2\tfrac{1}{2}}$

35 $\dfrac{\tfrac{9}{10}}{\tfrac{5}{6}}$

36 $\tfrac{2}{3}\times\tfrac{6}{7}-\tfrac{5}{8}\div 1\tfrac{1}{4}$

37 $\dfrac{\tfrac{7}{8}}{3\tfrac{1}{2}-\tfrac{2}{3}}$

38 $\dfrac{3\tfrac{1}{4}}{2\tfrac{3}{5}}$

39 $\dfrac{\tfrac{2}{3}\times\tfrac{3}{4}}{\tfrac{5}{6}\times\tfrac{3}{10}}$

Tip Remember: multiplication before subtraction.

Tip Remember: division before addition.

Tip Work out the brackets first.

Tip Get a single fraction on top and a single fraction on the bottom before dividing the top by the bottom.

Decimal fractions

Long after meeting common fractions, we learn that we can represent fractions of an object by placing a point after the units and continuing to add figures to the right. The first figure after the decimal point is the number of tenths, the next figure is the number of hundredths, and so on.

For example, 0.75 is 7 tenths plus 5 hundredths.

Fractions written this way are called *decimal fractions*.

Usually we refer to common fractions simply as fractions and to decimal fractions simply as decimals.

Interchanging decimals and fractions

Express 0.705 as a fraction.

$$0.705 = \frac{7}{10} + \frac{5}{1000} \qquad \text{(this step is usually omitted)}$$

$$= \frac{\cancel{705}^{141}}{\cancel{1000}_{200}} \qquad \text{(divide top and bottom by 5)}$$

$$= \frac{141}{200}$$

Express the following decimals as fractions:

1 0.35	**3** 0.204	**5** 0.03	**7** 0.005	**9** 0.11	**11** 1.104
2 0.216	**4** 1.36	**6** 0.012	**8** 1.01	**10** 2.05	**12** 0.0001

Express $\frac{7}{8}$ as a decimal.

To express a fraction as a decimal, divide the top by the bottom.

$$\frac{7}{8} = 0.875$$

$$\begin{array}{r} 0.875 \\ 8\overline{)7.000} \end{array} \qquad \text{Add zeros for the extra decimal places.}$$

Express the following fractions as decimals:

13 $\frac{3}{20}$

14 $\frac{1}{8}$

15 $\frac{3}{5}$

16 $\frac{6}{25}$

17 $\frac{1}{16}$

18 $\frac{27}{50}$

19 $1\frac{3}{4}$

20 $\frac{5}{32}$

21 $\frac{4}{25}$

22 $\frac{5}{16}$

23 $2\frac{3}{8}$

24 $\frac{1}{500}$

> **Tip** Change to an improper fraction first.

Recurring decimals

If we try to change $\frac{1}{6}$ to a decimal, i.e. $6\overline{)1.0000\ldots}$ with quotient $0.1666\ldots$ we discover that

a we cannot write $\frac{1}{6}$ as an exact decimal

b from the second decimal place, the 6 recurs for as long as we have the patience to continue the division.

Similarly if we convert $\frac{2}{11}$ to a decimal by dividing 2 by 11, we get $0.18181818\ldots$ and we see that

a $\frac{2}{11}$ cannot be expressed as an exact decimal

b the pair of figures '18' recurs indefinitely.

Decimals like these are called *recurring decimals*. To save time and space we place a dot over the figure that recurs. In the case of a group of figures recurring we place a dot over the first and last figure in the group.

Therefore   we write 0.166666 ...   as $0.1\dot{6}$

and   we write 0.181818 ...   as $0.\dot{1}\dot{8}$

Similarly   we write 0.316316316 ... as $0.\dot{3}1\dot{6}$

Exercise 2f

Use the dot notation to write the following fractions as decimals:

1 $\frac{1}{3}$		**3** $\frac{5}{6}$		**5** $\frac{1}{7}$		**7** $\frac{1}{11}$		**9** $\frac{5}{12}$		**11** $\frac{7}{30}$
2 $\frac{2}{9}$		**4** $\frac{1}{15}$		**6** $\frac{1}{12}$		**8** $\frac{1}{18}$		**10** $\frac{1}{14}$		**12** $\frac{1}{13}$

Addition and subtraction of decimals

Decimals are added and subtracted in the same way as whole numbers. It is sensible to write them in a column so that the decimal points are in a vertical line. This ensures that units are added to units, tenths are added to tenths, and so on.

Multiplication of decimals

To multiply decimals we can first convert them to fractions.

For example,
$$0.05 \times 1.04 = \frac{5}{100} \times \frac{104}{100}$$
(2 d.p.)   (2 d.p.)

$$= \frac{520}{10\,000}$$

$$= 0.0520$$
(4 d.p.)

From such examples we deduce the following rule.

First ignore the decimal points and multiply the numbers together. Then the sum of the number of decimal places in the original numbers gives the number of decimal places in the answer (including any zeros at the end).

Division by decimals

To divide a decimal by a whole number, proceed as with whole numbers, adding zeros after the point when necessary. For example, to find $3.14 \div 5$ we have

$$\begin{array}{r} 0.628 \\ 5\overline{)3.140} \end{array}$$

(Make sure that the decimal points are in a vertical line.)

Therefore $3.14 \div 5 = 0.628$

To divide by a decimal we use the fact that the top and bottom of a fraction can be multiplied by the same number without altering the value of the fraction. Division by a decimal can therefore be converted to division by a whole number.

For example,
$$3.14 \div 0.5 = \frac{3.14}{0.5}$$
$$= \frac{31.4}{5} \qquad \left(\frac{3.14 \times 10}{0.5 \times 10} \right)$$
$$= 6.28$$

Exercise 2g

Calculate, without using a calculator:

1 $1.26 + 3.75$

2 $12.4 + 6.7$

3 $5.82 + 0.35$

4 $0.04 + 8.76$

5 $1.8 + 0.02$

6 $25 + 1.36$

7 $4.002 + 0.83$

8 $0.016 + 1.09$

9 $0.00032 + 0.0017$

10 $5.3 - 2.1$

11 $8.2 - 4.9$

12 $0.16 - 0.08$

13 $1.3 - 0.09$

14 $1.07 - 0.58$

15 $24 - 0.98$

16 $0.37 - 0.009$

17 $2 - 0.17$

18 $0.0127 - 0.0059$

19 1.2×0.8

20 0.7×0.06

21 0.4×0.02

22 0.1×0.1

23 0.5×0.5

24 1.002×0.36

25 42×0.08

26 3.0501×1.1

27 0.0012×0.32

28 $2.8 \div 0.4$

Tip For addition and subtraction write the numbers as you would for whole numbers but with the decimal points one under the other.

29 $0.36 \div 1.2$

30 $1.08 \div 0.4$

31 $0.02 \div 2.5$

32 $0.018 \div 1.2$

33 $5.31 \div 0.9$

34 $0.1 \div 0.1$

35 $0.01 \div 0.5$

36 $0.0013 \div 1.3$

Evaluate:

 Tip Remember, do $\times$ and $\div$ before $+$ and $-$

37 $2.6 - 1.4 \div 0.7$

38 $200 \times 0.04 - 0.2$

39 $(1.2 - 0.8) \div 0.8$

40 $4.3 \times 2.1 \div 0.07$

41 $3.2(0.6 - 0.09) + 10.25$

42 $2.73 \div 0.9 \times 1.02$

43 $(20 \times 0.06) \div (3.1 - 1.9)$

44 $(2.5 + 1.3) \div (2.06 - 0.16)$

45 $\dfrac{2.4 + 0.98}{1.78 + 0.22}$

46 $\dfrac{0.04 \times 1.02}{3.4 \times 0.06}$

47 $\dfrac{1.2 \times 0.05}{0.07 + 0.08}$

48 $\dfrac{4.02 + 12.09}{0.9 \times 2}$

Relative sizes

Before we can compare the sizes of a set of numbers we must either change them all into decimals or change them all into fractions with the same denominator. Choose whichever method is easier.

Exercise 2h   Place > or < between the two numbers $\frac{3}{5}$, 0.67

(Remember that > means 'is greater than' and < means 'is less than'.)

(Convert each to a fraction or to a decimal. It is easier to convert $\frac{3}{5}$ to a decimal,

$$\frac{3}{5} = 0.6$$

Comparing 0.6 and 0.67 we see that 0.6 < 0.67

∴ $\frac{3}{5} < 0.67$

Place > or < between each of the following pairs of numbers:

1 $\frac{5}{8}$, 0.63

2 0.16, $\frac{3}{20}$

3 $\frac{9}{11}$, 0.9

4 $\frac{1}{16}$, 0.07

5 $\frac{2}{7}$, 0.16

6 0.48, $\frac{4}{9}$

7 0.33, $\frac{7}{25}$

8 $\frac{3}{11}$, 0.25

9 $\frac{2}{3}$, 0.66

Arrange the following numbers in ascending order (i.e. the smallest first): $\frac{1}{2}$, 0.52, $\frac{11}{20}$, 0.51

Writing them all as fractions with denominator 100 gives

$$\frac{1}{2} = \frac{50}{100}, \ 0.52 = \frac{52}{100}, \ \frac{11}{20} = \frac{55}{100}, \ 0.51 = \frac{51}{100}$$

Therefore the required order is $\frac{1}{2}$, 0.51, 0.52, $\frac{11}{20}$.

In each of the following questions arrange the numbers in ascending order of size:

10 $\frac{2}{3}$, 0.6, $\frac{4}{5}$

11 0.85, $\frac{4}{5}$, 0.79

12 $\frac{2}{7}$, 0.3, $\frac{1}{5}$

13 0.75, $\frac{5}{7}$, 0.875, $\frac{7}{9}$

14 0.16, $\frac{3}{20}$, 0.2, $\frac{6}{25}$

15 $1\frac{1}{5}$, 1.3, 1.24, $1\frac{1}{8}$

Tip   Convert $\frac{2}{3}$ and $\frac{4}{5}$ to decimals.

Indices

In the number 2^3, the 3 is called the *index* or *power* and 2^3 means $2 \times 2 \times 2$. Thus, indices are a kind of shorthand notation. All other forms of indices are derived from this.

Did you know that an index or power was first used by the Greek mathematician Hippocrates of Chios in the fifth century BC?

Exercise 2i   Find the value of   **a** $2^2 \times 3^3$ **b** $(3^3)^2$

   a $2^2 \times 3^3 = 2 \times 2 \times 3 \times 3 \times 3$ **b** $(3^3)^2 = (27)^2$

   $\qquad\qquad = 4 \times 27$ $\qquad\qquad = 27 \times 27$

   $\qquad\qquad = 108$ $\qquad\qquad = 729$

Find the value of:

1 5^2	**4** 5^3	**7** $8^2 \times 5^2$	**10** $7^3 \times 9^2$	**13** 0.072×10^4
2 3^4	**5** 4^3	**8** $6^3 \times 2^2$	**11** 3.25×10^2	**14** 1.102×10^3
3 2^5	**6** $2^4 \times 3^2$	**9** $4^4 \times 2^3$	**12** 8.01×10^3	**15** 1.1×10^6

Remember that we can multiply one number to a power by the *same* number to another power by adding the powers.

Where possible write as a single number in index form

   a $5^2 \times 5^4$ **b** $3^2 \times 2^3$

   a $5^2 \times 5^4 = 5^6$

         $(5^2 \times 5^4 = 5 \times 5 \times 5 \times 5 \times 5 \times 5$ i.e. you can add the indices)

   b $3^2 \times 2^3$ cannot be written as a single number in index form

            $(3^2 \times 2^3 = 3 \times 3 \times 2 \times 2 \times 2)$

Write the following, where possible, as a single number in index form:

16 $2^3 \times 2^4$	**21** $7^2 \times 7^5$
17 $3^2 \times 3^5$	**22** $4^2 \times 4^7$
18 $5^2 \times 3^5$	**23** $a^2 \times a^3$
19 $5^1 \times 5^3$	**24** $a^2 \times b^3$
20 2×2^4	

Tip To multiply powers of the same number you can add the indices.

We can also divide one number to a power by the *same* number to another power by subtracting the powers.

Where possible write as a single number in index form

   a $2^5 \div 2^2$ **b** $3^3 \div 5^3$

   a $2^5 \div 2^2 = 2^3$

   $\left(2^5 \div 2^2 = \dfrac{\cancel{2} \times \cancel{2} \times 2 \times 2 \times 2}{\cancel{2} \times \cancel{2}} = 2^3 \right)$

   b $3^3 \div 5^3$ cannot be simplified.

   $\left(3^3 \div 5^3 = \dfrac{3 \times 3 \times 3}{5 \times 5 \times 5} \right)$

Write the following, where possible, as a single number in index form:

25 $2^4 \div 2^2$ **27** $5^3 \div 3^2$ **29** $2^3 \div 3^2$ **31** $3^4 \div 3$ **33** $a^3 \div b^2$

26 $7^3 \div 7^2$ **28** $4^5 \div 4^2$ **30** $3^6 \div 3^2$ **32** $a^8 \div a^4$

Laws of indices

We already know the first two laws, namely $a^m \times a^n = a^{m+n}$

and $a^m \div a^n = a^{m-n}$

Now consider $(a^4)^2$; this means $a^4 \times a^4$ which, using the first law is a^8,

i.e. $(a^4)^2 = a^{4 \times 2}$.

This gives the third law: $(a^m)^n = a^{mn}$

Exercise 2j

1 Write as a single number in index form.

   a $(2^3)^2$ **b** $(3^2)^2$ **c** $(5^2)^2$

> **Tip** $(2^3)^2 = (2^3) \times (2^3)$

> Simplify $2(a^3b^2)^3$
>
> Notice that only the items in the bracket are cubed, so
> $$2(a^3b^2)^3 = 2 \times (a^3)^3 \times (b^2)^3$$
> $$= 2 \times a^9 \times b^6 = 2a^9b^6$$

2 Simplify

   a $3(t^3)^5$ **b** $(2d^5)^3$ **c** $2(a^2)^3$ **d** $(5p^3)^3$

3 Find the value of

   a $3(2^2)^3$ **b** $(5 \times 2^4)^2$

   c $10(3^2)^2$ **d** $2(3^2 - 2^2)^2$

> **Tip**  Read this carefully.

4 Simplify

   a $(3x^2y)^3$ **b** $5(ab^3)^4$ **c** $(4u^2v)^3$ **d** $2p^2(p^3q)^2$

Zero and negative indices

Consider $a^3 \div a^3$.

Subtracting indices gives $a^3 \div a^3 = a^0$

Dividing gives $a^3 \div a^3 = 1$

> $a^0 = 1$ i.e. (any number)$^0 = 1$

Now consider $a^3 \div a^5$.

Subtracting indices gives $\qquad a^3 \div a^5 = a^{-2}$

Dividing gives $\dfrac{a^3}{a^5} = \dfrac{\cancel{a} \times \cancel{a} \times \cancel{a}}{\cancel{a} \times \cancel{a} \times \cancel{a} \times a \times a} = \dfrac{1}{a^2}$

Therefore a^{-2} means $\dfrac{1}{a^2}$

A negative sign in front of the index means 'the reciprocal of'

i.e. $\qquad a^{-b} = \dfrac{1}{a^b}$

Exercise 2k

Find the value of 3^{-1}

$$3^{-1} = \frac{1}{3^1} = \frac{1}{3}$$

Find the value of:

1 2^{-1} **3** 5^{-1} **5** 8^{-1} **7** a^{-1}

2 10^{-1} **4** 7^{-1} **6** 4^{-1} **8** x^{-1}

Find the value of   **a** $\left(\frac{1}{2}\right)^{-1}$ **b** $\left(\frac{2}{5}\right)^{-1}$

a $\left(\frac{1}{2}\right)^{-1} = \left(\frac{2}{1}\right)^{1} = 2$

b $\left(\frac{2}{5}\right)^{-1} = \left(1 \div \frac{2}{5}\right) = \left(1 \times \frac{5}{2}\right) = \left(\frac{5}{2}\right)^{1} = 2\frac{1}{2}$

Find the value of:

9 $\left(\frac{1}{3}\right)^{-1}$ **11** $\left(\frac{1}{4}\right)^{-1}$ **13** $\left(\frac{1}{5}\right)^{-1}$ **15** $\left(\frac{1}{a}\right)^{-1}$

10 $\left(\frac{2}{3}\right)^{-1}$ **12** $\left(\frac{3}{4}\right)^{-1}$ **14** $\left(\frac{4}{5}\right)^{-1}$ **16** $\left(\frac{x}{y}\right)^{-1}$

Find the value of 3^{-2}

$$3^{-2} = \frac{1}{3^2} = \frac{1}{9}$$

Find the value of:

17 2^{-3} **19** 10^{-3} **21** 2^{-5} **23** 10^{-2}

18 5^{-2} **20** 6^{-2} **22** 10^{-4} **24** 4^{-3}

Find the value of $\left(\frac{1}{3}\right)^{-2}$

$$\left(\tfrac{1}{3}\right)^{-2} = \left(1 \div \tfrac{1}{3}\right)^2$$
$$= \left(1 \times \tfrac{3}{1}\right)^2 = 3^2 = 9$$

Find the value of:

25 $\left(\frac{1}{5}\right)^{-3}$ **27** $\left(\frac{1}{2}\right)^{-5}$ **29** $\left(\frac{1}{8}\right)^{-3}$ **31** $\left(\frac{1}{2}\right)^{-3}$

26 $\left(\frac{1}{4}\right)^{-2}$ **28** $\left(\frac{1}{3}\right)^{-4}$ **30** $\left(\frac{1}{10}\right)^{-4}$ **32** $\left(\frac{1}{6}\right)^{-2}$

Find the value of $\left(\frac{2}{5}\right)^{-3}$

$$\left(\frac{2}{5}\right)^{-3} = \left(\frac{5}{2}\right)^{3}$$

$$= \frac{5^3}{2^3} = \frac{125}{8} = 15\frac{5}{8}$$

Find the value of:

33 $\left(\frac{3}{4}\right)^{-2}$ **35** $\left(\frac{4}{9}\right)^{-2}$ **37** $\left(\frac{2}{3}\right)^{-4}$ **39** $\left(\frac{3}{10}\right)^{-4}$

34 $\left(\frac{2}{3}\right)^{-3}$ **36** $\left(\frac{2}{7}\right)^{-2}$ **38** $\left(\frac{3}{5}\right)^{-2}$ **40** $\left(\frac{5}{8}\right)^{-2}$

Exercise 2I

Find the value of:

1 $\left(\frac{1}{8}\right)^{-1}$ **5** $\left(\frac{1}{2}\right)^{0}$ **9** $\left(\frac{1}{2}\right)^{-4}$ **13** 5^0 **17** 12^{-1}

2 $\left(\frac{2}{5}\right)^{-2}$ **6** $\left(\frac{2}{3}\right)^{0}$ **10** 6^0 **14** $\left(\frac{7}{10}\right)^{-3}$ **18** 9^3

3 4^{-2} **7** 5^3 **11** $\left(\frac{3}{4}\right)^{-3}$ **15** 2^{-2} **19** $\left(\frac{1}{4}\right)^{-3}$

4 8^2 **8** 9^{-1} **12** $\left(\frac{2}{7}\right)^{-1}$ **16** $\left(\frac{4}{5}\right)^{3}$ **20** $\left(\frac{3}{7}\right)^{0}$

Puzzle

Complete this addition so that every digit from 1 to 9 is used once.

```
  * 6 *
  * 1 * +
  * 8 *
```

Standard form (scientific notation)

Very large or very small numbers are more briefly written in standard form. It is easier to compare sizes of numbers written in standard form.

Standard form is a number between 1 and 10 multiplied by the appropriate power of 10.

Exercise 2m

The following numbers are given in standard form. Write them as ordinary numbers:

1 3.45×10^2

2 1.2×10^3

3 5.01×10^{-2}

4 4.7×10^{-3}

5 2.8×10^2

6 7.3×10^{-1}

7 9.02×10^5

8 6.37×10^{-4}

9 8.72×10^6

> **Tip** 10^2 means $10 \times 10 = 100$

> **Tip** 10^{-2} means $\frac{1}{10^2} = \frac{1}{100}$

Write the following numbers in standard form:

a 3840

b 0.0025

(First, write the given figures as a number between 1 and 10 and then decide what power of 10 to multiply the number by to bring it back to the correct size.)

a $3840 = 3.84 \times 10^3$ (The decimal point has been moved 3 places to the left so the index of 10 is 3.)

b $0.0025 = 2.5 \times 10^{-3}$ (The decimal point has been moved 3 places to the right so the index of 10 is −3.)

Write the following numbers in standard form:

10 265

11 0.18

12 3020

13 0.019

14 76 700

15 390 000

16 0.000 85

17 7000

18 0.004

19 58 700

20 2600

21 450 000

22 0.000 007

23 0.8

> **Tip** Check your answers by writing them as ordinary numbers.

24 0.000 56

25 24 000

26 39 000 000

27 0.000 000 000 08

If you have a scientific calculator, it will display very large or very small numbers in scientific notation, but only the power of ten is given; 10 itself does not appear in the display.

Try this: enter 0.000 05, then press $\boxed{x^2}$.

The display will read $\boxed{2.5 \quad -09}$. This means 2.5×10^{-9}.

Now try entering 50 000 and then pressing $\boxed{x^2}$.

The display will read $\boxed{2.5 \quad 09}$. This means 2.5×10^9.

28 Use your calculator to find the value of

a $(250\,000)^2$

b $(25\,700)^2$

c $(0.000\,08)^2$

d $(0.000\,007)^2$

Approximations (decimal places and significant figures)

We have seen that it is sometimes unnecessary and often impossible to give exact values. In the case of measurements this is particularly true. However, we do need to know the degree of accuracy of an answer. For example, if a manufacturer is asked to make screws that are about $12\frac{1}{2}$ cm long, he does not know what is acceptable as being 'about $12\frac{1}{2}$ cm long'! But if he is asked to make them 12.5 cm long correct to one decimal place, he knows what tolerances to work to.

Decimal places

To correct 0.078 22 to *two* decimal places (d.p.) we look at the *third* decimal place. If it is 5 or larger, we add 1 to the figure in the second decimal place. If it is less than 5, we do not alter the figure in the second decimal place.

In this example the figure in the third decimal place is 8,

so $\qquad$ 0.07|822 = 0.08 $\qquad$ correct to 2 d.p.

Significant figures

To determine the 1st, 2nd, 3rd, ... significant figure (s.f.) in a number we can change it into standard form. Then the figure to the left of the decimal point is the 1st s.f., the next figure is the 2nd s.f., ... and so on.

For example, $\qquad$ $0.0\ 7\ 0\ 2\ 5 = 7.0\ 2\ 5 \times 10^{-2}$

1st s.f. — 2nd s.f. — 3rd s.f. — $\qquad$ 1st s.f. — 2nd s.f. — 3rd s.f. —

To correct 0.078 22 to two significant figures we look at the third significant figure; in this case it is 2.

Therefore $\qquad$ 0.078 22 = 0.078 $\quad$ correct to 2 s.f.

$\quad$ Give 0.070 25 correct to $\quad$ **a** 3 d.p. $\quad$ **b** 3 s.f.

a $\quad$ 0.070|25 = 0.070 $\,$ correct to 3 d.p. (2 is less than 5 so the zero after the 7 is unchanged.)

(We leave the zero at the end of the corrected answer to indicate that the third decimal place is zero.)

b $\quad$ 0.070 2|5 = 7.02|5 $\times 10^{-2}$ $\,$ (this step can be omitted)

$\qquad\qquad$ = 0.070 3 $\,$ correct to 3 s.f.

Give each of the following numbers correct to

a three decimal places   b three significant figures:

1 2.7846	4 0.073 25	7 0.7801	10 0.000 925 8
2 0.1572	5 0.150 76	8 3.2994	11 7.8196
3 3.2094	6 0.020 39	9 254.1627	12 0.009 638

Estimate a rough value for 0.826×38.3 by correcting each number to 1 significant figure. Then use your calculator to give the answer correct to 3 significant figures.

$0.826 \times 38.3 \approx 0.8 \times 40 = 32 \approx 30$ (For 0.826 the first figure is 8 and the second figure is less than 5 so the 8 is not changed. For 38.3 the second figure is greater than 5 so the first figure is increased by 1.)

$0.826 \times 38.3 = 31.6$ correct to 3 s.f.

First estimate a rough value for each of the following calculations, then use your calculator to give the answer correct to three significant figures:

13 0.035×1.098	20 $204 \div 942$	27 $0.532 \times 3.621 \times 36$
14 258×184	21 $0.827 \div 0.093$	28 $(0.32)^3$
15 0.0932×0.48	22 $0.0026 \div 0.0378$	29 $\dfrac{2.27 \times 3.84}{5.01}$
16 18.27×3.82	23 $16.97 \div 3.702$	30 $\dfrac{0.016 \times 5.82}{2.31}$
17 0.295×0.732	24 $3502 \div 651$	31 $\dfrac{2.93 \times 0.037\,2}{1.84 \times 0.562}$
18 1250×532	25 $(3.827)^3$	
19 $86.27 \div 39.8$	26 $\dfrac{3.257}{83.6}$	

Investigation

On the last day of a sale a retailer sold two anoraks at $48 each. By doing this he made a 25% profit on the one but a loss of 20% on the other. He believed that he had made a profit overall. Investigate.

Mixed exercises

Exercise 2p

1   Find the LCM of   a   2, 5 and 6   b   3, 7 and 14

2   Find the reciprocal of   a $\frac{3}{4}$ b $\dfrac{x}{y}$

3 Calculate   **a** $1\frac{1}{3} \div \frac{8}{9}$ **b** $1\frac{1}{3} - \frac{8}{9}$

4 Calculate $\left(3\frac{1}{4} + 2\frac{1}{2}\right) \times \frac{2}{5}$

5 Find, without using a calculator: **a** $3.27 + 0.09$ **b** 3.27×0.09 **c** $3.27 \div 0.03$

6 Find the value of   **a** 2^4 **b** 2^0 **c** 2^{-4}

7 Write as a single number in index form   **a** $5^6 \div 5^4$ **b** $5^{10} \times 5^2$

8 Write the following numbers in standard form   **a** 2560 **b** $0.000\,256$

Exercise **2q**

1 Find the LCM of   **a** $3, 8$ and 24 **b** $2, 3$ and 5

2 Find the reciprocal of   **a** $\frac{1}{5}$ **b** $1\frac{1}{2}$

3 Calculate   **a** $\frac{3}{5} \times 1\frac{1}{4}$ **b** $\frac{3}{5} + 1\frac{1}{4}$

4 Calculate $2\frac{1}{2} \times \frac{7}{10} + 1\frac{1}{3}$

5 Find, without using a calculator:

   a $2.5 - 1.05$ **b** 2.5×1.05 **c** $1.05 \div 2.5$

6 Find the value of   **a** $\left(\frac{1}{2}\right)^2$ **b** $\left(\frac{1}{2}\right)^0$ **c** $\left(\frac{1}{2}\right)^{-2}$

7 Write in standard form   **a** $570\,000$ **b** 0.057

Investigation

How accurate are calculators

a Use your calculator to perform these instructions.

Enter 5×10^{-20} on your calculator. Now add 2. Next subtract 1. Multiply the result by 10^{20}

b Repeat the instruction given in part **a** without using a calculator. Comment on the results.

c For this part, you will need to use a calculator on which you can enter more than 8 figures. (A graphics calculator will cope with this.)

Find the value of $\sqrt{100\,000\,000} - \sqrt{99\,999\,999}$

How many significant figures are there in your answer?

It is possible to use such a calculator to evaluate $\sqrt{100\,000\,000} - \sqrt{99\,999\,999}$ to 11 significant figures. Can you discover how to do this?

d A basic scientific calculator, on which you cannot enter more than eight figures, can be used to enter the numbers $\sqrt{100\,000\,000} - \sqrt{99\,999\,999}$ by first writing them in another form. How can this be done? Try it and comment on the result.

MATHS IS OUT THERE

Pythagoras once defined a friend as 'One who is the other I, such as 220 and 284'.

These two numbers are AMICABLE, because the proper divisors of one number adds up to the other number.

Divisors of 284 are $\{1, 2, 4, 71, 142\}$. Their sum, $1 + 2 + 4 + 71 + 142 = 220$.

Divisors of 220 are $\{1, 2, 4, 5, 10, 11, 20, 22, 44, 55, 110\}$.

Their sum is $1 + 2 + 4 + 5 + 10 + 11 + 20 + 22 + 44 + 55 + 110 = 284$.

The Hebrews believed that such numbers were good omens.

Today more than 600 pairs of friendly numbers are known.

In 1866 a sixteen year old Italian boy, Nicole Paganini, discovered the amicable numbers 1184 and 1210. Can you discover any?

IN THIS CHAPTER...

you have seen that:

- fractions can be added and subtracted by expressing them as equivalent fractions with the same denominator

- fractions can be multiplied together by multiplying the numerators together and multiplying the denominators together

- the reciprocal of a number is 1 divided by that number, e.g. the reciprocal of $\frac{5}{7}$ is $\frac{7}{5}$ and the reciprocal of 6 is $\frac{1}{6}$

- to divide by a fraction you multiply by its reciprocal, e.g. $\frac{2}{3} \div \frac{4}{7} = \frac{2}{3} \times \frac{7}{4}$

- before multiplying or dividing by fractions, mixed numbers must be changed to improper fractions

- brackets are always worked out first, then multiplication and division and lastly addition and subtraction

- some fractions, e.g. $\frac{1}{7} = 0.1428571428\ldots$ and $\frac{3}{11} = 0.272727\ldots$, cannot be expressed as exact decimals. They are called recurring decimals. The first can be written $0.\dot{1}4285\dot{7}$ and the second $0.\dot{2}\dot{7}$

- numbers can be compared by converting all of them either into decimals or into fractions

- indices provide a shorthand notation for multiplying and dividing numbers.

  An index can be positive, negative or zero. Remember that $a^0 = 1$ and $a^{-2} = \frac{1}{a^2}$

  The laws of indices are $a^m \times a^n = a^{m+n}$, $a^m \div a^n = a^{m-n}$, $(a^m)^n = a^{mn}$

- any number can be written as a number between 1 and 10 multiplied by the appropriate power of 10. This is called standard form, e.g. $26700 = 2.67 \times 10^4$ and $0.0086 = 8.6 \times 10^{-3}$.

3 DIRECTED NUMBERS

AT THE END OF THIS CHAPTER...

you should be able to:

1   Show the positions of positive and negative numbers on a number line.

2   Perform operations of addition, subtraction, multiplication and division on directed numbers.

3   Illustrate on a number line the range of values for a given number to a stated number of decimal places.

Did you know that the Hindus were probably the first people to understand the concept of negative numbers? Brahmagupta (598–670) gave the rules for using them.

The Greeks had no concept of negative numbers – in fact Diophantus (about 200–284) called any equation with negative solutions an absurd equation.

BEFORE YOU START

you need to know:
✓ how to add, subtract, multiply and divide whole numbers
✓ the meaning of decimal places
✓ the meaning of a^2

KEY WORDS

corrected number, diameter, directed number, negative number, number line, positive number, range of values, the symbols $\geqslant$, $<$ and $\leqslant$

Positive and negative numbers

We use numbers to describe quantities. For example, we may talk about $\frac{2}{3}$ of a cake, 10.5 cm of wire, 35 apples, and so on.

These numbers, $\frac{2}{3}$, 10.5 and 35, are examples of *positive numbers*.

We cannot however use positive numbers to describe temperatures below 0° C (the freezing point of water), or any other quantity that can fall below a zero level.

To do this we need *negative numbers*.

Positive numbers are written, for example, as +2 or simply as 2.

Negative numbers are written, for example, as −2.

Positive numbers and negative numbers are together known as *directed numbers*.

Using a number line

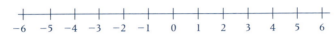

On the number line, positive numbers are in positions to the right of zero and negative numbers are in positions to the left of zero.

If a number a is to the *right* of a number b, then

$\qquad$ a is greater than b

e.g. $\qquad$ $3 > 1$ and $-1 > -3$

If a number c is to the *left* of a number d then

$\qquad$ c is less than d

e.g. $\qquad$ $2 < 6$ and $-6 < -4$

Exercise **3a**

Insert > or < between each of the following pairs of numbers.

1 $\quad$ 4 $\quad$ **2** $\quad$ 3 $\quad$ −1 $\quad$ −4 $\quad$ **5** $\quad$ 2 $\quad$ −1 $\quad$ **7** $\quad$ −5 $\quad$ −6 $\quad$ **9** $\quad$ 5 $\quad$ −5 $\quad$ **11** $\quad$ −3 $\quad$ 0

2 $\quad$ 3 $\quad$ 5 $\quad$ **4** $\quad$ −5 $\quad$ −2 $\quad$ **6** $\quad$ −4 $\quad$ 3 $\quad$ **8** $\quad$ 0 $\quad$ −4 $\quad$ **10** $\quad$ 4 $\quad$ −2 $\quad$ **12** $\quad$ 0 $\quad$ 6

Adding and subtracting directed numbers

Adding directed numbers can be interpreted as adding steps to the left or right,

e.g. $\qquad$ $+(+2)$ can mean 'add 2 steps to the right'

therefore $\qquad$ $+(+2) = +2$

Also $\qquad$ $+(-2)$ can mean 'add 2 steps to the left'

therefore $\qquad$ $+(-2) = -2$

Adding and subtracting directed numbers

Subtracting directed numbers can be interpreted as taking away steps.

e.g. $-(+2)$ can mean 'take away 2 steps to the right'

i.e.   'go 2 steps to the left'

therefore $-(+2) = -2$

Also $-(-2)$ can mean 'take away 2 steps to the left'

i.e.   'go 2 steps to the right'

therefore $-(-2) = +2$

> i.e.   subtracting a negative number has the same effect as adding a positive number.

Exercise 3b   Find $+2+(-4)$

$$+2+(-4) = +2-4 = -2$$

$(+2-4$ means 'go 2 steps to the right and then 4 steps to the left')

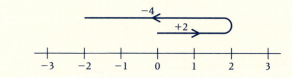

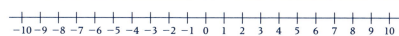

Use this number line, if it helps, to calculate:

1 $-2+(-3)$	**7** $-5+(-4)$	**13** $-3+(-2)$	**19** $-4-(-8)$
2 $+4-(+6)$	**8** $3-(-3)$	**14** $+4-(-2)$	**20** $+6+(-9)$
3 $-2+(-4)$	**9** $-3-(+3)$	**15** $-3-(+2)$	**21** $-5-(+7)$
4 $-8+(+6)$	**10** $+2-(-5)$	**16** $+2-(+5)$	**22** $+4+(-10)$
5 $-3-(-4)$	**11** $-6+(-4)$	**17** $+7+(-4)$	**23** $-8-(-3)$
6 $+2-(-6)$	**12** $-2+(+8)$	**18** $+8-(+10)$	**24** $+10-(+6)$

Remember that the + sign is often left out, i.e. $2+8-4$ means $+2+(+8)-(+4)$

Calculate $4-8+7$

$$4-8+7 = -4+7 = 3$$

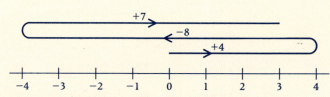

Calculate:

25 $3-4+6$	**28** $-4+2-(-4)$	**31** $-7+9-2$	**34** $5+6-4$
26 $2+6-4$	**29** $5-3+9$	**32** $3+(-2)-3$	**35** $-3-7+9$
27 $-3+2-4$	**30** $3-(-2)+2$	**33** $5-9+3$	**36** $2-7-(-6)$

Multiplying and dividing directed numbers

$(+2)\times(+3)$ means 'add 2 lots of +3'

therefore $\qquad (+2)\times(+3)=+(+6)=+6$

$(-2)\times(-3)$ means 'take away 2 lots of −3'

therefore $\qquad (-2)\times(-3)=-(-6)=+6$

In the same way $(+2)\times(-3)=+(-6)=-6$

and $\qquad (-2)\times(+3)=-(+6)=-6$

Dividing by, say, -2 is the same as multiplying by $-\frac{1}{2}$

e.g. $\qquad 12\div(-2)=12\times\left(-\frac{1}{2}\right)=-6$

From this we can deduce the general rule that

> when multiplying or dividing,
> like signs give a positive answer
> unlike signs give a negative answer

Exercise **3c** Find **a** $(-4)^2$ **b** $6\div(-2)$

a $(-4)^2=(-4)\times(-4)=+16$

b $6\div(-2)=-3$

Find:

1 $(3)\times(-2)$	**7** $(+3)\times(-4)$	**13** $(+6)\div(+3)$	**19** $(+9)\div(+3)$
2 $(-2)\times(-6)$	**8** $(-2)\times(-7)$	**14** $(+6)\div(-3)$	**20** $(-5)\div(-1)$
3 7×8	**9** $(-2)^2$	**15** $(-6)\div(+3)$	**21** $4\div2$
4 $(-3)\times(-7)$	**10** $(-5)\times(+3)$	**16** $(-6)\div(-3)$	**22** $(-12)\div(+4)$
5 $(-2)\times(+4)$	**11** $(-3)^2$	**17** $(-8)\div(+4)$	**23** $(-3)\div3$
6 5×6	**12** $(+3)\times(-6)$	**18** $(+10)\div(-2)$	**24** $15\div(-5)$

Puzzle

Wilhelm Fliess was convinced that all nature could be explained in terms of the numbers 23 and 28.

A table can be made listing all the numbers from 1 to 28 as sums of multiples of 23 and 28.

For example, $1 = (23 \times 11) - (28 \times 9)$.

Can you find 2 in this way?

Range of values for a corrected number

Suppose we are told that, correct to the nearest 10, 250 people boarded a particular train. People are counted in whole numbers only. Hence in this case, 245 is the lowest number that gives 250 when corrected to the nearest 10 and 254 is the highest number that can be corrected to 250. We can therefore say that the actual number of people who boarded the train is any whole number from 245 to 254.

Now suppose that we are given a nail and are told that its length is 25 mm correct to the nearest millimetre. The length may be a whole number of millimetres, but may also be any part of a millimetre.

Look at this magnified section of a measuring gauge:

The lowest number that can be rounded up to 25 is 24.5. The highest number that can be rounded down to 25 is not so easy to determine. All we can say is that any number up to, but not including, 25.5 can be rounded down to 25.

The length of the nail is therefore in the range from 24.5 mm up to, but not including, 25.5 mm.

If l mm is the length of the nail, we can therefore write

$$24.5 \leqslant l < 25.5$$

where $\leqslant$ means 'is less than or equal to' and $<$ means 'is less than'.

To illustrate this on the diagram we use a solid circle to show that 24.5 is included in the range and an open circle to show that 25.5 is not included in the range.

Exercise **3d** Illustrate on a number line the range of values of x given by $0.1 < x \leqslant 0.8$

The open circle means that 0.1 is not in the range. The solid circle means that 0.8 is in the range.

Use a number line like this for questions **1** to **10**:

In each case illustrate the range on your number line:

1 $5 \leqslant x \leqslant 10$ **3** $-2 \leqslant x \leqslant 6$ **5** $0 < x < 10$ **7** $4 \leqslant x \leqslant 12$ **9** $-5 < x < 8$

2 $0 < x \leqslant 15$ **4** $5 \leqslant x < 15$ **6** $-5 < x \leqslant 5$ **8** $3 < x \leqslant 18$ **10** $-1 < x \leqslant 11$

Use a number line like this for questions **11** to **20**. In each case illustrate the range on your number line:

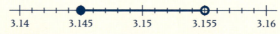

11 $0 < x < 0.1$ **14** $0.08 < x \leqslant 0.16$ **17** $0.05 \leqslant x \leqslant 0.2$

12 $0.1 \leqslant x < 0.2$ **15** $0.02 < x < 0.08$ **18** $0.07 < x \leqslant 0.14$

13 $0.05 \leqslant x \leqslant 0.15$ **16** $0.03 < x \leqslant 0.13$ **19** $0.10 < x < 0.19$

> A number is given as 3.15 correct to 2 d.p. Illustrate on a number line the range in which this number lies.
>
> (3.15 is between 3.14 and 3.16 so we will use just that part of the number line.)
>
>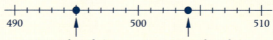
>
> 3.155 would be rounded up to 3.16, so 3.155 is not included in the range and we indicate this with an open circle.

Illustrate on a number line the range of possible values for each of the following corrected numbers: The numbers in questions **20** to **29** are correct to 1 d.p.

20 1.5   **22** 0.1   **24** 2.0   **26** 1.3   **28** 8.0

21 0.2   **23** 4.8   **25** 0.6   **27** 6.2   **29** 12.9

The numbers in questions **30** to **39** are correct to 2 d.p.

30 0.25   **32** 12.26   **34** 3.10   **36** 6.89   **38** 8.50

31 1.15   **33** 0.05   **35** 0.52   **37** 26.35   **39** 0.70

Problems

Exercise 3e   It is stated that a packet of pins contains 500 pins to the nearest 10. Find the range in which the actual number of pins lies.

(There must be a whole number of pins in the packet.)

If n is the number of pins in the packet then n is a whole number such that $495 \leqslant n \leqslant 504$.

A copper tube is sold as having an internal bore (diameter) of 10 mm to the nearest millimetre. Find the range in which the actual bore lies.

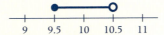

If d mm is the diameter of the tube then $9.5 \leqslant d < 10.5$

In some of the following questions you are asked to find a range of values for a quantity that can only have whole number values. When this is the case, it *must* be clearly stated in your answer:

> **Tip** Read the question carefully to decide whether the quantity can take any value in the range or whether it can have whole number values only.

1 The weight, w kg, of a bag of sand is given as 5.6 kg correct to 1 decimal place. Find the range of values in which the actual weight lies.

2 A shop is said to make a profit of $2500 a month. If this figure is given correct to the nearest $100, find the range in which the actual monthly figure, x, lies.

3 Correct to one decimal place, the length of a room is given as 2.8 m. Find the range in which the actual length of the room, x m, lies.

4 On a certain model of bicycle, the brake pads have to be 12.5 mm thick to work efficiently. Find the range in which the thickness, x mm, can lie, if the figure given is correct to 1 decimal place.

5 Referring to a football match, a newspaper headline proclaimed '7000 watch Barbados win'. If this figure for the number of spectators is correct to the nearest 1000, find the range in which x, the number of people who actually attended the match, lies.

6 One of the component parts of a metal hinge is a pin. In order to work properly this pin must have a diameter of 1.25 mm correct to 2 decimal places. Find the range in which the diameter, d mm, must lie.

7 One hundred people were asked if they ate cornflakes for breakfast that morning. To the nearest 10, 40 people said they did. Of the 100 people interviewed, what is the largest possible number who did *not* eat cornflakes?

8 An advertisement says 'For about $250 you can buy this automatic watch'. If this price is correct to the nearest $10, what is the most that you would expect to have to pay for such a watch?

9 The length of a car is given as 255 cm. If this figure is correct to the nearest 5 cm, find the range in which the actual length lies.

10 Knitting yarn is sold by weight. It is found that 10 g of double knitting pure wool has a length of 20 m correct to the nearest metre. What is the minimum length of yarn you would expect in a 50 g ball of double knitting pure wool?

Investigation

You will need access to a pair of scales that give weights in kilograms and grams. They do not need to be able to measure small masses accurately; a set of kitchen scales is ideal.

You also need a bag of uncooked rice.

a Try weighing one grain of rice and report on the result.

b Describe a method by which it is possible to estimate the mass of one grain of rice in grams to 2 decimal places.

c Use your method to estimate the mass of one grain.

d Suggest a way to judge the accuracy of your estimate.

Mixed exercise

Exercise 3f

Find:

1	$2 + 6 - 5$	**7**	$2 - (-3)$	**13**	$8 - 7 - 3$	**19**	$(-10) \div 5$
2	$(+3) \times (-4)$	**8**	$(-6)^2$	**14**	$(-4) \times 7$	**20**	$7 + (-3) - 4$
3	$(-5)^2$	**9**	$3 - (-4) + 6$	**15**	$(-9)^2$	**21**	$(-2) + 6 - 3$
4	$2 + 7 - 9 + 2$	**10**	$(-5) \times 6$	**16**	$8 \div (-2)$	**22**	$15 \div (-3)$
5	$(-3) \div (-1)$	**11**	$12 \div (-3)$	**17**	$5 - 7 - (-4)$	**23**	$7 \times (-2)$
6	$2 \times (-6)$	**12**	$2 - (-5) - 4$	**18**	4×5	**24**	$(-14) \div (-2)$

25 The diameter of a ball bearing is given as 1.5 mm correct to 1 decimal place. Find the range in which the diameter lies.

26 To the nearest 10, 70 children brought a packed lunch to school. Find the range in which the actual number of children lies.

27 A box contains 450 tacks to the nearest 10. Find the range in which the actual number of tacks lies.

28 To fit properly, the diameter of a bicycle tyre has to be 0.75 m correct to 2 decimal places. Find the range in which the diameter can lie.

MATHS IS OUT THERE

Look up the meaning of each of the following words and write it in your notebook:

(i) GEMATRIA   (ii) NUMEROLOGY   (iii) TRISKAIDEKAPHOBIA.

IN THIS CHAPTER...

you have seen that:

- directed numbers can be added and subtracted using the rules that $-(-a)$ gives $+a$ and $-(+a)$ or $+(-a)$ gives $-a$

- directed numbers can be multiplied using the rules that $(+a) \times (+b)$ and $(-a) \times (-b)$ both give a positive answer whereas $(+a) \times (-b)$ and $(-a) \times (+b)$ both give a negative answer

- when you are giving the range in which a corrected number lies, you need to decide whether the quantity can have any value within the range or only whole number values. You also need to decide whether the end values are included or not.

4 EQUATIONS AND FORMULAE

AT THE END OF THIS CHAPTER...

you should be able to:

1 Simplify algebraic expressions.

2 Solve simple equations.

3 Construct formulae from data.

4 Substitute numbers into formulae.

5 Change the subject of a simple formula.

MATHS IS OUT THERE

Did you know that equations first appeared in cuneiform texts of the Babylonians and go back to the third millennium BC? They are among the first mathematical achievements of mankind.

BEFORE YOU START

you need to know:
✓ how to remove brackets in algebraic expressions
✓ how to use index notation
✓ how to work with positive and negative numbers

KEY WORDS

expression, equation, index (plural indices), subject of a formula

Simplifying algebraic expressions

If an algebraic expression contains two terms each with the same combination of letters and powers then we can usually add them.

For example, $x + x$ can be written as $2x$

and $2x + 3x$ can be written as $5x$

$$(2x + 3x = x + x + x + x + x)$$

However, if the two terms contain different combinations of letters or different powers then, as we do not know what numbers the letters stand for, we cannot add them.

For example, $x + y$ cannot be simplified, neither can $x - x^2$.

Exercise 4a Simplify where possible **a** $x - (-y)$ **b** $x + 2y - 3x$

a $x - (-y) = x + y$
(Like signs together give a +)

b $x + 2y - 3x = 2y - 2x$
(We usually write the positive term first.)

Simplify where possible:

1 $p + q$ **3** $2s + 3t$ **5** $5x - 3x$ **7** $x + 3x - 2y$ **9** $a + 3b - 2a$

2 $a + a$ **4** $3v + 4v$ **6** $p - (-q)$ **8** $3u - (-2u)$ **10** $c + 2d - (-3c)$

When terms are multiplied together we can write them in a slightly shorter form by omitting the multiplication sign.

For example

$a \times b$ can be written as ab

$p \times p$ can be written as p^2 (using index notation)

$5a \times 2b$ can be written as $10ab$ (multiplication can be done in any order
so 5 can be multiplied by 2)

But we cannot simplify ab or $\dfrac{x}{y}$.

Exercise 4b Simplify **a** $4x \div (-2y)$ **b** $2p \times q$

a $4x \div (-2y) = \dfrac{{}^2\cancel{4}x}{-\cancel{2}y}_1$

$= -\dfrac{2x}{y}$ (A + number divided by a − number gives a − number)

b $2p \times q = 2pq$

Simplify where possible:

1 $x \times y$	**7** $a \div a$	**13** $a - 3a$	**19** $q - 5q$
2 $a \times a$	**8** $6b \div 2c$	**14** $p \times 4p^2$	**20** $r - (-4s)$
3 $2s \times 3s$	**9** $s + t$	**15** $8u \div 4w$	**21** $8p \div 2q$
4 $4x \times 3x$	**10** $m - n$	**16** $2z - 3y$	**22** $(-3s) \times (-2t)$
5 $u \div v$	**11** $m \times (-n)$	**17** $3s \times 2t$	**23** $(b) \times (-2b)$
6 $a \div (-b)$	**12** $3s - 2t$	**18** $p \times 2p$	**24** $(-x) \div (-y)$

Simplify $a(a - b) - (b - a)$

Remember that $-(b - a) = -1 \times (b - a)$

$$a(a - b) - (b - a) = a^2 - ab - b - (-a)$$
$$= a^2 - ab - b + a$$

Simplify $3(a - b) - 2(a + b)$

$$3(a - b) - 2(a + b) = 3a - 3b - 2a - 2b$$

(Everything in the first bracket is multiplied by 3 and everything in the second bracket is multiplied by -2)

$$= a - 5b$$

Simplify:

25 $a - 3(a - b)$	**30** $4(x + y) + 2(x + z)$	**35** $x(x - y) + y(y - x)$
26 $2a + a(a - 3)$	**31** $3(p + q) - 2(p + r)$	**36** $4(b - c) - 2(b + c)$
27 $2(a - b) - (b - a)$	**32** $2x - (x + y)$	**37** $4(p - q) + 2(q - p)$
28 $4(a - c) + 2(a - b)$	**33** $2(p + q) - 3(p - q)$	**38** $w(w + x) - x(w - x)$
29 $y - 2(y - z)$	**34** $a(a + b) - 2(a - b)$	**39** $3(m + n) - 5(m - n)$

Solving equations

Consider the statement

$$3x + 6 = 2 - x$$

This is an equation. It means that the left hand side is *equal* to (i.e. has the same value as) the right hand side.

We can think of the two sides of an equation as the contents of the two pans on a pair of scales which are exactly balanced. The equality will remain true provided that we always do the same thing to both sides.

Solving an equation means finding the number that the letter stands for so that the two sides *are* equal.

When solving equations like these it is sensible to proceed in the following order:

1. remove any brackets
2. collect any like terms on each side
3. collect the letter terms on one side (choose the side with the greater number to start with remembering that, say, $-2x > -3x$)
4. collect the number terms on the other side.

Exercise 4c Solve the equation $3x + 6 = 2 - x$

$$3x + 6 = 2 - x$$

(Aim to get all the xs on the LHS and all the numbers on the RHS.)

Add x to each side $\qquad 4x + 6 = 2$

Take 6 from each side $\qquad 4x = -4$

Divide each side by 4 $\qquad x = -1$

(Always check that your value of x satisfies the original equation.)

Check: when $x = -1$, LHS $= 3(-1) + 6 = -3 + 6 = 3$

$\qquad\qquad\qquad$ RHS $= 2 - (-1) = 2 + 1 = 3$

$\therefore \qquad\qquad\qquad x = -1$

Solve the following equations:

1 $2p + 5 = 3 - p$ $\qquad$ **6** $3y + 2 = 7 - 2y$

2 $3 - s = 4 - 3s$ $\qquad$ **7** $4 - x = 8 - 3x$

 3 $x + 5 = 3x - 2$ $\qquad$ **8** $a + 7 = 4a - 5$

4 $2a + 3 = 4 - 3a$ $\qquad$ **9** $2x + 5 = 7 - 2x$

5 $3 - 2x = 8x - 7$ $\qquad$ **10** $3 - 2x = 9 - 5x$

Tip Subtract x from each side, then add 2 to each side.

Solve the equation $3 - 2(4 + x) = x - 6$

$$3 - 2(4 + x) = x - 6$$

Expand bracket $\qquad\qquad 3 - 8 - 2x = x - 6$

Collect like terms $\qquad\qquad -5 - 2x = x - 6$

Add $2x$ to both sides $\qquad\qquad -5 = 3x - 6$

Add 6 to both sides $\qquad\qquad 1 = 3x$

Divide both sides by 3 $\qquad\qquad \frac{1}{3} = x$ or $x = \frac{1}{3}$

Solve:

11 $3 - 2(x - 2) = 8$

12 $x - 4(x + 3) = 3$

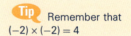

13 $4x = 2 - 3(x + 1)$

14 $2y = 8 - (y - 2)$

15 $3(2 - x) = 4(x - 3)$

16 $5 - 3(x - 4) = 6$

17 $a - 2(a + 3) = 5$

18 $3p - 2 = 4 - 3(p + 2)$

19 $4w = w - (w - 8)$

20 $7(5 - x) = 3(x - 5)$

21 $5(x - 2) - 3(x + 1) = 0$

22 $3 - 5(2x + 1) = 2x$

23 $3(2x + 1) - 2(4x - 1) = 0$

24 $4 - 2(3 - 2x) = 5$

25 $3x - 4(1 - 3x) = 2x - (x + 1)$

26 $7(b - 4) - 5(b + 2) = 0$

27 $12 - 3(4x - 1) = 5$

28 $5(3x - 1) - 4(3x - 2) = 0$

29 $8 - 5(2 - x) = 8$

30 $4x - 2(1 - 3x) = 3 - (2x - 1)$

 Puzzle

A supermarket offers an extra tin of baked beans in exchange for the labels from three tins on their own brand baked beans. Cherie buys 15 tins. How many extra tins can she claim?

 Investigation

Are the two orange lines straight?

Check your answer.

Investigate other optical illusions.

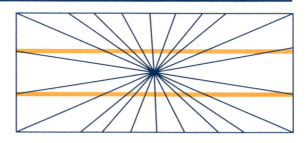

Constructing formulae

Electricity bills are presented every quarter. They are made up of a fixed standing charge plus the cost of the number of units used in the quarter.

By using letters for the unknown quantities, we can construct a formula for a quarterly electricity bill.

If $C is the total bill, $R is the standing charge, the cost of one unit is x cents and N units are used, then

the cost of the units is Nx cents or $\dfrac{Nx}{100}$

therefore $C = R + \dfrac{Nx}{100}$

Notice that the units are the same throughout, i.e. the cost of the units was converted from cents to dollars so that we added dollars to dollars, not dollars to cents.

A number p is equal to the sum of a number q and twice a number r. Write down a formula for p in terms of q and r.

p equals q plus $2 \times r$

i.e. $\qquad\qquad\qquad p = q + 2r$

Read each question carefully. Be sure that you understand what you are being asked.

In questions 1 to 12 write down a formula connecting the given letters:

1 A number a is equal to the sum of two numbers b and c.

2 A number m is equal to twice the sum of two numbers n and p.

3 A number z is equal to the product of two numbers x and y.

4 A number a is equal to twice the product of two numbers b and c.

5 A number v is equal to the square of a number n.

6 A number d is equal to the difference of two numbers e and f, where e is greater than f.

7 A number x is equal to half a number y.

8 A number a is equal to a number b divided by twice a number c.

9 A number k is equal to the sum of twice a number u and three times a number v.

10 A number x is equal to twice a number y, minus a number z.

11 A number n is equal to the sum of a number p and its square.

12 A number v is equal to the sum of a number u and the product of the numbers a and t.

13 Cloth is sold at $p per metre.
The cost of N metres is $R.
Find a formula for R in terms of N and p.

> **Tip** The cost of 1 metre is $p so the cost of N metres is N times as much.

14 Oranges are sold at x cents each.
The cost of n oranges is y cents.
Find a formula for y in terms of n and x.

15 A ship moving with constant speed takes x minutes to cover one nautical mile. It takes X minutes to cover y nautical miles. Find a formula for X in terms of x and y.

16 A shop sells two brands of baked beans. It has N tins of baked beans altogether; y of them are one brand and z of them are the other brand. Find a formula for N in terms of y and z.

17 A rectangle is x centimetres wide and y centimetres long and its perimeter is P metres. Find a formula for P in terms of x and y.

18 Fertiliser is applied at the rate of a grams per square metre. It takes b kilograms to cover a field of area c square metres. Find a formula for b in terms of a and c.

19 A money box contains l one-cent coins and m five-cent coins. The total value of money in the money box is n cents. Find a formula for n in terms of l and m.

20 A bag of coins contains x ten-cent coins and y five-cent coins. The total value of the coins is $\$R$. Find a formula for R in terms of x and y.

Substituting numbers into formulae

Exercise **4e** Given that $s = ut - \frac{1}{2}gt^2$, find s when $u = 8$, $t = 6$ and $g = -10$

$$s = ut - \tfrac{1}{2}gt^2$$

When $u = 8$, $t = 6$ and $g = -10$

$$s = (8)(6) - (\tfrac{1}{2})(-10)(6)^2$$

(Notice that we have put each number in brackets; this is particularly important in the case of negative numbers.)

$$= 48 - (-5)(36)$$
$$= 48 - (-180)$$
$$= 48 + 180 = 228$$

1 Given that $p = r - nt$, find p when $r = 40$, $n = 8$ and $t = 4$

2 Given that $v = \dfrac{u - t}{3}$, find v when $u = 4$ and $t = -2$

3 Given that $z = \dfrac{1}{x} + \dfrac{1}{y}$, find z when $x = 2$ and $y = 4$

4 Given that $a = bc + d$, find a when $b = 3$, $c = 2$ and $d = -4$

5 If $x = y^2$, find x when $y = 5$

6 If $C = rt$, find C when $r = -3$ and $t = -10$

7 If $x = rt - v$, find x when $r = 2$, $t = 10$ and $v = -4$

8 If $p = x + x^2$, find p when $x = 2$

9 Given that $s = \frac{1}{2}(a + b + c)$, find s when $a = 6$, $b = 9$ and $c = 5$

10 Given that $v = ab + bc + cd$, find v when $a = 2$, $b = \frac{1}{2}$, $c = 4$ and $d = -2$

11 If $p = r(2t - s)$, find p when $r = \frac{1}{2}$, $t = 3$ and $s = -2$

12 If $a = (b + c)^2$, find a when $b = 8$ and $c = -5$

13 If $r = \dfrac{2}{s+t}$, find r when $s = \frac{1}{2}$ and $t = \frac{1}{4}$

Tip Find the value of $s + t$ first.

14 If $n = \dfrac{r}{p-q}$, find n when $r = 4$, $p = 3$ and $q = -5$

15 Given that $a = bc - \frac{1}{2}dc$, find a when $b = 3$, $c = -4$ and $d = 7$

16 Given that $V = \frac{1}{2}(X - Y)^2$, find V when $X = 3$ and $Y = -5$

17 Given that $P = 2Q + 5RT$, find P when $Q = 8$, $R = -2$ and $T = -\frac{1}{2}$

18 Given that $a = (b - c)(c - d)$, find a when $b = 2$, $c = 4$ and $d = 7$

Changing the subject of a formula

Consider the formula

$$A = B + C$$

Taking C from each side gives

$$A - C = B$$

i.e. $\qquad\qquad B = A - C$

We have changed the subject of this formula from A to B. Changing the subject of a formula is very similar to solving equations.

Exercise 4f

Make the letter in brackets the subject of the formula:

1 $p = s + r$ (s) **11** $x = 2y$ (y)

2 $x = 3 + y$ (y) **12** $v = \frac{1}{2}t$ (t)

3 $a = b - c$ (b) **13** $a = bc$ (b)

4 $X = Y - Z$ (Y) **14** $t = \dfrac{u}{3}$ (u)

5 $r = s + 2t$ (s) **15** $l = \dfrac{m}{k}$ (m)

6 $k = l + m$ (m) **16** $a = 3b$ (b)

7 $u = v - 5$ (v) **17** $X = \frac{1}{10}N$ (N)

8 $z = x + y$ (y) **18** $v = ut$ (u)

9 $N = P - Q$ (P) **19** $z = \dfrac{w}{100}$ (w)

10 $v = u + 10t$ (u) **20** $n = \dfrac{p}{q}$ (p)

Tip [1] Subtract r from each side.

Tip [2] Add c to both sides.

Tip [11] Divide both sides by 2.

Tip [12] Multiply both sides by 2.

Make t the subject of the formula $v = u + 2t$

Think of this as solving an equation in t.

$$v = u + 2t$$

Take u from each side $\qquad v - u = 2t$

$\therefore \qquad\qquad\qquad\qquad 2t = v - u$

Divide each side by 2 $\qquad\qquad t = \dfrac{(v - u)}{2}$

Make the letter in brackets the subject of the formula:

21 $p = 2s + r$ (s) **25** $x = 2w - y$ (w) **29** $x = \dfrac{3y}{4}$ (y) **33** $V = \dfrac{2R}{l}$ (R)

22 $v = u - 3t$ (t) **26** $l = k + 4t$ (t) **30** $u = v + 5t$ (t) **34** $p = 2r - w$ (r)

23 $a = b - 4c$ (c) **27** $w = x - 6y$ (y) **31** $A = P + \frac{1}{10}l$ (l) **35** $a = b + \frac{1}{2}c$ (c)

24 $V = 2v + 3u$ (v) **28** $N = lt - 2s$ (s) **32** $z = x - \dfrac{y}{3}$ (y) **36** $p = q - \dfrac{r}{5}$ (r)

37 Make u the subject of the formula $v = u + at$
Find u when $v = 80$, $a = -10$ and $t = 6$

38 Make B the subject of the formula $A = \dfrac{C}{100} + B$

Find B when $A = 20$ and $C = 250$

39 Make C the subject of the formula $P = \dfrac{C}{N}$

Find C when $N = 20$ and $P = \frac{1}{2}$

40 Make x the subject of the formula $z = \frac{1}{2}x - 3t$
Find x when $z = 4$ and $t = -3$

41 A number a is equal to the sum of a number b and twice a number c

a Find a formula for a in terms of b and c

b Find a when $b = 8$ and $c = -2$

c Make b the subject of the formula.

42 A number x is equal to the product of a number z and twice a number y

a Find a formula for x in terms of z and y

b Find x when $z = 3$ and $y = 2$

c Make y the subject of the formula.

43 A number d is equal to the square of a number e plus twice a number f

a Find a formula for d in terms of e and f

b Make f the subject of the formula

c Find f when $d = 10$ and $e = 3$

44 A retailer sells bags of crisps at x cents each. The cost of n dozen bags of crisps is R.

    a Find a formula for R in terms of x and n

    b Find R when $x = 10$ and $n = 4$

Mixed exercises

Exercise 4g

1 Evaluate   **a** $8 \times (-\frac{1}{2})$ **b** $5 + 2 - 8$ **c** $8 \div (-4)$

2 Simplify   **a** $x + 3x$ **b** $2b - (-4b)$ **c** $(x) \times (x) \times (-3x)$

3 Simplify   **a** $2a - (a - b)$ **b** $3(a + b) - 2(a - b)$

4 Solve   **a** $x - 3 = 2 - 3x$ **b** $3(x - 2) - 8 = 0$

5 Make r the subject of the formula   **a** $v = 4r + u$ **b** $s = \dfrac{5r}{p}$

6 Find P, given that $P = \dfrac{100l}{RT}$, when

    a $l = 3$, $R = 4$ and $T = 2$ **b** $l = 6$, $R = 5$, $T = 3$

Exercise 4h

1 Evaluate   **a** $3 - (-10)$ **b** $(-\frac{1}{2}) \times (-4)$ **c** $(-20) \div (-5)$

2 Simplify   **a** $4a - 3b + 6a$ **b** $x + x^2 + 3x$ **c** $3a \times 4b$

3 Simplify   **a** $x - 3(x - y)$ **b** $2(x + 3y) - 4(2x + y)$

4 Solve   **a** $4 - 3a = 5 - 2a$ **b** $2(x + 4) = 3(5 - 2x)$

5 Make d the subject of the formula   **a** $C = \pi d$ **b** $a = 7d - s$

6 Find u, given that $u = v - gt$, when

    a $v = 16$, $g = -10$ and $t = 4$ **b** $v = -6$, $g = 10$ and $t = 8$

Did you find the meaning of 'GEMATRIA'?

Gematria has been defined as a method of interpretation of the Hebrew scriptures based upon the numerical value of the letters in the words.

It was thought that a person whose numerical 'value' exceeded another's was superior.

Example: Assign the numbers 1 to 26 to the letters of the alphabet as follows:

$$A = 1, B = 2, C = 3, \text{etc.}$$

Using the above, John's numerical value is $10 + 15 + 8 + 14 = 47$.

Jane's numerical value is $10 + 1 + 14 + 5 = 30$.

This shows that John is superior to Jane.

What is your numerical value?

In the Bible, the number 666 is referred to as the 'number of the beast'. Because 666 was associated with the Devil, any person with a gematria of 666 was thought to be the Antichrist. The enemies of Martin Luther, Pope Leo X, and Napoleon identified them as being 666's.

IN THIS CHAPTER...

you have seen that:

- like terms can be collected together but unlike terms cannot

- simple linear equations can be solved by collecting the terms in the unknown on one side and number terms on the other side

- you can change the subject of a formula by treating the formula as an equation in the letter you wish to isolate. Then solve the equation for this letter.

5 PERCENTAGES

Paying interest probably started when money itself was first introduced. Working out interest on a per cent basis probably comes from the Romans who worked out taxes as hundredths.

49

BEFORE YOU START	you need to know:
	✓ how to work with fractions and decimals
	✓ the basic meaning of percentage
	✓ how to solve equations involving fractions
	✓ how to find one quantity as a fraction of another

KEY WORDS	amount, compound interest, cost price, loss, multiplying factor, percentage, percentage decrease, percentage increase, principal, profit, rate per cent, selling price, simple interest

Meaning of percentages

We first met percentages in Book 1, and we begin this chapter by reminding ourselves of the work we did then:

$$65\% \text{ means } 65 \text{ out of } 100$$

Therefore, as a fraction $65\% = \frac{65}{100} = \frac{13}{20}$, i.e. to change a percentage to a fraction, put it over 100 and simplify

and as a decimal $65\% = \frac{65}{100} = 0.65$, i.e. to change a percentage to a decimal, divide it by 100.

Similarly the fraction $\frac{7}{20} = \frac{7}{20} \times 100\% = 35\%$, i.e. to change a fraction to a percentage, multiply it by 100.

and $0.35 = 0.35 \times 100\% = 35\%$, i.e. to change a decimal to a percentage, multiply it by 100.

Exercise 5a

Complete the following tables:

	Fraction	Percentage	Decimal
1	$\frac{3}{5}$		
2		40%	
3			0.55
4	$\frac{17}{20}$		
5		54%	

	Fraction	Percentage	Decimal
6			0.24
7			0.92
8		84%	
9	$\frac{37}{40}$		
10	$\frac{2}{3}$		

If 20% of the children in a class play the violin, what percentage do not?

(All the children (i.e. 100%) either do, or do not, play the violin.) The percentage who do play the violin is 20% and the percentage who do not play the violin is

$$100\% - 20\% = 80\%$$

11 Seventy-six per cent of the passengers on an aeroplane are Jamaican. What percentage of them are not?

12 Deductions from a man's wage were: income tax 21%, pension scheme 8%, other deductions 7%. What percentage did he keep?

13 In the third year at Stanley School, 35% take woodwork only, 25% take metalwork only and 20% take both subjects. What percentage of the third year pupils study neither?

Express 45 cm as a percentage of 2 m

(First express both lengths in the same unit. We will express 2 m in cm.)

$$2\,m = 2 \times 100\,cm = 200\,cm$$

Then 45 cm as a percentage of 2 m is

$$\frac{45}{200} \times 100\% = 22\tfrac{1}{2}\%$$

Express the first quantity as a percentage of the second:

> **Tip** Put the first quantity over the second then multiply by 100. Make sure that the units are the same.

14 20 cm, 50 cm	**17** 15 cm, 3 m	**20** 54 mm, 20 cm
15 $3\,cm^2$, $12\,cm^2$	**18** $210\,mm^2$, $84\,cm^2$	**21** 800 g, 2 kg
16 $1.53, $4.50	**19** $200\,mm^3$, $10\,cm^3$	**22** $4500\,cm^3$, 2 litres

Find 24% of 7.5 m

$$24\% \text{ of } 7.5\,m = \frac{24}{100} \times 7.5\,m$$

$$= 1.8\,m$$

Alternatively: 24% of 7.5 m = 0.24 × 7.5 m = 1.8 m

Find the value of:

23 30% of 250	**25** $12\tfrac{1}{2}\%$ of 4.88 cm	**27** 84% of 225 g
24 5% of $18.40	**26** 12% of 4.5 km	**28** $33\tfrac{1}{3}\%$ of $126\,m^2$

A pupil measures the length of a line as 8.2 cm. If its actual length is 8 cm find the percentage error.

Percentage error means the actual error as a percentage of the actual value.

$$\text{Error} = 0.2 \text{ cm}$$

$$\text{Percentage error} = \frac{\text{error}}{\text{actual value}} \times 100\%$$

$$= \frac{0.2}{8} \times 100\% = 2.5\%$$

Therefore the percentage error is 2.5%

Find the percentage error for each of the following values:

29 Measured length 12.3 m, actual length 12 m.

30 Measured area 147 cm^2, actual area 150 cm^2.

31 Measured volume 456.75 cm^3, actual volume 450 cm^3.

32 Measured weight 975 g, actual weight 1000 g.

33 Estimated cost $25.60, actual cost $25.

In Treetown 1680 out of a workforce of 12 000 are unemployed. What percentage is this?

This means we need to find 1680 as a percentage of 12 000.

$$\text{Percentage unemployed} = \frac{1680}{12\,000} \times 100\% = 14\%$$

34 Paul scored 27 out of a possible 45 in a science test. What percentage was this?

35 In a form of 30 pupils, 21 of them are boys. What percentage are girls?

36 A department store employs 250 female staff. If 26 of them have long hair, what percentage of the female staff do not have long hair?

If 72% of the 25 pupils in a class are good at mathematics how many are not good at it?

If 72% are good at mathematics

$(100 - 72)\% = 28\%$ are not good at it

$$28\% \text{ of class} = \frac{28}{100} \times 25 = 7$$

Therefore 7 pupils are not good at mathematics.

37 There are 1460 pupils in my school and 35% are non-swimmers. How many pupils can swim?

38 All the students at a college live either in hall or in lodgings. If 62% of the 2650 students live in lodgings, how many live in hall?

39 Sixty-two per cent of the audience of 1650 at a concert were females. How many males attended?

Increase 250 by 42%

The new value is $250 + 42\%$ of 250, i.e. 100% of $250 + 42\%$ of $250 = 142\%$ of 250.

i.e.   new value $= \frac{142}{100} \times 250 = 355$

Note that $\frac{142}{100}$ is called the multiplying factor.

40 Increase 300 by 27%	**44** Decrease 200 by 14%
41 Increase 44 by 12%	**45** Decrease 84 by 23%
42 Increase 240 by 45%	**46** Decrease 350 by 16%
43 Increase 23.4 by 35%	**47** Decrease 16.4 by 65%

48 John is 8% heavier now than he was two years ago. If he weighed 55 kg then, how much does he weigh now?

49 The price of a car that cost $18 000 last year was increased by 7.5% on 1st January this year. What is its present price?

50 A school employs 60 teachers. Next year they must reduce their teaching staff by 15%. How many teachers will there be next year?

Percentage increase and percentage decrease

Percentage increase or decrease arises in many different areas of life today. We read that certain workers are to receive an increase in their wages of 8%; sales tax may be increased from 15% to 18%; that the basic rate of income tax should be reduced from 30% to 27%; or that all the items in a sale are offered at a discount of 20%.

Changes are expressed in percentage terms, because this makes it easier to calculate the actual change in a particular case and to compare one change with another.

If a wage of $100 per week is increased by 8% then the new wage is

$$\frac{108}{100} \times \$100 = \$108$$

If an article costs $55 plus sales tax at 15%, then the full cost is

$$\frac{115}{100} \times \$55 = \$63.25$$

If a woman earns $550 and has to pay tax on it at the rate of 30% she actually receives

$$\frac{70}{100} \times \$550 = \$385$$

If a discount of 25% is offered in a sale, a piece of furniture, originally marked at $760, will cost

$$\frac{75}{100} \times \$760 = \$570$$

Profit and loss are normally given as a percentage of the *original* price. If a store makes a profit of 50% on an article it buys for $100 (called the cost price, CP), its profit is

$$\frac{50}{100} \times \$100 = \$50$$

and the selling price (SP) is

$$\frac{150}{100} \times \$100 = \$150$$

Exercise 5b   A second-hand car dealer bought a car for $3500 and sold it for $4340. Find his percentage profit.

$$\text{Profit} = \text{SP} - \text{CP}$$

$$= \$4340 - \$3500 = \$840$$

$$\% \text{ profit} = \frac{\text{profit}}{\text{CP}} \times 100\%$$

$$= \frac{\$840}{\$3500} \times 100\% = 24\%$$

Therefore the percentage profit is 24%

Find the percentage profit:

1   CP $12, profit $3   **3**   CP $16, profit $4

2   CP $28, profit $8.40   **4**   CP $55, profit $5.50

A retailer bought a leather chair for $375 and sold it for $285. Find his percentage loss.

$$\text{Loss} = \text{CP} - \text{SP}$$

$$= \$375 - \$285 = \$90$$

$$\% \text{ loss} = \frac{\text{loss}}{\text{CP}} \times 100\%$$

$$= \frac{\$90}{\$375} \times 100\% = 24\%$$

Therefore the percentage loss is 24%

Find the percentage loss:

5   CP $20, loss $4   **7**   CP $64, loss $9.60

6   CP $125, loss $25   **8**   CP $160, loss $38.40

An article costing $30 is sold at a profit of 25%. Find the selling price.

Method 1 Find the profit then add it to the cost.

$$\text{Profit} = 25\% \text{ of } \$30 = \frac{25}{100} \times \$30 = \$7.50$$

$$\text{SP} = \$30 + \$7.50$$

Therefore the selling price is $37.50

Method 2 First find the SP as a percentage of the CP

$$\text{SP} = 125\% \text{ of CP,}$$

$$\therefore \qquad \text{SP} = \frac{125}{100} \times \$30$$

Therefore the selling price is $37.50

Find the selling price:

9   CP $50, profit 12%   11   CP $29, profit 110%   13   CP $75, loss 64%

10   CP $64, profit $12\frac{1}{2}$%   12   CP $36, loss 50%   14   CP $128, loss $37\frac{1}{2}$%

> **Tip** Read these questions carefully so that you know exactly what you are being asked to find.

Find the weekly cash increase for each of the following employees:

15   Ian Dickenson earning $80 p.w. receives a rise of 10%

16   Nairn Williams earning $120 p.w. receives a rise of 12%

17   Sylvia Smith earning $150 p.w. receives a rise of 8%

18   Lyn Wyman earning $180 p.w. receives a rise of 9%

19   Joe Bright earning $200 p.w. receives a rise of 7%

Which is the better cash pay rise, and by how much?

20   a   12% on a weekly pay of $100, or b   8% on a weekly pay of $250

21   a   7% on a weekly pay of $90, or b $3\frac{1}{2}$% on a weekly pay of $200

22   a   4% on a weekly pay of $300, or b   3% on a weekly pay of $400

Find the purchase price of the given object, assuming that the rate of sales tax is 15%:

23   An electric iron marked $120 plus sales tax.

24   A food mixed marked $320 plus sales tax.

25   A calculator marked $84 plus sales tax.

26   A car tyre costing $152 plus sales tax.

27   A living room suite costing $2400 plus sales tax.

28   A ticket for a concert costing $8 plus sales tax.

29   Two chisels each marked $20.80 plus sales tax.

30   Three metres of material costing $5 per metre plus sales tax.

Income tax and sale reductions

Exercise **5c**

Assuming that the basic rate of income tax is 30%, find the yearly tax on the following taxable incomes:

1   $5000   3 $6500   5 $6450

2   $8000   4 $12 500   6 $8260

> **Tip** The tax due is 30% of the taxable income.

Find the yearly income tax due on a taxable income of:

7 $10 000 if the basic tax rate is 33%   **9** $16 000 if the basic tax rate is 25%

8 $8000 if the basic tax rate is 28%   **10** $24 000 if the basic tax rate is 32%

Use the following details to find the income tax due in each case:

> **Tip** Take the allowances away from the gross income to find the taxable income.

	Name	Gross Income	Allowances	Basic Tax Rate
11	Miss Deats	$24 000	$6000	30%
12	Mr Evans	$30 000	$9000	30%
13	Mrs Khan	$45 000	$9600	30%
14	Mr Amos	$27 000	$7800	33%
15	Miss Eyles	$60 000	$14 250	28%

During the January sales, a department store offers a discount of 10% off marked prices. What is the purchase price of

a a dinner service marked $338.00   **b** a pair of jeans marked $65.20?

Method 1 First find the discount then subtract this from the marked price.

a Discount of 10% on a marked price of $338.00 is

$$\frac{10}{100} \times \$338 = \$33.80$$

∴   Purchase price $= \$338 - \$33.80$

$$= \$304.20$$

b Discount of 10% on a marked price of $65.20 is

$$\frac{10}{100} \times \$65.20 = \$6.52$$

∴   Purchase price $= \$65.20 - \$6.52$

$$= \$58.68$$

Method 2 Find the purchase price as a percentage of the marked price.

a If the discount is 10%, the cash price is 90% of the marked selling price

i.e. purchase price of dinner service is

$$\frac{90}{100} \times \$338 = \$304.20$$

b Similarly, purchase price of jeans is

$$\frac{90}{100} \times \$65.20 = \$58.68$$

In a sale, a shop offers a discount of 20%. What would be the cash price for each of the following articles?

16 A dress marked $105

17 A lawn mower marked $460

18 A pair of shoes marked $64

19 A set of garden tools priced $290.00

20 Light fittings marked $168 each

In a sale, a department store offers a discount of 50% on the following articles. Find their sale price:

21 A pair of curtains marked $153

22 A leather football marked $129.20

23 A boy's jacket marked $85.80

24 A girl's dress marked $64.50

25 In order to clear a large quantity of goods a shopkeeper puts them on sale at a discount of $33\frac{1}{3}$%. Find the cash price of

    a a shirt marked $36.60

    b a skirt marked $66.60

Finding the original quantity

Sometimes we are given an increased or decreased quantity and we want to find the original quantity. For example, if the cost of a chair including sales tax at 15% is $172.50, we might wish to find the price of the chair before the tax was added.

Exercise **5d**

An article is sold for $252. If this gives a profit of 5% find the cost price.

If there is a profit of 5% and the cost price is $$x$

then SP = 105% of CP,

i.e.
$$SP = \frac{105}{100} \text{ of the CP}$$

so
$$252 = \frac{105}{100} \times x$$

$$100 \times 252 = \cancel{100} \times \frac{105}{\cancel{100}} \times x$$

∴
$$252 \times \frac{100}{105} = x$$

i.e.
$$x = 240$$

Therefore the cost price is $240

Find the cost price:

1 SP $98, profit 40%

2 SP $64, profit 60%

3 SP $28, profit 75%

4 SP $12, profit 100%

5 SP $4.50, profit $66\frac{2}{3}$%

6 SP $40, profit 25%

7 SP $920, profit 15%

8 SP $1008, profit 12%

9 SP $888, profit 11%

A book is sold for $12.60 at a loss of 30%. Find the cost price.

If there is a loss of 30% and the cost price is x cents

then SP $= 70\%$ of CP, i.e. SP $= \frac{70}{100}$ of the CP

so $\qquad\qquad\qquad 1260 = \frac{70}{100}x$

$\therefore \qquad\qquad 1260 \times \frac{100}{70} = x$

i.e. $\qquad\qquad\qquad x = 1800$

The cost price of the book is 1800 c, or $18

Find the cost price:

10   SP $30, loss 25%   **13**   SP $12, loss $33\frac{1}{3}$%   **16**   SP $120, loss 25%

11   SP $56, loss 30%   **14**   SP $8.16, loss 40%   **17**   SP $8.50, loss 50%

12   SP $70, loss 65%   **15**   SP $45, loss 10%   **18**   SP $64, loss 60%

After a pay rise of 5%, Peter's weekly pay is $273. How much did he earn before the rise?

If Peter's original pay was x and he receives a 5% pay rise his new pay will be 105% of his original pay.

But his new pay is $273,

$\therefore \qquad\qquad\qquad\qquad 273 = \frac{105}{100} \times x$

i.e. $\qquad\qquad\qquad\qquad x = \frac{273 \times 100}{105} = 260$

Peter's original weekly pay was $260

The following table shows the weekly wage of a number of employees after percentage increases as shown. Find the original weekly wage of each employee:

	Name	% increase in pay	Weekly wage after increase
19	George Black	10%	$264
20	Anne Reed	8%	$270
21	John Rowlands	15%	$598
22	Beryl Lewis	7%	$196.88

The purchase price of a watch is $276. If this includes sales tax at 15%, find the price before sales tax was added.

If the cost of an article is C,

the cost including sales tax at 15% will be 115% of $C = \frac{115}{100} \times \$C$

i.e. $\qquad\qquad\qquad \frac{115}{100} \times C = 276$

i.e. $\qquad\qquad\qquad C = 276 \times \frac{100}{115} = 240$

The price before sales tax was added was $240

23 The purchase price of a hairdryer is $82.80. If this includes sales tax at 15%, find the price before sales tax was added.

24 I paid $2242.50 for a dining table and four chairs. If the price includes sales tax at 15%, find the price before sales tax was added.

25 John's income last week was $224 after income tax had been deducted at 30%. Calculate his pay before the tax was deducted.

26 The stretched length of an elastic string is 31 cm. If this is 24% more than its unstretched length, find its unstretched length.

Problems involving percentage increase and decrease

Exercise **5e**

Modifications were made to the engine of a racing car which increased the car's top speed by 4% to 260 km/h. What was the top speed of the car before the modifications?

We need to find the top speed before the modifications: let this be x km/h

After the modifications, the top speed is 260 km/h which is 104% of the top speed before the modifications.

So 260 km/h = 104% of x km/h,

i.e.
$$260 = \tfrac{104}{100} \times x$$

∴
$$260 \times \tfrac{100}{104} = x$$

$$x = 250$$

The top speed before modifications was 250 km/h.

Tip Remember that a percentage increase or decrease is always calculated as a percentage of the original quantity, that is the quantity *before* any change takes place.

1 A house is bought for $112 000 and sold at a profit of 14%. Find the selling price.

2 Carpets that had been bought for $55.50 per square metre were sold at a loss of 26%. Find the selling price per square metre.

3 Potatoes bought at $45 per 50 kg are sold at $1.20 per kg. Find the percentage profit.

4 A shopkeeper buys 80 articles for $600 and sells them for $10.50 each. Find his percentage profit.

5 An art dealer sold a picture for $1980 thus making a profit of 65%. What did she pay for it?

6 My present average weekly grocery bill is $121.50, which is 8% more than the same goods cost me, on average, each week last year. What was my average weekly grocery bill last year?

7 When water freezes, its volume increases by 4%. What volume of water is required to make $221\,\text{cm}^3$ of ice?

8 What is Fred Proctor's monthly wage, if, after paying deductions of 35%, he is left with $2236?

9 Engine modifications were made to a particular model of car. As a result the number of kilometres it travels on one litre of petrol increases by 8%. If the new petrol consumption is 16.2 km/l what was the previous value?

10 Between two elections, the size of the electorate in a constituency fell by 16%. For the second election, 37 191 people were entitled to vote. How many could have voted at the first election?

11 When the rate of sales tax is 15% an LP record costs $13.80. What will it cost if the rate of sales tax is

    a increased to 20%   **b** decreased to 10%?

12 The purchase price of a diamond ring increases by $60 when the rate of sales tax is increased from 12% to 20%. Find the original purchase price of the ring.

 Puzzle

> Pauline manages an Art Gallery. She buys a picture for £2000 and sells it at a profit of 50%. Six months later she buys it back for 50% less than she sold it for, but soon sells it again at a profit of 50%. This process of buying back at a loss of 50% of her selling price, and selling the picture again soon after, at a profit of 50% is repeated on a further two occasions. How much profit (or loss) did she make altogether?

Simple interest

If you borrow money, you will normally have to pay for it, i.e. you must repay more than you borrow.

The difference between what you borrow and what you repay is called the *interest*. The sum of money borrowed (or lent) is called the *principal*. If the interest is an agreed percentage of this sum it is called *simple interest*.

In Book 1, we developed the formula

$$I = \frac{PRT}{100}$$

where I is the simple interest, P the principal, R the rate per cent each year, and T the time in years for which the principal is borrowed or invested.

Hence $\qquad 100I = PRT$

This formula enables us to find any one of four quantities, if the other three are known.

The sum of the principal (P) and the interest (I) is called the amount. This is denoted by A, i.e. $A = P + I$

Exercise 5f

Find the simple interest on:

 1 $200 invested for 2 years at 8% p.a.

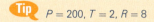

Tip $P = 200, T = 2, R = 8$

2 $160 invested for 5 years at 14% p.a.

3 $730 invested for 3 years at 13% p.a.

4 $337 invested for 7 years at 8% p.a.

5 $650 invested for 4 years at $6\frac{1}{4}$ p.a.

6 $420 invested for 3 years at 10% p.a.

7 $280 invested for 6 years at 12% p.a.

8 $834 invested for 5 years at 9% p.a.

9 $500 invested for 8 years at $12\frac{1}{2}$% p.a.

Tip Read the question carefully. Substitute the values you know into the formula then 'solve' it to find the unknown value.

10 $726 invested for 3 years at $7\frac{3}{4}$% p.a

11 What sum of money invested for 5 years at 12% p.a. gives $264 simple interest?

12 How long must $3700 be invested at 9% p.a. simple interest to give interest of $2331?

13 What annual rate of simple interest is necessary to give interest of $416 on a principal of $800 invested for 8 years?

14 What sum of money earns $312 simple interest if invested for 8 years at 13%?

15 Find the annual rate per cent that earns $234 simple interest when $900 is invested for $6\frac{1}{2}$ years.

16 Find the amount if:

Tip Remember that the amount = principal + interest.

a $120 is invested for 3 years at 14% p.a. simple interest.

b $280 is invested for 5 years at 8% p.a. simple interest.

c $530 is invested for 6 years at 9% p.a. simple interest.

17 What length of time is required if:

Tip First find the interest, then substitute the known quantities into the formula.

a $500 amounts to $635 when invested at 9% p.a. simple interest?

b $380 amounts to $570 when invested at 10% p.a. simple interest?

c $144 amounts to $192.96 when invested at $8\frac{1}{2}$% p.a. simple interest?

Compound percentage problems

There are many occasions when a percentage increase or decrease happens more than once. Suppose that a plot of land is bought for $20 000 and increases in value (appreciates) by 10% of its value each year.

After one year, its value will be 110% of its initial value, i.e.

$$\frac{110}{100} \times \$20\,000 = \$22\,000$$

The next year it will increase by 10% of the $22 000 it was worth at the beginning of the year, i.e. its value after two years will be

$$\frac{110}{100} \times \$22\,000 = \$24\,200$$

While some things increase in value year after year, many things decrease in value (depreciate) each year. If you buy a car or a motorcycle it will probably depreciate in value more quickly than anything else you buy.

If you invest money in a Savings Account at a fixed rate of interest and do not spend the interest, your money will increase by larger amounts each year.

This kind of interest is called *compound interest*.

Exercise 5g

In this exercise give all answers that are not exact correct to the nearest cent.

Find the compound interest on $2600 invested for 2 years at 8% p.a.

We need to find the interest for the first year, then add this to $2600 to find the principal for the second year.

$$\text{Simple interest for first year at 8\%} = \frac{8}{100} \times \$2600$$
$$= \$\frac{20\,800}{100}$$
$$= \$208$$

$$\text{New principal} = \text{original principal} + \text{interest}$$
$$= \$2600 + \$208 = \$2808$$

Now we can find the interest for the second year.

Simple interest for second year at 8%

$$= \frac{8}{100} \times \$2808 = \$224.64$$

The compound interest is the sum of the interests for the two years.

∴ total of interest for the two years is

$$\$208 + \$224.64 = \$432.64$$

i.e. compound interest on $2600 for 2 years at 8% is $432.64

Find the compound interest on:

1 $200 for 2 years at 10% p.a.

2 $300 for 2 years at 12% p.a.

3 $400 for 3 years at 8% p.a.

4 $650 for 3 years at 9% p.a.

5 $520 for 2 years at 13% p.a.

6 $690 for 2 years at 14% p.a.

7 $624 for 3 years at 12% p.a.

Tip Read these questions carefully to make sure that you know what you are being asked to find.

8 A house is bought for $80 000 and appreciates at 10% each year. What will it be worth in 2 years' time?

9 A postage stamp increases in value by 15% each year. If it is bought for $50, what will it be worth in 3 years' time?

10 A motorcycle bought for $6000 depreciates in value by 10% each year. Find its value after 3 years.

11 A motor car bought for $20 000 depreciates in any one year by 20% of its value at the beginning of that year. Find its value after 2 years.

Puzzle

This puzzle was posed by Fibonacci (1170–1250):

A man saves one denarius at an interest rate such that after five years he has two denarii, and every five years after his money doubles.

How many denarii will the man gain from his one denarius in 100 years?

Numerologists believe that events in our lives can be explained by using numbers.

Ask your history teacher to tell you about the following people: Churchill, Hitler, Mussolini, Roosevelt and Stalin. Some information on them is shown below:

Name	Year born	Age in 1944	Year he took office	No. of years in office	Total
Churchill	1874	70	1940	4	3888
Hitler	1889	55	1933	11	3888
Mussolini	1883	61	1921	23	3888
Roosevelt	1882	62	1933	11	3888
Stalin	1879	65	1924	20	3888

1944 was considered to be the year of the turning point of World War II.

$$1944 \times 2 = 3888\,!!$$

IN THIS CHAPTER...

you have seen that:

- a percentage increase or a percentage decrease is calculated on the quantity before any change takes place

- taxable income is gross income less any allowances; tax is charged on taxable income

- to find a quantity before a percentage change, let the original quantity be x then form an equation

- simple interest can be found from the formula $I = \dfrac{PRT}{100}$

- compound interest occurs when the simple interest for each period is added to the principal to form a new principal for the next period.

6 SIMULTANEOUS EQUATIONS

AT THE END OF THIS CHAPTER...

you should be able to:

1 Solve a pair of simultaneous equations by

   a the method of elimination

   b a graphical method

   c the substitution method.

2 Solve problems using simultaneous equations.

3 Identify a pair of equations which have

   a no solution

   b an infinite number of solutions.

MATHS IS OUT THERE

Did you know that equations containing several unknown quantities were known to the ancient Egyptians, Greeks and Indians? The Hindus used the names of colours to distinguish the unknown quantities.

BEFORE YOU START

you need to know:

✓ how to solve a linear equation in one unknown

✓ the rules for working with negative numbers

✓ the angle sum of a triangle

✓ the meaning of perimeter

✓ how to draw the graph of a straight line from its equation

KEY WORDS

elimination, equation, infinite, perimeter, proportion, ratio, simultaneous, substitution, triangle

Equations with two unknown quantities

Up to now the equations we have solved have had only one unknown quantity, but there can be more.

Looking at the equation $2x + y = 8$, we can see that there are many possible values which will fit, for instance $x = 2$ and $y = 4$, or $x = 1$ and $y = 6$. We could also have $x = -1$ and $y = 10$ or even $x = 1.681$ and $y = 4.638$. Indeed, there is an infinite set of pairs of solutions.

If however we are *also* told that $x + y = 5$, then we shall find that not every pair of numbers which satisfies the first equation also satisfies the second. While $x = 2$, $y = 4$ satisfies the first equation, it does not satisfy $x + y = 5$. On the other hand $x = 3$, $y = 2$ satisfies both equations.

These two equations together form a pair of *simultaneous* equations. 'Simultaneous' means that the two equations are both satisfied by the same values of x and y; that is, when x has the same value in both equations then y has the same value in both.

There are several different methods for solving simultaneous equations. We start with the simplest.

Elimination method

Whenever we meet a new type of equation, we try to reorganise it so that it is similar to equations we have already met.

Previous equations have had only one unknown quantity, so we try to *eliminate* one of the two unknowns.

Consider the pair of equations

$$2x + y = 8 \qquad [1]$$

$$x + y = 5 \qquad [2]$$

In this case, if we try subtracting the second equation from the first we find that the y term disappears but the x term does not:

i.e. $\qquad$ [1] − [2] gives $\qquad x = 3$

Then, substituting 3 for x in equation [2], we see that $3 + y = 5$ so $y = 2$

We can check that $x = 3$ and $y = 2$ also satisfy equation [1].

Notice that it is essential to number the equations and to say that you are subtracting them.

Sometimes it is easier to subtract the first equation from the second rather than the second equation from the first. (In this case we would write equation [1] again, underneath equation [2].)

Sometimes we can eliminate x rather than y.

Exercise 6a

Solve the equations

$$x + y = 5$$
$$3x + y = 7$$

$$x + y = 5 \qquad \text{[1]}$$
$$3x + y = 7 \qquad \text{[2]}$$
$$x + y = 5 \qquad \text{[1]}$$

[2] − [1] gives

$$2x = 2$$
$$x = 1$$

(To find y choose the simpler equation, i.e. the first.)

Substituting 1 for x in [1] gives

$$1 + y = 5$$
$$y = 4$$

(Check in the equation *not* used for finding y.)

Check in [2]   left-hand side $= 3 + 4$

$$= 7 = \text{right-hand side}$$

Therefore the solution is $x = 1$, $y = 4$

Solve the following pairs of equations:

1 $x + y = 5$
     $4x + y = 14$

2 $5x + y = 14$
     $3x + y = 10$

3 $2a + b = 11$
     $4a + b = 17$

4 $2x + 3y = 23$
     $x + 3y = 22$

5 $5x + 2y = 14$
     $7x + 2y = 22$

6 $x + 2y = 12$
     $x + y = 7$

7 $4p + 3q = -5$
     $7p + 3q = -11$

8 $12x + 5y = 65$
     $9x + 5y = 50$

9 $3x + 4y = 15$
     $3x + 2y = 12$

10 $9c + 2d = 54$
     $c + 2d = 6$

11 $2x + 3y = -8$
     $2x + y = -4$

12 $9x + 5y = 45$
     $4x + 5y = 45$

Not all pairs of simultaneous equations can be solved by subtracting one from the other.

Consider $4x + y = 6$ [1]

             $2x - y = 0$ [2]

If we subtract we get $2x + 2y = 6$ which is no improvement.

On the other hand, if we add we get $6x = 6$ which eliminates y.

If the signs in front of the letter to be eliminated are the same we should *subtract*; if the signs are different we should *add*.

Solve the equations

$$x - 2y = 1$$
$$3x + 2y = 19$$

$$x - 2y = 1 \qquad [1]$$
$$3x + 2y = 19 \qquad [2]$$

$[1] + [2]$ gives

$$4x = 20$$
$$x = 5$$

(It is easier to use the equation with the $+$ sign to find y.)

Substitute 5 for x in [2] \qquad $15 + 2y = 19$

Take 15 from both sides \qquad $2y = 4$

$$y = 2$$

Check in [1] \qquad left-hand side $= 5 - 4 = 1 =$ right-hand side

Therefore the solution is $x = 5$, $y = 2$

Solve the following pairs of equations:

1 $\quad x - y = 2$
$\quad 3x + y = 10$

2 $\quad 2x - y = 6$
$\quad 3x + y = 14$

3 $\quad p + 2q = 11$
$\quad 3p - 2q = 1$

4 $\quad 3a - b = 10$
$\quad a + b = 2$

5 $\quad 6x + 2y = 19$
$\quad x - 2y = 2$

6 $\quad 4x + y = 37$
$\quad 2x - y = 17$

7 $\quad x + y = 2$
$\quad 2x - y = 10$

8 $\quad 5p + 3q = 5$
$\quad 4p - 3q = 4$

9 $\quad 3x - 4y = -24$
$\quad 5x + 4y = 24$

To solve the following equations, first decide whether to add or subtract:

10 $\quad 3x + 2y = 12$
$\quad x + 2y = 8$

11 $\quad x - 2y = 6$
$\quad 4x + 2y = 14$

12 $\quad x + 3y = 12$
$\quad x + y = 8$

13 $\quad 9x + 2y = 48$
$\quad x - 2y = 2$

14 $\quad 4x + y = 19$
$\quad 3x + y = 15$

15 $\quad 2x + 3y = 13$
$\quad 2x + 5y = 21$

16 $\quad 5x - 2y = 24$
$\quad x + 2y = 0$

17 $\quad x + 3y = 0$
$\quad x - y = -4$

18 $\quad 5p - 3q = 9$
$\quad 4p + 3q = 9$

Solve the equations

$$4x - y = 10$$
$$x - y = 1$$

$$4x - y = 10 \qquad [1]$$
$$x - y = 1 \qquad [2]$$

(The signs in front of the y terms are the same so we subtract: remember that $-y - (-y) = -y + y = 0$)

$[1] - [2]$ gives

$$3x = 9$$
$$x = 3$$

Substitute 3 for x in [2] $3 - y = 1$

Add y to both sides $3 = 1 + y$

Take 1 from both sides $2 = y$

Check in [1]   left-hand side $= 12 - 2 = 10 =$ right-hand side

Therefore the solution is $x = 3$, $y = 2$

Solve the following pairs of equations:

19 $2x - y = 4$
   $x - y = 1$

20 $2p - 3q = -7$
   $4p - 3q = 1$

21 $x - y = 3$
   $3x - y = 9$

22 $6x - y = 7$
   $2x - y = 1$

23 $5x - 2y = -19$
   $x - 2y = -7$

24 $2x - 3y = 14$
   $2x - y = 10$

25 $3x - 2y = 14$
   $x + 2y = 10$

26 $3p - 5q = -3$
   $4p - 5q = 1$

27 $3p + 5q = 17$
   $4p + 5q = 16$

28 $3p - 5q = 7$
   $4p + 5q = -14$

29 $3p + 5q = 35$
   $4p - 5q = 0$

30 $3x - y = 10$
   $x + y = -2$

Harder elimination

Equations are not always as simple as the ones we have had so far.

Consider $2x + 3y = 4$ [1]

$4x + y = -2$ [2]

Whether we add or subtract neither letter will disappear, so it is necessary to do something else first. If we multiply the second equation by 3 to give $12x + 3y = -6$, we have the same number of ys in each equation. Then we can use the same method as before:

[2] × 3 $12x + 3y = -6$ [3]

$2x + 3y = 4$ [1]

[3] − [1] gives $10x = -10$

$x = -1$

Substitute -1 for x in [2] $-4 + y = -2$

Add 4 to both sides $y = 2$

Therefore the solution is $x = -1$, $y = 2$

Exercise 6c

Solve the equations $3x - 2y = 1$
$4x + y = 5$

$3x - 2y = 1$ [1]

$4x + y = 5$ [2]

[2] × 2 gives $8x + 2y = 10$ [3]

$3x - 2y = 1$ [1]

$[1] + [3]$ gives $\qquad\qquad 11x = 11$

$\qquad\qquad\qquad\qquad\qquad\qquad\qquad x = 1$

Substitute 1 for x in [2] $\qquad 4 + y = 5$

Take 4 from both sides $\qquad\qquad y = 1$

Check in [1] $\qquad\qquad$ left-hand side $= 3 - 2$

$\qquad\qquad\qquad\qquad\qquad\qquad\qquad = 1 =$ right-hand side

Therefore the solution is $x = 1$, $y = 1$

Solve the following pairs of equations:

1 $\quad 2x + y = 7$
$\qquad 3x + 2y = 11$

2 $\quad 5x - 4y = -3$
$\qquad 3x + y = 5$

3 $\quad 9x + 7y = 10$
$\qquad 3x + y = 2$

4 $\quad 5x + 3y = 21$
$\qquad 2x + y = 3$

5 $\quad 6x - 4y = -4$
$\qquad 5x + 2y = 2$

6 $\quad 4x + 3y = 25$
$\qquad x + 5y = 19$

Solve the following pairs of equations:

7 $\quad 5x + 3y = 11$
$\qquad 4x + 6y = 16$

8 $\quad 2x - 3y = 1$
$\qquad 5x + 9y = 19$

9 $\quad 2x + 5y = 1$
$\qquad 4x + 3y = 9$

10 $\quad 9x + 5y = 15$
$\qquad\ \ 3x - 2y = -6$

11 $\quad 4x + 3y = 1$
$\qquad\ \ 16x - 5y = 21$

12 $\quad 7p + 2q = 22$
$\qquad\ \ 3p + 4q = 11$

> **Tip** Multiplying the top equation by 2 will give $6y$ in both equations.

Exercise **6d**

Solve the equations $\quad\begin{aligned}3x + 5y = 6 \\ 2x + 3y = 5\end{aligned}$

Method 1 $\qquad\qquad\qquad\qquad\qquad 3x + 5y = 6 \qquad\qquad [1]$

$\qquad\qquad\qquad\qquad\qquad\qquad\qquad\ 2x + 3y = 5 \qquad\qquad [2]$

We can choose to either get the same number of xs ([1] $\times$ 2 and [2] $\times$ 3 does this) or the same number of ys, which is what we will do.

$[1] \times 3$ gives $\qquad\qquad\qquad\qquad 9x + 15y = 18 \qquad\ [3]$

$[2] \times 5$ gives $\qquad\qquad\qquad\qquad 10x + 15y = 25 \qquad [4]$

$\qquad\qquad\qquad\qquad\qquad\qquad\qquad 9x + 15y = 18 \qquad\ [3]$

$[4] - [3]$ gives $\qquad\qquad\qquad\qquad\qquad\qquad x = 7$

Substitute 7 for x in [2] $\qquad\qquad 14 + 3y = 5$

Take 14 from both sides $\qquad\qquad\qquad 3y = -9$

Divide both sides by 3 $\qquad\qquad\qquad\ y = -3$

Check in [1] $\qquad\qquad$ left-hand side $= 21 - 15$

$\qquad\qquad\qquad\qquad\qquad\qquad\qquad = 6 =$ right-hand side

Therefore the solution is $x = 7$, $y = -3$

Method 2

Another method of solving linear equations is by *ratio and proportion*:

$$3x + 5y = 6 \qquad [1]$$

$$2x + 3y = 5 \qquad [2]$$

$$\frac{3x + 5y}{2x + 3y} = \frac{6}{5}$$

$$\therefore \qquad 5(3x + 5y) = 6(2x + 3y)$$

$$\therefore \qquad 15x + 25y = 12x + 18y$$

$$\therefore \qquad 3x = -7y$$

$$\therefore \qquad \frac{x}{y} = -\frac{7}{3}$$

Therefore $x = -7K$ and $y = 3K$, where K is a constant

Substituting these values in [1], we have

$$3(-7K) + 5(3K) = 6$$

$$\therefore \qquad -21K + 15K = 6$$

$$-6K = 6$$

$$\therefore \qquad K = -1$$

$$\therefore \qquad x = -7(-1) = 7$$

$$y = 3(-1) = -3$$

Therefore the solution is $x = 7$, $y = -3$

Solve the following pairs of equations:

1 $2x + 3y = 12$	**8** $9x + 8y = 17$	**15** $6x + 5y = 8$	**22** $3x + 8y = 56$
$5x + 4y = 23$	$2x - 6y = -4$	$3x + 4y = 1$	$5x - 6y = 16$
2 $3x - 2y = -7$	**9** $9x - 2y = 14$	**16** $7x - 3y = 20$	**23** $7x + 3y = -9$
$4x + 3y = 19$	$7x + 3y = 20$	$2x + 4y = -4$	$2x + 5y = 14$
3 $2x - 5y = 1$	**10** $5x + 4y = 11$	**17** $10x + 3y = 12$	**24** $7x + 6y = 0$
$5x + 3y = 18$	$2x + 3y = 3$	$3x + 5y = 20$	$5x - 8y = 43$
4 $6x + 5y = 9$	**11** $4x + 5y = 26$	**18** $6x - 5y = 4$	**25** $2x + 6y = 30$
$4x + 3y = 6$	$5x + 4y = 28$	$4x + 2y = -8$	$3x + 10y = 49$
5 $14x - 3y = -18$	**12** $2x - 6y = -6$	**19** $5x + 3y = 8$	**26** $4x - 3y = -7$
$6x + 2y = 12$	$5x + 4y = -15$	$3x + 5y = 8$	$3x + 2y = 16$
6 $6x - 7y = 25$	**13** $5x - 6y = 6$	**20** $7x + 2y = 23$	**27** $17x - 2y = 47$
$7x + 6y = 15$	$2x + 9y = 10$	$3x - 5y = 4$	$5x - 3y = 9$
7 $5x + 4y = 21$	**14** $3p + 4q = 5$	**21** $6x - 5y = 17$	**28** $8x + 3y = -17$
$3x + 6y = 27$	$2p + 10q = 18$	$5x + 4y = 6$	$7x - 4y = 5$

Mixed questions

Exercise **6e**

Solve the following pairs of equations:

1 $x + 2y = 9$
    $2x - y = -2$

2 $x + y = 4$
    $x + 2y = 9$

3 $2x + 3y = 0$
    $3x + 2y = 5$

4 $3x - y = -10$
    $4x - y = -4$

5 $5x + 2y = 16$
    $2x - 3y = -5$

6 $3x + 2y = -5$
    $3x - 4y = 1$

7 $x + y = 6$
    $x - y = 1$

8 $3x - 5y = 13$
    $2x + 5y = -8$

9 $7x + 3y = 35$
    $2x - 5y = 10$

10 $9x + 2y = 8$
    $7x + 3y = 12$

11 $2x - 5y = 1$
    $3x + 4y = 13$

12 $3x - 2y = -2$
    $5x - y = -15$

Sometimes the equations are arranged in an awkward fashion and need to be rearranged before solving them.

Exercise **6f**

Solve the equations $x = 4 - 3y$
                $2y - x = 1$

$$x = 4 - 3y \qquad [1]$$

$$2y - x = 1 \qquad [2]$$

(We must first arrange the equations so that the letters are in the same corresponding positions in both equations.

By adding $3y$ to both sides, equation [1] can be written $3y + x = 4$)

$$3y + x = 4 \qquad [3]$$

$$2y - x = 1 \qquad [2]$$

[3] + [2] gives $5y = 5$

$$y = 1$$

Substitute 1 for y in [1] $x = 4 - 3$

$$x = 1$$

Check in [2]   left-hand side $= 2 - 1 = 1 =$ right-hand side

Therefore the solution is $x = 1$, $y = 1$

Solve the following pairs of equations:

1 $y = 6 - x$
    $2x + y = 8$

2 $x - y = 2$
    $2y = x + 1$

3 $3 = 2x + y$
    $4x + 6 = 10y$

4 $9 + x = y$
    $x + 2y = 12$

5 $2y = 16 - x$
    $x - 2y = -8$

6 $3x + 4y = 7$
    $2x = 5 - 3y$

As long as the x and y and number terms are in corresponding positions in the two equations, they need not be in the order we have had so far.

Solve the equations $\begin{aligned} y &= x + 5 \\ y &= 7 - x \end{aligned}$

$$y = x + 5 \qquad\qquad [1]$$
$$y = 7 - x \qquad\qquad [2]$$

Rewrite [1] as $\qquad\qquad y = 5 + x \qquad\qquad [3]$

[2] + [3] gives $\qquad\qquad 2y = 12$

$$y = 6$$

Substitute 6 for y in [1] $\qquad 6 = x + 5$

$$x = 1$$

Check in [2] $\qquad$ left-hand side $= 6$

$\qquad\qquad$ right-hand side $= 7 - 1 = 6$

Therefore the solution is $x = 1$, $y = 6$

Solve the following pairs of equations:

7 $\quad y = 9 + x$
$\quad\;\; y = 11 - x$

8 $\quad x = 3 + y$
$\quad\;\; 2x = 4 - y$

9 $\quad y = 4 - x$
$\quad\;\; y = x + 6$

10 $\quad 2y = 4 + x$
$\quad\;\;\; y = x + 8$

11 $\quad x + 4 = y$
$\quad\;\;\; y = 10 - 2x$

12 $\quad x + y = 12$
$\quad\;\;\; y = 3 + x$

Special cases

Some pairs of equations have no solution and some have an infinite number of solutions.

Exercise **6g**

Try solving the following pairs of equations. Comment on why the method breaks down:

1 $\quad x + 2y = 6$
$\quad\;\; x + 2y = 7$

2 $\quad 3x + 4y = 1$
$\quad\;\; 6x + 8y = 2$

3 $\quad y = 4 + 2x$
$\quad\;\; y - 2x = 6$

4 $\quad 9x = 3 - 6y$
$\quad\;\; 3x + 2y = 1$

5 $\quad$ Make up other pairs of equations which either have no solution or have an infinite set of solutions.

Substitution method

This is a method for solving simultaneous equations that avoids adding or subtracting equations. We start with one of the equations and rearrange it to make one unknown the subject of that equation.

For example, for the equations $\qquad 2x - y = 9 \qquad$ [1]

and $\qquad\qquad\qquad\qquad\qquad 3y + x = 1 \qquad$ [2]

we can choose [2] to 'solve' for x: $\qquad x = 1 - 3y \quad$ (subtracting $3y$ from both sides)

The next step is to substitute $1 - 3y$ for x in equation [1].

This gives $\qquad\qquad\qquad\qquad 2(1 - 3y) - y = 9$

Solving this equation gives $\qquad 2 - 6y - y = 9 \quad$ (expanding the bracket)

$$2 - 7y = 9 \quad \text{(collecting like terms)}$$

$$2 = 9 + 7y \quad \text{(adding } 7y \text{ to both sides)}$$

$$-7 = 7y \quad \text{(taking 9 from both sides)}$$

$$-1 = y \text{ or } y = -1$$

Then using $x = 1 - 3y$ and substituting -1 for y gives $x = 1 - 3(-1)$

$$= 1 + 3 = 4$$

So the solution is $x = 4$ and $y = -1$

Exercise 6h

Use the substitution method to solve the following pairs of equations.

  1 $\quad 2x + y = 4$
$\qquad 5x - 2y = 1$

2 $\quad 3a - 2b = 1$
$\qquad 2a = 5 - b$

3 $\quad 3x - y = 8$
$\qquad y = 2 - 2x$

4 $\quad 2x + 3y = 4$
$\qquad 4x = 1 - y$

  5 $\quad 3s - 2t = 1$
$\qquad 3t = 4s$

6 $\quad 2a = b - 3$
$\qquad 2a - 5b = 1$

> **Tip** Start by choosing the equation with a single letter and 'solve' it for that letter, i.e. make that letter the subject of the 'formula'.

> **Tip** Use the second equation to find t in terms of s. Be careful with the fractions.

Use the substitution method to solve questions **7** to **12** of Exercise 6f.

Puzzle

There are 12 identical coins. One of them is a forgery and its weight is different from the others. It is not known if the forged coin is heavier or lighter than the genuine coins. How can you find the forged coin by three weighings on a simple balance?

Problems

Use either elimination or substitution to solve the equations.

I think of two numbers. If I add three times the smaller number to the bigger number I get 14. If I subtract the bigger number from twice the smaller number I get 1. Find the two numbers.

First allocate letters to the unknown numbers.

Let the smaller number be x and the bigger number be y.

Now interpret the information in terms of these letters.

Second sentence $\Rightarrow$	$3x + y = 14$	[1]
Third sentence $\Rightarrow$	$2x - y = 1$	[2]
[1] + [2] gives	$5x = 15$	
	$x = 3$	
Substitute 3 for x in [1]	$9 + y = 14$	
Take 9 from both sides	$y = 5$	

Therefore, the two numbers are 3 and 5

(Check by reading the original statements to see if the numbers fit.)

Solve the following problems by forming a pair of simultaneous equations:

1 The sum of two numbers is 20 and their difference is 4. Find the numbers.

2 The sum of two numbers is 16 and they differ by 6. What are the numbers?

3 I think of two numbers. If I double the first and add the second I get 18. If I double the first and subtract the second I get 14. What are the numbers?

4 Three times a number added to a second number is 33. The first number added to three times the second number is 19. Find the two numbers.

5 Find two numbers such that twice the first added to the second is 26 and the first added to three times the second is 28.

6 Find the two numbers such that twice the first added to the second gives 27 and twice the second added to the first gives 21.

A shop sells bread rolls. If five brown rolls and six white rolls cost 98 c while three brown rolls and four white rolls cost 62 c, find the cost of each type of roll.

First allocate letters to the unknown quantities.

Let one brown roll cost x c and one white roll cost y c.

Now interpret the information in terms of x and y.

$$5x + 6y = 98 \qquad [1]$$

$$3x + 4y = 62 \qquad [2]$$

[1] × 2 gives	$10x + 12y = 196$	[3]
[2] × 3 gives	$9x + 12y = 186$	[4]
[3] − [4] gives	$x = 10$	
Substitute 10 for x in [1]	$50 + 6y = 98$	
Take 50 from both sides	$6y = 48$	
	$y = 8$	

Therefore one brown roll costs 10 c and one white roll costs 8 c.

7 I buy x choc ices and y orange ices and spend \$2.30. I buy ten ices altogether. The choc ices cost 30 c each and the orange ices cost 20 c each. How many of each do I buy?

8 x is bigger than y.
The difference between x and y is 18.
Find x and y.

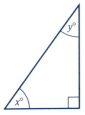

Tip There is information in the diagram that you can use.

9 A cup and saucer cost \$3.15 together. A cup and two saucers cost \$4.50. Find the cost of a cup and of a saucer.

10 The cost of two Roti is the same as the cost of three Petties. One Roti and one Pettie together cost \$3.50. What do they each cost?

11 In a test, the sum of Harry's marks and Adam's marks is 42. Sam has twice as many marks as Adam, and the sum of Harry's and Sam's marks is 52. What are the marks of each of the three boys?

12 The perimeter of triangle ABC is 14 cm. AB is 2 cm longer than AC. Find x and y.

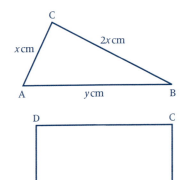

13 The perimeter of the rectangle is 31 cm. The difference between the lengths of AB and BC is $3\frac{1}{2}$ cm. Find the lengths of AB and BC.

14 The equation of a straight line is $y = mx + c$. When $x = 1$, $y = 6$ and when $x = 3$, $y = 10$. Form two equations for m and c and hence find the equation of the line.

 **Puzzle**

Andy wants to invest some money now that will give him enough to pay a deposit on a house in five years' time. He sees an advertisement offering a bond that states that after 5 years it would be worth an amount equivalent to earning simple interest paid at 5% per annum.

How much does he need to invest so that the bond will be worth $60 000 when he withdraws it?

Graphical solutions of simultaneous equations

We saw in Book 2 that when we are given an equation we can draw a graph. Any of the equations which occur in this chapter give us a straight line. Two equations give us two straight lines which usually cross one another.

Consider the two equations $x + y = 4$, $y = 1 + x$

Suppose we know that the x coordinate of the point of intersection is in the range $0 \leqslant x \leqslant 5$: this means that we can plot these lines in that range.

$x + y = 4$

x	0	4	5
y	4	0	−1

$y = 1 + x$

x	0	2	5
y	1	3	6

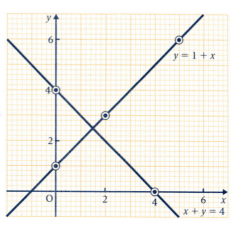

At the point where the two lines cross, the values of x and y are the same for both equations, so they are the solutions of the pair of equations.

From the graph we see that the solution is $x = 1\frac{1}{2}$, $y = 2\frac{1}{2}$.

Solve the following equations graphically. In each case draw axes for x and y and use values in the ranges indicated, taking 2 cm to 1 unit:

1 $x + y = 6$ $0 \leqslant x \leqslant 6$, $0 \leqslant y \leqslant 6$
$y = 3 + x$

2 $x + y = 5$ $0 \leqslant x \leqslant 6$, $0 \leqslant y \leqslant 6$
$y = 2x + 1$

3 $y = 4 + x$ $0 \leqslant x \leqslant 6$, $0 \leqslant y \leqslant 6$
$y = 1 + 3x$

4 $x + y = 1$ $-3 \leqslant x \leqslant 2$, $-2 \leqslant y \leqslant 4$
$y = x + 2$

5 $2x + y = 3$ $0 \leqslant x \leqslant 3$, $-3 \leqslant y \leqslant 3$
$x + y = 2\frac{1}{2}$

6 $y = 5 - x$ $0 \leqslant x \leqslant 5$, $0 \leqslant y \leqslant 7$
$y = 2 + x$

77

7 $3x + 2y = 9$ $0 \leqslant x \leqslant 4$, $-2 \leqslant y \leqslant 5$ **9** $x + 3y = 6$ $0 \leqslant x \leqslant 5$, $0 \leqslant y \leqslant 5$
   $2x - 2y = 3$ $3x - y = 6$

8 $2x + 3y = 4$ $-2 \leqslant x \leqslant 2$, $0 \leqslant y \leqslant 4$ **10** $x = 2y - 3$ $-2 \leqslant x \leqslant 3$, $0 \leqslant y \leqslant 4$
   $y = x + 2$ $y = 2x + 1$

Investigation

Try to solve the following equations graphically. Why do you think the method breaks down?

1 $x + y = 9$ $0 \leqslant x \leqslant 9$ **2** $y = 2x + 3$ $0 \leqslant x \leqslant 4$

   $x + y = 4$ $0 \leqslant y \leqslant 9$ $y = 2x - 1$ $-1 \leqslant y \leqslant 11$

Did you know that Rene Descartes was nicknamed 'The philosopher'? He was born into a rich French family. He was so small and sickly that nobody thought that he would live long, but he surprised everyone and lived fifty-three years. He studied law to please his father but law did not really interest him. He turned to gambling and was very lucky. He did serious thinking about life and mathematics. He is famous for his statement 'I think, therefore I am'. He is remembered for inventing a branch of mathematics known as analytical geometry.

I think therefore I am

IN THIS CHAPTER...

you have seen that:

- two linear equations in two unknowns usually have one solution, that is one value for each unknown

- you can solve two simultaneous equations by adding or subtracting the equations when they both have the same number of one of the unknowns: this eliminates that unknown

- you may have to multiply one or both equations by numbers in order to get the same number of one unknown

- you can also solve two simultaneous equations by using one equation to express one letter as an expression in terms of the other letter, then substituting that expression into the other equation; this gives one equation in one unknown

- you can solve two simultaneous equations graphically by drawing the lines that the equations represent; the point of intersection gives the solution.

7 MATRICES

AT THE END OF THIS CHAPTER...

you should be able to:

1 Arrange a set of data in a rectangular array called a matrix.

2 State the size of a given matrix.

3 Write down the entry in a matrix, given its row and column.

4 Add and subtract two or more matrices.

5 Multiply a matrix by a constant.

6 Multiply two matrices.

7 Determine if two matrices are compatible for multiplication.

Did you know that the symbol '=', which stands for 'is equal to', was first proposed by a Royal Court physician named Robert Recorde (1510–1558) in a book published in 1558 entitled *The Whetstone of Witte?* However, it was a long time before it was generally adopted.

BEFORE YOU START

you need to know:
✓ how to work with directed numbers

KEY WORDS

compatibility, entry, matrix (plural matrices), matrix multiplication, power of a matrix, rectangular array, row–column multiplication, vector

Shopping lists

Mrs Smith and Mrs Jones go shopping for oranges, lemons and grapefruit. Mrs Smith buys six oranges, two lemons and three grapefruit while Mrs Jones buys five oranges, one lemon and two grapefruit. We can arrange this information in a table:

	Oranges	Lemons	Grapefruit
Mrs Smith	6	2	3
Mrs Jones	5	1	2

We can write this information more briefly in the form $\begin{pmatrix} 6 & 2 & 3 \\ 5 & 1 & 2 \end{pmatrix}$

All the numerical information is given, though we have missed out the descriptions of what the various rows and columns mean.

A *rectangular array* of numbers is called a *matrix* (plural *matrices*). It is held together by curved brackets. Each item is called an *entry* or *element*. A matrix can have any number of *rows* and *columns*; rows go across and columns go down. If we had the shopping lists of three people buying four different sorts of fruit we would have a matrix with three rows and four columns.

e.g. $\begin{pmatrix} 4 & 3 & 1 & 2 \\ 1 & 2 & 0 & 4 \\ 9 & 0 & 1 & 1 \end{pmatrix}$

We use the number of rows and the number of columns to describe the size of a matrix. This matrix is a 3×4 matrix (number of rows first, then the number of columns).

Exercise 7a

a Give the size of the matrix $\begin{pmatrix} 1 & 4 & 3 & 2 \\ 7 & 0 & 6 & 5 \end{pmatrix}$

b What is the entry in the second row, third column?

a The matrix has 2 rows and 4 colums so it is a 2×4 matrix

b The entry in the second row, third column is 6

Give the sizes of the matrices in questions **1** to **6**:

1 $\begin{pmatrix} 2 & 4 \\ 1 & 2 \end{pmatrix}$

2 $\begin{pmatrix} 2 & 3 & 4 \\ 1 & 1 & 4 \end{pmatrix}$

3 $\begin{pmatrix} 1 \\ 2 \end{pmatrix}$

4 $(1 \ 2 \ 3)$

5 (2)

6 $\begin{pmatrix} 2 & 3 \\ 4 & 5 \\ 6 & 7 \end{pmatrix}$

7 In the matrix $\begin{pmatrix} 1 & 2 & 3 \\ 4 & 5 & 6 \\ 7 & 8 & 9 \end{pmatrix}$ give the entry in

a the second row, third column

b the third row, second column

c the first row, second column

d the third row, first column

8 Write down the second row and the third column of the matrix $\begin{pmatrix} 5 & 2 & 4 \\ 3 & 1 & 7 \\ 9 & 6 & 2 \end{pmatrix}$

    a Give the entry in the second row of the third column.

    b Give the entry in the third row of the second column.

    c Give the entry in the first row of the third column.

9 Write down the matrix with three rows and three columns in which the first row is a row of zeros, the second row is a row of ones and the third row is a row of twos.

10 Write down the 3×2 matrix in which the first column is a column of threes and the second column is a column of ones.

Addition of matrices

Mrs Smith's and Mrs Jones' first shopping lists were represented by the matrix $\begin{pmatrix} 6 & 2 & 3 \\ 5 & 1 & 2 \end{pmatrix}$

The next week, Mrs Smith buys three oranges, no lemons and one grapefruit and Mrs Jones buys four oranges, two lemons and one grapefruit. We can arrange this in a table:

	Oranges	Lemons	Grapefruit
Mrs Smith	3	0	1
Mrs Jones	4	2	1

and represent it by the matrix $\begin{pmatrix} 3 & 0 & 1 \\ 4 & 2 & 1 \end{pmatrix}$

We see that over the two weeks Mrs Smith buys nine oranges, two lemons and four grapefruit and Mrs Jones buys nine oranges, three lemons and three grapefruit. This information can be written briefly in the form

$$\begin{pmatrix} 6 & 2 & 3 \\ 5 & 1 & 2 \end{pmatrix} + \begin{pmatrix} 3 & 0 & 1 \\ 4 & 2 & 1 \end{pmatrix} = \begin{pmatrix} 9 & 2 & 4 \\ 9 & 3 & 3 \end{pmatrix} \qquad \begin{array}{l} \text{Add the corresponding} \\ \text{elements e.g. } 6 + 3 = 9 \end{array}$$

Each of these matrices is a 2×3 matrix.

We have added the entries in corresponding positions, e.g. in the first row, third column, $3 + 1 = 4$

In the third week, Mrs Smith buys two oranges and one lemon and Mrs Jones buys four oranges and one lemon.

The matrix showing this information is $\begin{pmatrix} 2 & 1 \\ 4 & 1 \end{pmatrix}$

This is a 2×2 matrix.

We do not know how many grapefruit were bought in the third week. This makes it impossible to add the matrices for the second and third weeks

i.e. $\qquad \begin{pmatrix} 3 & 0 & 1 \\ 4 & 2 & 1 \end{pmatrix} + \begin{pmatrix} 2 & 1 \\ 4 & 1 \end{pmatrix}$ cannot be worked out.

We can add matrices if they are the same size, but not if they are of different sizes.

Exercise 7b　　Find, where possible

a $\begin{pmatrix} 6 & 3 \\ 1 & 2 \end{pmatrix} + \begin{pmatrix} 4 & 1 \\ 3 & 2 \end{pmatrix}$　　b $\begin{pmatrix} 1 \\ 2 \end{pmatrix} + \begin{pmatrix} 4 & 1 \\ 0 & 3 \end{pmatrix}$

a $\begin{pmatrix} 6 & 3 \\ 1 & 2 \end{pmatrix} + \begin{pmatrix} 4 & 1 \\ 3 & 2 \end{pmatrix} = \begin{pmatrix} 10 & 4 \\ 4 & 4 \end{pmatrix}$

b $\begin{pmatrix} 1 \\ 2 \end{pmatrix} + \begin{pmatrix} 4 & 1 \\ 0 & 3 \end{pmatrix}$ not possible since the matrices are not the same size.

Find, where possible:

1 $\begin{pmatrix} 2 \\ 3 \end{pmatrix} + \begin{pmatrix} 10 \\ 12 \end{pmatrix}$

6 $\begin{pmatrix} 7 & 8 \\ 9 & 1 \end{pmatrix} + \begin{pmatrix} 4 & 3 \\ 2 & 4 \end{pmatrix}$

2 $\begin{pmatrix} 9 & 2 \\ 4 & 1 \end{pmatrix} + \begin{pmatrix} 6 & 2 \\ 3 & 0 \end{pmatrix}$

7 $(2 \ 1 \ 4) + (3 \ 2 \ 1)$

3 $\begin{pmatrix} 1 \\ 2 \\ 3 \end{pmatrix} + \begin{pmatrix} 3 & 6 \\ 4 & 7 \\ 5 & 8 \end{pmatrix}$

8 $\begin{pmatrix} 1 \\ 2 \end{pmatrix} + (3 \ 4)$

4 $(6 \ 1) + (3 \ 4)$

9 $\begin{pmatrix} 1 & 2 \\ 3 & 4 \\ 5 & 6 \end{pmatrix} + \begin{pmatrix} 5 & 6 \\ 4 & 3 \\ 2 & 1 \end{pmatrix}$

5 $\begin{pmatrix} 4 & 1 & 2 \\ 3 & 5 & 6 \end{pmatrix} + \begin{pmatrix} 7 & 1 & 0 \\ 3 & 2 & 1 \end{pmatrix}$

10 $(6 \ 3) + (4 \ 5)$

Negative numbers can be used.

Find $\begin{pmatrix} 4 & -3 \\ 4 & -1 \end{pmatrix} + \begin{pmatrix} -3 & 2 \\ 1 & 4 \end{pmatrix}$

$\begin{pmatrix} 4 & -3 \\ 4 & -1 \end{pmatrix} + \begin{pmatrix} -3 & 2 \\ 1 & 4 \end{pmatrix} = \begin{pmatrix} 4+(-3) & -3+2 \\ 4+1 & -1+4 \end{pmatrix} = \begin{pmatrix} 1 & -1 \\ 5 & 3 \end{pmatrix}$

Find, where possible:

11 $\begin{pmatrix} 2 & 4 \\ -1 & 4 \end{pmatrix} + \begin{pmatrix} -1 & 4 \\ -3 & 3 \end{pmatrix}$

15 $\begin{pmatrix} -2 \\ -3 \\ 4 \end{pmatrix} + \begin{pmatrix} -1 \\ 0 \\ 2 \end{pmatrix}$

12 $\begin{pmatrix} 1 \\ -4 \end{pmatrix} + \begin{pmatrix} -3 \\ 6 \end{pmatrix}$

16 $(1 \ -6 \ 1) + (-4 \ 1)$

13 $\begin{pmatrix} 4 & 3 \\ -2 & 1 \end{pmatrix} + \begin{pmatrix} -6 & 4 \\ -3 & 2 \end{pmatrix}$

17 $\begin{pmatrix} 3 & 6 \\ 2 & -5 \end{pmatrix} + \begin{pmatrix} -1 & 4 \\ 3 & 2 \end{pmatrix}$

14 $(1 \ 2) + (3 \ 4)$

18 $\begin{pmatrix} 1 & -2 \\ -2 & 1 \end{pmatrix} + \begin{pmatrix} 4 & -3 \\ 5 & -1 \end{pmatrix}$

19 $\begin{pmatrix} 3 & 2 \\ -1 & -4 \end{pmatrix} + \begin{pmatrix} -3 & 6 \\ 9 & 2 \end{pmatrix}$

20 $\begin{pmatrix} 4 \\ -2 \end{pmatrix} + (4 \quad -2)$

21 $\begin{pmatrix} 3 & 2 & 0 \\ 1 & 4 & 3 \end{pmatrix} + \begin{pmatrix} 2 & 4 \\ -3 & 4 \\ 1 & 2 \end{pmatrix}$

22 $(1 \quad 4 \quad -3) + (0 \quad 2 \quad 0)$

23 $\begin{pmatrix} 3 & -1 & 1 \\ 5 & 6 & -7 \end{pmatrix} + \begin{pmatrix} -1 & 4 & 3 \\ 0 & -6 & -5 \end{pmatrix}$

Multiples of matrices

If Mrs Smith and Mrs Jones each have the same shopping list for three weeks running we can see that

$$\begin{pmatrix} 3 & 1 & 4 \\ 1 & 2 & 1 \end{pmatrix} + \begin{pmatrix} 3 & 1 & 4 \\ 1 & 2 & 1 \end{pmatrix} + \begin{pmatrix} 3 & 1 & 4 \\ 1 & 2 & 1 \end{pmatrix} = 3\begin{pmatrix} 3 & 1 & 4 \\ 1 & 2 & 1 \end{pmatrix} = \begin{pmatrix} 9 & 3 & 12 \\ 3 & 6 & 3 \end{pmatrix}$$

In the same way $\quad 5\begin{pmatrix} 1 & 4 \\ 3 & -2 \end{pmatrix} = \begin{pmatrix} 5 & 20 \\ 15 & -10 \end{pmatrix}$

When we multiply a matrix by five, we multiply *every* entry by five.

Subtraction of matrices

We can subtract matrices if they are the same size.

Two weeks' shopping First week's shopping Second week's shopping

$$\begin{pmatrix} 7 & 9 & 2 \\ 6 & 4 & 1 \end{pmatrix} - \begin{pmatrix} 3 & 2 & 1 \\ 3 & 3 & 1 \end{pmatrix} = \begin{pmatrix} 4 & 7 & 1 \\ 3 & 1 & 0 \end{pmatrix}$$

Exercise 7c

Find, where possible:

1 $3\begin{pmatrix} 1 \\ 2 \\ 4 \end{pmatrix}$

2 $2\begin{pmatrix} 1 & 4 & 0 \\ 2 & -1 & 3 \end{pmatrix}$

3 $\frac{1}{2}\begin{pmatrix} 2 & 4 \\ 1 & 6 \\ 3 & 8 \end{pmatrix}$

4 $6\begin{pmatrix} 1 & 4 \\ 3 & -2 \end{pmatrix}$

5 $6\begin{pmatrix} -1 & -5 \\ 1 & 2 \end{pmatrix}$

6 $\frac{2}{3}\begin{pmatrix} 6 & 0 \\ 1 & 2 \\ 3 & 5 \end{pmatrix}$

7 $\begin{pmatrix} 3 & 2 \\ 1 & 4 \end{pmatrix} - \begin{pmatrix} 1 & 4 \\ 0 & 1 \end{pmatrix}$

8 $\begin{pmatrix} 1 \\ 3 \end{pmatrix} - \begin{pmatrix} 2 & 3 \\ 4 & 1 \end{pmatrix}$

Tip Multiply every entry by 3.

Tip Multiply every entry by 2.

Tip Subtract each entry in the second matrix from the corresponding entry in the first matrix.

Tip Are these the same size?

9 $\begin{pmatrix} 1 \\ 2 \\ 3 \end{pmatrix} - \begin{pmatrix} 4 \\ -1 \\ 3 \end{pmatrix}$

11 $(1 \quad 6 \quad 2) - (4 \quad 3)$

10 $\begin{pmatrix} 4 & 5 & 3 \\ 1 & -2 & 1 \end{pmatrix} - \begin{pmatrix} 2 & 1 & 1 \\ 4 & 1 & 2 \end{pmatrix}$

12 $\begin{pmatrix} 1 & 2 & 3 \\ 4 & 5 & 6 \\ 7 & 8 & 9 \end{pmatrix} - \begin{pmatrix} 4 & 3 & 1 \\ -5 & 0 & 2 \\ 6 & -3 & 4 \end{pmatrix}$

Mixed questions

Exercise **7d**

Find, where possible:

1 $\begin{pmatrix} 1 & 4 \\ 3 & 2 \end{pmatrix} + \begin{pmatrix} -2 & 4 \\ 3 & -1 \end{pmatrix}$

7 $2\begin{pmatrix} 1 & 4 \\ 3 & 2 \end{pmatrix}$

2 $\begin{pmatrix} 1 & 4 \\ 3 & 2 \end{pmatrix} - \begin{pmatrix} -2 & 4 \\ 3 & -1 \end{pmatrix}$

8 $\frac{1}{2}\begin{pmatrix} -2 & 4 \\ 3 & -1 \end{pmatrix}$

3 $(1 \quad 2 \quad 4) - (3 \quad 2)$

9 $\begin{pmatrix} 1 \\ 2 \\ 3 \end{pmatrix} + \begin{pmatrix} 4 \\ 5 \\ -1 \end{pmatrix} + \begin{pmatrix} 3 \\ 2 \\ 1 \end{pmatrix}$

4 $\begin{pmatrix} 3 & 2 \\ 4 & -1 \end{pmatrix} + \begin{pmatrix} 4 & -3 \\ 1 & 0 \end{pmatrix}$

10 $4\begin{pmatrix} 6 & 2 & -1 \\ 4 & 3 & 4 \end{pmatrix}$

5 $\frac{1}{3}\begin{pmatrix} 4 \\ 5 \\ -1 \end{pmatrix}$

11 $\begin{pmatrix} 6 & 2 & 1 \\ 4 & 3 & 4 \end{pmatrix} + \begin{pmatrix} -2 & 4 \\ 3 & -1 \end{pmatrix}$

6 $\begin{pmatrix} 1 \\ 2 \\ 3 \end{pmatrix} - \begin{pmatrix} 4 \\ 5 \\ 6 \end{pmatrix}$

12 $\begin{pmatrix} 6 & 2 & -1 \\ 4 & 3 & 4 \end{pmatrix} - \begin{pmatrix} -2 & 4 & 1 \\ 3 & -1 & 0 \end{pmatrix}$

 Puzzle

The large square is divided into sixteen smaller squares.

Each small square is coloured red (R), white (W), yellow (Y) or blue (B).

Divide the large square into four pieces so that each piece is made up of four small squares with four different colours.

R	R	W	Y
W	Y	R	B
B	R	W	Y
Y	W	B	B

Use of letters

In Book 2, we saw that we could use a single small letter in heavy type to denote a vector,

e.g. $\qquad \mathbf{a} = \begin{pmatrix} 1 \\ 2 \end{pmatrix}$

In the same way we can represent a matrix by giving it a capital letter in heavy type:

e.g. $\qquad \mathbf{A} = \begin{pmatrix} 1 & 2 \\ 4 & 1 \end{pmatrix}$ and $\mathbf{B} = \begin{pmatrix} 4 & 1 \\ 3 & 0 \end{pmatrix}$

then $\qquad \mathbf{A} + \mathbf{B} = \begin{pmatrix} 1 & 2 \\ 4 & 1 \end{pmatrix} + \begin{pmatrix} 4 & 1 \\ 3 & 0 \end{pmatrix} = \begin{pmatrix} 5 & 3 \\ 7 & 1 \end{pmatrix}$

and $\qquad 2\mathbf{A} = 2\begin{pmatrix} 1 & 2 \\ 4 & 1 \end{pmatrix} = \begin{pmatrix} 2 & 4 \\ 8 & 2 \end{pmatrix}$

We cannot *write* $\mathbf{A}$, so we write $\underline{A}$.

Exercise 7e

The questions in this exercise refer to the following matrices:

$$\mathbf{A} = \begin{pmatrix} 4 & 3 & 1 \\ 1 & 2 & 3 \end{pmatrix} \qquad \mathbf{B} = \begin{pmatrix} 4 \\ 1 \end{pmatrix} \qquad \mathbf{C} = \begin{pmatrix} 2 & 3 \\ 1 & -2 \end{pmatrix} \qquad \mathbf{D} = \begin{pmatrix} 6 & 2 \\ 1 & 4 \end{pmatrix}$$

$$\mathbf{E} = (6 \quad -1 \quad 2) \qquad \mathbf{F} = (3 \quad 2) \qquad \mathbf{G} = \begin{pmatrix} 5 & 1 & 3 \\ 6 & -1 & 4 \end{pmatrix}$$

Give the size of $\mathbf{A}$.

$\mathbf{A}$ is a 2×3 matrix (It has 2 rows and 3 columns.)

Find, where possible (a) $\mathbf{C} - \mathbf{D}$ (b) $\mathbf{B} + \mathbf{F}$.

a $\quad \mathbf{C} - \mathbf{D} = \begin{pmatrix} 2 & 3 \\ 1 & -2 \end{pmatrix} - \begin{pmatrix} 6 & 2 \\ 1 & 4 \end{pmatrix} = \begin{pmatrix} -4 & 1 \\ 0 & -6 \end{pmatrix}$

b $\quad \mathbf{B} + \mathbf{F} = \begin{pmatrix} 4 \\ 1 \end{pmatrix} + (3 \quad 2)$ not possible since B and F are not the same size.

1 Give the sizes of the matrices $\mathbf{B}$ to $\mathbf{G}$.

Find, where possible:

2 $\mathbf{A} + \mathbf{G}$ **5** $3\mathbf{A}$ **8** $\mathbf{G} + \mathbf{B}$ **11** $\frac{3}{4}\mathbf{D}$

3 $\mathbf{B} + \mathbf{C}$ **6** $\mathbf{E} + \mathbf{F}$ **9** $\mathbf{G} - \mathbf{A}$ **12** $\mathbf{C} + \mathbf{D} + \mathbf{G}$

4 $\mathbf{D} - \mathbf{C}$ **7** $\frac{1}{2}\mathbf{F}$ **10** $6\mathbf{B}$ **13** $\mathbf{F} - \mathbf{E}$

Matrix multiplication

We can use the idea of shopping lists to find an operation that we call matrix multiplication.

Mrs Smith buys six bananas and one lemon and Mrs Jones buys four bananas and two lemons.

The matrix showing this information is $\begin{pmatrix} 6 & 1 \\ 4 & 2 \end{pmatrix}$

The bananas cost 12 c each and the lemons 8 c and this information can be written as

$$\begin{pmatrix} 12 \\ 8 \end{pmatrix}$$

We can work out that Mrs Smith spends 80 c and Mrs Jones 64 c. This can be written in a table

	Amount spent in cents
Mrs Smith	80
Mrs Jones	64

or as a matrix $\begin{pmatrix} 80 \\ 64 \end{pmatrix}$

We can say $\begin{pmatrix} 6 & 1 \\ 4 & 2 \end{pmatrix}\begin{pmatrix} 12 \\ 8 \end{pmatrix} = \begin{pmatrix} 80 \\ 64 \end{pmatrix}$

This is a sort of multiplication, though different from any multiplication we have done so far.

It is called *matrix multiplication* or *row–column multiplication*.

Notice how the numbers have been paired:

$$\begin{pmatrix} 6 & 1 \end{pmatrix}\begin{pmatrix} 12 \\ 8 \end{pmatrix} = \begin{pmatrix} 80 \end{pmatrix} \qquad 6 \times 12 \quad + \quad 1 \times 8 = 80$$

and $\begin{pmatrix} 4 & 2 \end{pmatrix}\begin{pmatrix} 12 \\ 8 \end{pmatrix} = \begin{pmatrix} 64 \end{pmatrix} \qquad 4 \times 12 \quad + \quad 2 \times 8$

Exercise 7f

Find $\begin{pmatrix} 4 & 3 \\ 6 & 1 \end{pmatrix}\begin{pmatrix} 3 \\ 2 \end{pmatrix}$

$$\begin{pmatrix} 4 & 3 \\ 6 & 1 \end{pmatrix}\begin{pmatrix} 3 \\ 2 \end{pmatrix} = \begin{pmatrix} 18 \\ 20 \end{pmatrix} \qquad \begin{matrix} 4 \times 3 & + & 3 \times 2 = 18 \\ 6 \times 3 & + & 1 \times 2 = 20 \end{matrix}$$

Find the following products:

1 $\begin{pmatrix} 1 & 4 \\ 3 & 2 \end{pmatrix}\begin{pmatrix} 5 \\ 6 \end{pmatrix}$

2 $\begin{pmatrix} 6 & 2 \\ 3 & 5 \end{pmatrix}\begin{pmatrix} 2 \\ 1 \end{pmatrix}$

3 $\begin{pmatrix} 3 & 2 \\ 4 & 3 \end{pmatrix}\begin{pmatrix} 1 \\ 1 \end{pmatrix}$

4 $\begin{pmatrix} 1 & 3 \\ 0 & 1 \end{pmatrix}\begin{pmatrix} 3 \\ 2 \end{pmatrix}$

5 $\begin{pmatrix} 5 & 2 \\ 1 & 2 \end{pmatrix}\begin{pmatrix} 1 \\ 2 \end{pmatrix}$

6 $\begin{pmatrix} 6 & 2 \\ 1 & 4 \end{pmatrix}\begin{pmatrix} 2 \\ 3 \end{pmatrix}$

7 $\begin{pmatrix} 5 & 4 \\ 3 & 1 \end{pmatrix}\begin{pmatrix} 2 \\ 4 \end{pmatrix}$

8 $\begin{pmatrix} 6 & 4 \\ 1 & 2 \end{pmatrix}\begin{pmatrix} 5 \\ 7 \end{pmatrix}$

9 $\begin{pmatrix} 11 & 12 \\ 10 & 9 \end{pmatrix}\begin{pmatrix} 4 \\ 1 \end{pmatrix}$

Mrs Smith bought six bananas and one lemon and Mrs Jones bought four bananas and two lemons. Bananas cost 12 c each and lemons 8 c each.

This gives the matrix multiplication $\begin{pmatrix} 6 & 1 \\ 4 & 2 \end{pmatrix}\begin{pmatrix} 12 \\ 8 \end{pmatrix} = \begin{pmatrix} 80 \\ 64 \end{pmatrix}$

If they chose to do their shopping at a second shop where the bananas cost 11 c each and the lemons 9 c, the matrix multiplication would be $\begin{pmatrix} 6 & 1 \\ 4 & 2 \end{pmatrix}\begin{pmatrix} 11 \\ 9 \end{pmatrix} = \begin{pmatrix} 75 \\ 62 \end{pmatrix}$

The table for the bills is as follows:

	First shop	Second shop
Mrs Smith	80	75
Mrs Jones	64	62

This gives the 2 × 2 matrix $\begin{pmatrix} 80 & 75 \\ 64 & 62 \end{pmatrix}$

We can combine the previous two matrix multiplications into one:

$$\begin{array}{c} \text{Mrs Smith} \\ \text{Mrs Jones} \end{array} \begin{pmatrix} 6 & 1 \\ 4 & 2 \end{pmatrix} \times \begin{pmatrix} 12 & 11 \\ 8 & 9 \end{pmatrix} = \begin{pmatrix} 80 & 75 \\ 64 & 62 \end{pmatrix}$$

First shop, Second shop

Notice that we pair the entries in the *first row* of the first matrix with the entries in the *first column* of the second matrix to give the entry in the *first row, first column* of the final matrix.

$$\begin{pmatrix} 6 & 1 \end{pmatrix}\begin{pmatrix} 12 \\ 8 \end{pmatrix} = \begin{pmatrix} 80 \end{pmatrix} \qquad 6 \times 12 + 1 \times 8 = 80$$

Then the entries in the *first row* of the first matrix paired with the entries in the *second column* of the second matrix give the entry in the *first row, second column* of the final matrix.

$$\begin{pmatrix} 6 & 1 \end{pmatrix}\begin{pmatrix} 11 \\ 9 \end{pmatrix} = \begin{pmatrix} 75 \end{pmatrix} \qquad 6 \times 11 + 1 \times 9 = 75$$

Then we use the second row in the same way.

Exercise 7g

Find the missing entries in the matrix multiplication

$$\begin{pmatrix} 2 & 4 \\ 3 & 2 \end{pmatrix}\begin{pmatrix} 4 & 1 \\ 3 & 2 \end{pmatrix} = \begin{pmatrix} 20 & \\ & 7 \end{pmatrix}$$

$$\begin{pmatrix} 2 & 4 \\ 3 & 2 \end{pmatrix}\begin{pmatrix} 4 & 1 \\ 3 & 2 \end{pmatrix} = \begin{pmatrix} 20 & 10 \\ 18 & 7 \end{pmatrix}$$

(1st row, 2nd column $2 \times 1 + 4 \times 2 = 10$) (2nd row, 1st column $3 \times 4 + 2 \times 3 = 18$)

Find the missing entries in the following matrix multiplications:

1 $\begin{pmatrix} 1 & 4 \\ 3 & 1 \end{pmatrix}\begin{pmatrix} 3 & 2 \\ 1 & 4 \end{pmatrix} = \begin{pmatrix} & 18 \\ & 10 \end{pmatrix}$

2 $\begin{pmatrix} 3 & 2 \\ 4 & 6 \end{pmatrix}\begin{pmatrix} 4 & 1 \\ 1 & 3 \end{pmatrix} = \begin{pmatrix} & 9 \\ 22 & \end{pmatrix}$

3 $\begin{pmatrix} 9 & 1 \\ 0 & 2 \end{pmatrix}\begin{pmatrix} 1 & 4 \\ 3 & 1 \end{pmatrix} = \begin{pmatrix} 12 & \\ 6 & \end{pmatrix}$

4 $\begin{pmatrix} 5 & 6 \\ 7 & 9 \end{pmatrix}\begin{pmatrix} 1 & 2 \\ 3 & 1 \end{pmatrix} = \begin{pmatrix} & \\ 34 & 23 \end{pmatrix}$

$$\underline{5} \quad \begin{pmatrix} 1 & 2 \\ 3 & 1 \end{pmatrix}\begin{pmatrix} 5 & 6 \\ 7 & 9 \end{pmatrix} = \begin{pmatrix} & 24 \\ & 27 \end{pmatrix}$$

$$\underline{7} \quad \begin{pmatrix} 3 & 9 \\ 2 & 1 \end{pmatrix}\begin{pmatrix} 1 & 1 \\ 1 & 1 \end{pmatrix} = \begin{pmatrix} 12 \\ 3 \end{pmatrix}$$

$$\underline{6} \quad \begin{pmatrix} 5 & 1 \\ 0 & 6 \end{pmatrix}\begin{pmatrix} 4 & 3 \\ 2 & 1 \end{pmatrix} = \begin{pmatrix} 22 \\ & 6 \end{pmatrix}$$

$$\underline{8} \quad \begin{pmatrix} 5 & 3 \\ 1 & 1 \end{pmatrix}\begin{pmatrix} 1 & 2 \\ 4 & 3 \end{pmatrix} = \begin{pmatrix} & \\ & 5 \end{pmatrix}$$

Find the following products:

$$\underline{9} \quad \begin{pmatrix} 4 & 6 \\ 1 & 3 \end{pmatrix}\begin{pmatrix} 1 & 4 \\ 3 & 6 \end{pmatrix}$$

$$\underline{11} \quad \begin{pmatrix} 5 & 1 \\ 3 & 2 \end{pmatrix}\begin{pmatrix} 3 & 2 \\ 1 & 4 \end{pmatrix}$$

$$\underline{13} \quad \begin{pmatrix} 6 & 4 \\ 5 & 1 \end{pmatrix}\begin{pmatrix} 2 & 6 \\ 8 & 1 \end{pmatrix}$$

$$\underline{10} \quad \begin{pmatrix} 4 & 8 \\ 1 & 1 \end{pmatrix}\begin{pmatrix} 5 & 6 \\ 3 & 1 \end{pmatrix}$$

$$\underline{12} \quad \begin{pmatrix} 1 & 4 \\ 3 & 6 \end{pmatrix}\begin{pmatrix} 4 & 6 \\ 1 & 3 \end{pmatrix}$$

$$\underline{14} \quad \begin{pmatrix} 3 & 2 \\ 1 & 4 \end{pmatrix}\begin{pmatrix} 5 & 1 \\ 3 & 2 \end{pmatrix}$$

Sometimes the entries are negative.

Find $\begin{pmatrix} 1 & -2 \\ 2 & 1 \end{pmatrix}\begin{pmatrix} 2 & 4 \\ -3 & 1 \end{pmatrix}$

$$\begin{pmatrix} 1 & -2 \\ 2 & 1 \end{pmatrix}\begin{pmatrix} 2 & 4 \\ -3 & 1 \end{pmatrix} = \begin{pmatrix} 8 & 2 \\ 1 & 9 \end{pmatrix}$$ (1st row, 2nd column
$1 \times 4 + (-2) \times 1 = 4 - 2 = 2$)

Find the following products:

$$\underline{15} \quad \begin{pmatrix} 3 & 2 \\ 1 & 4 \end{pmatrix}\begin{pmatrix} -2 & 4 \\ 3 & 1 \end{pmatrix}$$

$$\underline{17} \quad \begin{pmatrix} 5 & 2 \\ 3 & 4 \end{pmatrix}\begin{pmatrix} -1 & -2 \\ 4 & 3 \end{pmatrix}$$

$$\underline{19} \quad \begin{pmatrix} 4 & 3 \\ -1 & 4 \end{pmatrix}\begin{pmatrix} 3 & 2 \\ 3 & 1 \end{pmatrix}$$

$$\underline{16} \quad \begin{pmatrix} 3 & 4 \\ -1 & 2 \end{pmatrix}\begin{pmatrix} 1 & 4 \\ 3 & 2 \end{pmatrix}$$

$$\underline{18} \quad \begin{pmatrix} 5 & -2 \\ 4 & 3 \end{pmatrix}\begin{pmatrix} -4 & -3 \\ 2 & 1 \end{pmatrix}$$

$$\underline{20} \quad \begin{pmatrix} 4 & -3 \\ 2 & -1 \end{pmatrix}\begin{pmatrix} -1 & -2 \\ 4 & -3 \end{pmatrix}$$

Order of multiplication

We can see from the answers to questions 4 and 5 and others in the previous exercise, that the order in which we write the matrices *does* matter when we multiply. For example,

if $\qquad \mathbf{A} = \begin{pmatrix} 3 & 1 \\ 2 & 4 \end{pmatrix}$ and $\mathbf{B} = \begin{pmatrix} 5 & 6 \\ 1 & 1 \end{pmatrix}$

then $\qquad \mathbf{AB} = \begin{pmatrix} 16 & 19 \\ 14 & 16 \end{pmatrix}$ but $\mathbf{BA} = \begin{pmatrix} 27 & 29 \\ 5 & 5 \end{pmatrix}.$

That is, $\qquad \mathbf{AB} \neq \mathbf{BA}.$

Occasionally, with a few special matrices, the order does not affect the answer, but in general matrix multiplication is not commutative.

Exercise 7h

The questions in this exercise refer to the following matrices:

$$\mathbf{A} = \begin{pmatrix} 4 & 3 \\ 2 & 1 \end{pmatrix} \qquad \mathbf{B} = \begin{pmatrix} 2 & 1 \\ 4 & 3 \end{pmatrix} \qquad \mathbf{C} = \begin{pmatrix} 7 & 8 \\ 1 & 1 \end{pmatrix} \qquad \mathbf{D} = \begin{pmatrix} 2 & 0 \\ 0 & 2 \end{pmatrix}$$

Find:

| 1 | **AB** | 3 | **AC** | 5 | **AD** | 7 | **BC** | 9 | **BD** | 11 | **CD** |
| 2 | **BA** | 4 | **CA** | 6 | **DA** | 8 | **CB** | 10 | **DB** | 12 | **DC** |

13 In some cases in questions 1 to 12 the order mattered and in some cases it did not. What is special about the matrices for which the order did not matter?

14 Make up a set of matrices similar to the ones used in this exercise and see whether the same results occur when you multiply them together in pairs.

Multiplication of matrices of different sizes

We can multiply matrices of sizes other than 2×2. If Mrs Smith and Mrs Jones are buying bananas, lemons and grapefruit then we might have the information in the following tables:

Shopping lists	Bananas	Lemons	Grapefruit
Mrs Smith	4	0	1
Mrs Jones	2	3	2

Costs in cents	First shop	Second shop
Bananas	12	11
Lemons	8	9
Grapefruit	15	12

Then we would have the matrix multiplication

$$\begin{pmatrix} 4 & 0 & 1 \\ 2 & 3 & 2 \end{pmatrix} \begin{pmatrix} 12 & 11 \\ 8 & 9 \\ 15 & 12 \end{pmatrix}$$

Again we pair entries in a row in the first matrix with the entries in a column in the second matrix:

$$\begin{pmatrix} 4 & 0 & 1 \end{pmatrix} \begin{pmatrix} 12 \\ 8 \\ 15 \end{pmatrix} = \begin{pmatrix} 63 \end{pmatrix} \qquad 4 \times 12 \ + \ 0 \times 8 \ + \ 1 \times 15 = 63$$

This gives us Mrs Smith's bill at the first shop; she would spend 63 c.

$$\begin{pmatrix} 4 & 0 & 1 \end{pmatrix} \begin{pmatrix} 11 \\ 9 \\ 12 \end{pmatrix} = \begin{pmatrix} 56 \end{pmatrix} \qquad 4 \times 11 \ + \ 0 \times 9 \ + \ 1 \times 12 = 56$$

Mrs Smith would spend 56 c at the second shop.

The complete multiplication is as follows:

$$\begin{pmatrix} 4 & 0 & 1 \\ 2 & 3 & 2 \end{pmatrix} \begin{pmatrix} 12 & 11 \\ 8 & 9 \\ 15 & 12 \end{pmatrix} = \begin{pmatrix} 63 & 56 \\ 78 & 73 \end{pmatrix}$$

For example, pairing the entries in the *second row* of the first matrix with the entries in the *first column* of the second matrix gives the entry in the *second row, first column* of the resulting matrix,

i.e. $2 \times 12 + 3 \times 8 + 2 \times 15 = 78$

Exercise 7i   Find **a** $\begin{pmatrix} 3 & 1 & 4 \\ 4 & 0 & 3 \end{pmatrix}\begin{pmatrix} 1 \\ 2 \\ 3 \end{pmatrix}$ **b** $(1 \ \ 2)\begin{pmatrix} 1 & 3 \\ 4 & 2 \end{pmatrix}$

a $\begin{pmatrix} 3 & 1 & 4 \\ 4 & 0 & 3 \end{pmatrix}\begin{pmatrix} 1 \\ 2 \\ 3 \end{pmatrix} = \begin{pmatrix} 17 \\ 13 \end{pmatrix}$

(first row, first column
$3 \times 1 + 1 \times 2 + 4 \times 3 = 17$)

(second row, first column
$4 \times 1 + 0 \times 2 + 3 \times 3 = 13$)

b $(1 \ \ 2)\begin{pmatrix} 1 & 3 \\ 4 & 2 \end{pmatrix} = (9 \ \ 7)$

(first row, first column
$1 \times 1 + 2 \times 4 = 9$)

(first row, second column
$1 \times 3 + 2 \times 2 = 7$)

Find:

1 $\begin{pmatrix} 1 & 4 \\ 3 & 1 \end{pmatrix}\begin{pmatrix} 3 \\ 1 \end{pmatrix}$

2 $\begin{pmatrix} 1 & 4 & 1 \\ 3 & 2 & 4 \end{pmatrix}\begin{pmatrix} 6 \\ 1 \\ 3 \end{pmatrix}$

3 $(1 \ \ 2)\begin{pmatrix} 4 \\ 3 \end{pmatrix}$

4 $\begin{pmatrix} 3 & 2 & 1 \\ 3 & 4 & 8 \end{pmatrix}\begin{pmatrix} 2 & 1 \\ 4 & 3 \\ 6 & 1 \end{pmatrix}$

5 $\begin{pmatrix} 4 & 2 \\ 3 & 1 \end{pmatrix}\begin{pmatrix} 4 & 6 & 3 \\ 8 & 1 & 2 \end{pmatrix}$

6 $\begin{pmatrix} 1 & 4 \\ 2 & 5 \\ 3 & 6 \end{pmatrix}\begin{pmatrix} 4 \\ 5 \end{pmatrix}$

7 $(7 \ \ 3)\begin{pmatrix} 1 & 4 & 3 \\ 2 & 1 & 2 \end{pmatrix}$

8 $\begin{pmatrix} 4 & 1 & 2 \\ 3 & 8 & 1 \\ 5 & 6 & 2 \end{pmatrix}\begin{pmatrix} 1 & 2 \\ 4 & 3 \\ 1 & 0 \end{pmatrix}$

9 $\begin{pmatrix} 1 & 4 & 1 \\ 2 & 3 & 0 \end{pmatrix}\begin{pmatrix} 4 & 1 & 2 \\ 3 & 8 & 1 \\ 5 & 6 & 2 \end{pmatrix}$

10 $(1 \ \ 2 \ \ 3)\begin{pmatrix} 4 \\ 1 \\ 3 \end{pmatrix}$

Compatibility for multiplication

It was possible to find all the matrix products in the last exercise but, just as there are pairs of matrices that cannot be added together, there are pairs of matrices that we cannot multiply together.

Consider $\begin{pmatrix} 1 & 2 \\ 3 & 4 \end{pmatrix}\begin{pmatrix} 1 & 4 \\ 2 & 5 \\ 3 & 6 \end{pmatrix}$

If we try to multiply the entries in the first row of the first matrix with entries in the first column of the second matrix,

i.e.
$$\begin{pmatrix} 1 & 2 \end{pmatrix}\begin{pmatrix} 1 \\ 2 \\ 3 \end{pmatrix}$$

we find that the entries do not fit together in pairs.

We need to make sure that the number of entries in the row we use is the same as the number of entries in the column that goes with it. The easiest way to check this is to write the size below each matrix:

$$\begin{pmatrix} 3 & 1 & 4 \\ 4 & 0 & 3 \end{pmatrix}\begin{pmatrix} 1 \\ 2 \\ 3 \end{pmatrix} = \begin{pmatrix} 17 \\ 3 \end{pmatrix}$$

$$2 \times 3 \qquad 3 \times 1 \qquad 2 \times 1$$

There are three columns in the first matrix and three rows in the second, so these two matrices are compatible for multiplication. The two outer numbers give us the size of the resulting matrix.

Exercise 7j

Under each matrix write its size and find the product

$$\begin{pmatrix} 1 & 2 \end{pmatrix}\begin{pmatrix} 1 & 2 & 3 \\ 3 & 4 & 6 \end{pmatrix}$$

Give the size of this product.

$$\begin{pmatrix} 1 & 2 \end{pmatrix}\begin{pmatrix} 1 & 2 & 3 \\ 3 & 4 & 6 \end{pmatrix} = \begin{pmatrix} 7 & 10 & 15 \end{pmatrix}$$

$1 \times 1 + 2 \times 3 = 7$
$1 \times 2 + 2 \times 4 = 10$
$1 \times 3 + 2 \times 6 = 15$

$$1 \times 2 \qquad 2 \times 3 \qquad\qquad 1 \times 3$$
$$\uparrow$$

When these two numbers are the same the matrices can be multiplied together.

In questions **1** to **8**, under each matrix write its size and give the size of the product. Find the product. (The pairs of matrices in these questions are all compatible for multiplication.)

1 $\begin{pmatrix} 3 & 2 \\ 4 & 1 \end{pmatrix}\begin{pmatrix} 1 \\ 2 \end{pmatrix}$

5 $\begin{pmatrix} 3 & 2 \\ 4 & 5 \end{pmatrix}\begin{pmatrix} 1 & 2 \\ 4 & 7 \end{pmatrix}$

2 $\begin{pmatrix} 1 & 4 & 5 \\ 3 & 2 & 1 \end{pmatrix}\begin{pmatrix} 1 \\ 4 \\ 1 \end{pmatrix}$

6 $\begin{pmatrix} 1 \\ 2 \end{pmatrix}\begin{pmatrix} 3 & 4 \end{pmatrix}$

3 $\begin{pmatrix} 1 & 2 \end{pmatrix}\begin{pmatrix} 4 \\ 3 \end{pmatrix}$

7 $\begin{pmatrix} 9 & 4 \end{pmatrix}\begin{pmatrix} 1 & 4 \\ 3 & 0 \end{pmatrix}$

4 $\begin{pmatrix} 3 & 2 & 1 \\ 3 & 4 & 8 \end{pmatrix}\begin{pmatrix} 2 & 1 \\ 4 & 3 \\ 6 & 1 \end{pmatrix}$

8 $\begin{pmatrix} 1 \\ 2 \\ 3 \end{pmatrix}\begin{pmatrix} 4 & 5 & 6 \end{pmatrix}$

Find, if possible, $\begin{pmatrix} 4 & 3 \\ 1 & 2 \end{pmatrix} (3 \ \ 1 \ \ 4)$

$$\begin{pmatrix} 4 & 3 \\ 1 & 2 \end{pmatrix} (3 \ \ 1 \ \ 4)$$

$$\underset{\uparrow}{2 \times 2 \quad \underline{\quad} \quad 1 \times 3}$$

These numbers are different so it is not possible.

Find, where possible, the following products. Some of the matrices are not compatible for multiplication; in these cases, write 'not possible':

9 $\begin{pmatrix} 3 & 4 \\ 1 & 2 \end{pmatrix} \begin{pmatrix} 4 \\ 1 \end{pmatrix}$

10 $\begin{pmatrix} 2 & 4 \\ 1 & 0 \end{pmatrix} \begin{pmatrix} 2 & 3 \\ 1 & 4 \\ 1 & 5 \end{pmatrix}$

11 $\begin{pmatrix} 3 & 2 \\ 4 & 5 \end{pmatrix} \begin{pmatrix} 1 & 2 \\ 4 & 7 \end{pmatrix}$

12 $\begin{pmatrix} 7 & 2 & 3 \\ 4 & 1 & 0 \end{pmatrix} \begin{pmatrix} 1 & 2 \\ 4 & 5 \end{pmatrix}$

13 $\begin{pmatrix} 1 & 2 \\ 4 & 5 \end{pmatrix} \begin{pmatrix} 7 & 2 & 3 \\ 4 & 1 & 0 \end{pmatrix}$

14 $\begin{pmatrix} 1 & 4 \\ 0 & 6 \end{pmatrix} (3 \ \ 2)$

15 $(2 \ \ 5 \ \ 4) \begin{pmatrix} 3 \\ 4 \\ 1 \end{pmatrix}$

16 $\begin{pmatrix} 1 \\ 2 \end{pmatrix} \begin{pmatrix} 3 & 2 \\ 4 & 1 \end{pmatrix}$

17 $(3 \ \ 2) \begin{pmatrix} 1 & 4 \\ 0 & 6 \end{pmatrix}$

18 $\begin{pmatrix} 3 \\ 4 \\ 1 \end{pmatrix} (2 \ \ 4 \ \ 5)$

Matrices containing negative numbers

Exercise 7k   Find $\begin{pmatrix} 1 & -2 \\ -4 & 3 \end{pmatrix} \begin{pmatrix} 4 \\ -2 \end{pmatrix}$

$$\begin{pmatrix} 1 & -2 \\ -4 & 3 \end{pmatrix} \begin{pmatrix} 4 \\ -2 \end{pmatrix} = \begin{pmatrix} 8 \\ -22 \end{pmatrix} \quad \begin{array}{l} 1 \times 4 + (-2) \times (-2) = 8 \\ (-4) \times 4 + 3 \times (-2) = -22 \end{array}$$

$$2 \times 2 \quad\quad 2 \times 1 \quad\quad 2 \times 1$$

Find the following matrix products:

1 $\begin{pmatrix} 4 & -2 \\ 2 & -3 \end{pmatrix} \begin{pmatrix} 1 \\ -1 \end{pmatrix}$

> **Tip** Remember that the product of two negative numbers is positive.

2 $\begin{pmatrix} 4 & -3 & 0 \\ 1 & 4 & -3 \end{pmatrix} \begin{pmatrix} 1 \\ -2 \\ 4 \end{pmatrix}$

3 $(2 \quad -1 \quad 4) \begin{pmatrix} 1 & 2 \\ 0 & -2 \\ -1 & -3 \end{pmatrix}$

4 $\begin{pmatrix} 1 & -2 & 3 \\ -4 & 1 & 4 \\ 5 & 2 & 6 \end{pmatrix} \begin{pmatrix} -2 \\ -3 \\ -1 \end{pmatrix}$

5 $\begin{pmatrix} -4 & 2 \\ 2 & 5 \end{pmatrix} \begin{pmatrix} -3 & 4 \\ -2 & -5 \end{pmatrix}$

6 $(-6 \quad -2) \begin{pmatrix} 5 & -1 \\ 4 & -3 \end{pmatrix}$

<u>7</u> $(6 \quad 1 \quad -3) \begin{pmatrix} -4 \\ 1 \\ 1 \end{pmatrix}$

<u>8</u> $\begin{pmatrix} -4 \\ 1 \\ 1 \end{pmatrix} (6 \quad 1 \quad -3)$

<u>9</u> $\begin{pmatrix} 2 & 3 \\ -2 & -3 \end{pmatrix} \begin{pmatrix} 5 & 6 & -2 \\ -1 & 2 & 1 \end{pmatrix}$

<u>10</u> $\begin{pmatrix} 5 & -1 \\ 6 & 2 \\ -2 & 1 \end{pmatrix} \begin{pmatrix} 2 & 3 \\ -2 & -3 \end{pmatrix}$

Exercise 7l

The questions in this exercise refer to the following matrices:

$$A = \begin{pmatrix} 4 & 3 \\ 1 & 2 \end{pmatrix} \qquad B = \begin{pmatrix} 1 & -2 \\ 3 & 0 \end{pmatrix} \qquad C = \begin{pmatrix} 6 & 2 \\ -1 & 0 \end{pmatrix} \qquad D = \begin{pmatrix} 1 \\ 2 \end{pmatrix}$$

$$E = (3 \quad 4) \qquad F = \begin{pmatrix} 6 \\ 1 \\ -4 \end{pmatrix} \qquad G = \begin{pmatrix} 1 & 2 & 3 \\ 4 & -5 & 1 \\ 3 & 2 & 4 \end{pmatrix} \qquad H = (3)$$

Find, where possible, the following products:

1 AB	**4** AG	**7** AF	**10** ED	**13** EH
2 BA	**5** DE	**8** GA	**11** EC	**14** HE
3 AD	**6** AE	**9** AH	**12** EF	**15** HD

<u>16</u> Find the other possible products of pairs of matrices chosen from **A** to **G**.

Addition, subtraction and multiplication

Exercise 7m

The questions in this exercise refer to the following matrices:

$$A = \begin{pmatrix} 1 & 2 \\ -4 & 3 \end{pmatrix} \qquad B = \begin{pmatrix} 6 & 2 \\ 1 & 0 \end{pmatrix} \qquad C = \begin{pmatrix} 1 & -1 & 3 \\ 2 & 2 & 0 \end{pmatrix}$$

$$D = \begin{pmatrix} 1 & 2 \\ 3 & 4 \\ 0 & 1 \end{pmatrix} \qquad E = (1 \quad -2) \qquad F = (3 \quad 1)$$

Find, where possible:

1 AB	3 A − B	5 A + C	7 E − F	9 DC	11 BE
2 A + B	4 B − A	6 EF	8 CD	10 C − D	12 EB

Multiplying more than two matrices together

Consider $\quad \mathbf{A} = \begin{pmatrix} 1 & 4 \\ -1 & 0 \end{pmatrix} \qquad \mathbf{B} = \begin{pmatrix} 6 & 2 \\ 0 & 3 \end{pmatrix} \qquad$ and $\qquad \mathbf{C} = \begin{pmatrix} 1 & 2 \\ 3 & 4 \end{pmatrix}$

If we wish to find **ABC** we do this in two steps.

We can think of **ABC** either as (**AB**)**C** or as **A**(**BC**)

either $\qquad \mathbf{AB} = \begin{pmatrix} 1 & 4 \\ -1 & 0 \end{pmatrix} \begin{pmatrix} 6 & 2 \\ 0 & 3 \end{pmatrix} = \begin{pmatrix} 6 & 14 \\ -6 & -2 \end{pmatrix}$

so $\qquad \mathbf{ABC} = \begin{pmatrix} 6 & 14 \\ -6 & -2 \end{pmatrix} \begin{pmatrix} 1 & 2 \\ 3 & 4 \end{pmatrix} = \begin{pmatrix} 48 & 68 \\ -12 & -20 \end{pmatrix}$

or $\qquad \mathbf{BC} = \begin{pmatrix} 6 & 2 \\ 0 & 3 \end{pmatrix} \begin{pmatrix} 1 & 2 \\ 3 & 4 \end{pmatrix} = \begin{pmatrix} 12 & 20 \\ 9 & 12 \end{pmatrix}$

so $\qquad \mathbf{ABC} = \begin{pmatrix} 1 & 4 \\ -1 & 0 \end{pmatrix} \begin{pmatrix} 12 & 20 \\ 9 & 12 \end{pmatrix} = \begin{pmatrix} 48 & 68 \\ -12 & -20 \end{pmatrix}$

We must pay careful attention to the order of the matrices. For example, if we found **AC** and then multiplied by **B**, the result would *not* be **ABC**.

Powers of matrices

$\mathbf{A}^2$ means $\mathbf{A} \times \mathbf{A}$, $\quad \mathbf{A}^3$ means $\mathbf{A} \times \mathbf{A} \times \mathbf{A}$.

Exercise 7n

The questions in this exercise refer to the following matrices:

$$\mathbf{A} = \begin{pmatrix} 4 & -1 \\ 3 & 2 \end{pmatrix} \qquad \mathbf{B} = \begin{pmatrix} 2 & 3 \\ -4 & 2 \end{pmatrix} \qquad \mathbf{C} = \begin{pmatrix} 4 & 0 \\ 3 & -1 \end{pmatrix}$$

Find the following products. Make use of the answers and working from one question to help you with another:

1 AB	3 $\mathbf{A}^2$	5 CAB	7 ABA	9 ACA	11 $\mathbf{B}^3$
2 ABC	4 CBA	6 $\mathbf{A}^3$	8 $\mathbf{A}^2\mathbf{B}$	10 $\mathbf{AB}^2$	12 CBC

Where possible, check your answers by finding them in two different ways. For instance, in question 2, **ABC** can be found by thinking of it as (**AB**)**C** or as **A**(**BC**).

Puzzle

At a farmers' market Miranda bought 100 pieces of fruit for her shop for which she paid $80. If apples cost 50c, peaches 70c and pears 90c, how many more pears than apples did Miranda buy?

Mixed exercises

Exercise 7p

The questions in this exercise refer to the following matrices:

$$A = \begin{pmatrix} 5 & 2 \\ 1 & 4 \end{pmatrix} \qquad B = \begin{pmatrix} 3 & -1 \\ 4 & -3 \end{pmatrix} \qquad C = \begin{pmatrix} 1 \\ 4 \end{pmatrix}$$

1 What are the sizes of **A** and **C**?

2 Are **A** and **B** compatible for multiplication?

3 Are **A** and **C** compatible for multiplication?

4 Find **AB** if it is possible.   **6** Find **A** − **C** if it is possible.

5 Find **A²** and **C²** if it is possible.   **7** Find 3**B**.

8 Find 2**A** + **B** if it is possible.

9 What is the entry in the second row of the first column of **C**?

10 Which is possible to find, **BC** or **CB**?

Exercise 7q

The questions in this exercise refer to the following matrices:

$$P = \begin{pmatrix} 2 & 1 & -1 \\ 4 & 3 & 1 \end{pmatrix} \qquad Q = \begin{pmatrix} 3 & 2 \\ -1 & 4 \end{pmatrix} \qquad R = \begin{pmatrix} 3 \\ 4 \end{pmatrix}$$

1 Find 2**P**.   **3** Find **Q** + 2**R** if it is possible.

2 Find 2**P** + **Q** if it is possible.   **4** What are the sizes of **P** and **Q**?

5 Are **P** and **R** compatible for multiplication?

6 What is the entry in the first row of the first column of **R**?

7 What is the entry in the second row of the third column of **P**?

8 Which is it possible to find, **PQ** or **QP**?

9 Find **QR** if it is possible.

10 Find **Q²** and **P²** if it is possible.

Consider the matrix $\begin{pmatrix} 5 & -2 & 3 \\ 0 & 2 & 4 \\ 1 & 6 & -1 \end{pmatrix}$

Find the sum of the terms in each row, column and diagonal.

For example, sum in column 1 is $5 + 0 + 1 = 6$.

What do you notice about the sum in each case?

Now consider another matrix whose terms are 2^x, where x is the entry in the matrix above.

For example, the element corresponding to 5 above is $2^5 = 32$.

Now complete this matrix.

Now find the product of the terms in each row, in each column, in each diagonal.

What do you notice about the products?

If we call our first matrix an 'additive magic square', we may refer to this matrix as a 'multiplicative magic square'.

Do you notice any relationship between the sum of entries in the 'additive square' and the product of entries in the 'multiplicative square'?

Try to form other multiplicative magic squares using bases other than 2.

IN THIS CHAPTER...

you have seen that:

- matrices that have the same number of rows and the same number of columns can be added together by adding the corresponding entries, e.g.

$$\begin{pmatrix} 2 & 3 & 5 \\ 3 & -1 & 2 \end{pmatrix} + \begin{pmatrix} 1 & 3 & 2 \\ 5 & 4 & 3 \end{pmatrix} =$$

$$\begin{pmatrix} 3 & 6 & 7 \\ 8 & 3 & 5 \end{pmatrix}$$

- matrices that have the same number of rows and the same number of columns can be subtracted by subtracting the corresponding entries, e.g.

$$\begin{pmatrix} 2 & 3 & 5 \\ 3 & -1 & 2 \end{pmatrix} - \begin{pmatrix} 1 & 3 & 2 \\ 5 & 4 & 3 \end{pmatrix} =$$

$$\begin{pmatrix} 1 & 0 & 3 \\ -2 & -5 & -1 \end{pmatrix}$$

- two matrices can be multiplied together if the number of columns in the first matrix is equal to the number of rows in the second matrix, e.g.

$\begin{pmatrix} 2 & 3 \\ 1 & 2 \end{pmatrix} \begin{pmatrix} 2 \\ 3 \end{pmatrix}$ can be multiplied together.

They give $\begin{pmatrix} 2 \times 2 + 3 \times 3 \\ 1 \times 2 + 2 \times 3 \end{pmatrix} = \begin{pmatrix} 13 \\ 8 \end{pmatrix}$

But $\begin{pmatrix} 2 & 1 \\ 3 & 2 \\ 1 & 3 \end{pmatrix} \begin{pmatrix} 2 & 3 \\ 4 & 2 \end{pmatrix}$ cannot be

multiplied together.

- $\mathbf{A}^2$ means $\mathbf{A} \times \mathbf{A}$.

- $\mathbf{ABC}$ means $(\mathbf{AB})\mathbf{C}$ or $\mathbf{A}(\mathbf{BC})$.

8

INVERSE AND SQUARE MATRICES

Did you know that the beginnings of matrices and determinants appeared back in the second century BC? However, they did not reappear until towards the end of the 17th Century so they are generally thought of as a modern branch of mathematics. They do have a very modern application – they are central to some compression techniques for digital pictures.

Square matrices

A square matrix has the same number of rows and columns.

$$\begin{pmatrix} 1 & 2 \\ 3 & -1 \end{pmatrix} \text{ and } \begin{pmatrix} 1 & 6 & 3 \\ 1 & 0 & 1 \\ 4 & 4 & 3 \end{pmatrix} \text{ are square matrices.}$$

A square matrix has two diagonals: $\begin{pmatrix} 1 & 3 & 4 \\ 0 & 1 & 2 \\ 3 & 1 & 2 \end{pmatrix}$

The diagonal that goes from top left to bottom right is the *leading diagonal*: $\begin{pmatrix} 1 & 2 \\ 3 & 4 \end{pmatrix}$

Exercise 8a

State which of the two matrices $\begin{pmatrix} 1 \\ 2 \end{pmatrix}$ and $\begin{pmatrix} 4 & 1 \\ 3 & 2 \end{pmatrix}$ is a square matrix.

Give its size and mark its leading diagonal.

$\begin{pmatrix} 4 & 1 \\ 3 & 2 \end{pmatrix}$ is a 2×2 square matrix.

State whether or not each of the following matrices is a square matrix. If it is, give its size and mark its leading diagonal:

1 $\begin{pmatrix} 3 & 6 & 1 \\ 4 & 0 & 1 \\ 3 & 2 & 1 \end{pmatrix}$ **3** $\begin{pmatrix} 2 & 1 \\ 4 & 3 \end{pmatrix}$ **5** $\begin{pmatrix} -1 & 4 \\ 4 & -1 \end{pmatrix}$

2 $\begin{pmatrix} 4 & 3 & 2 \\ 1 & 3 & 1 \end{pmatrix}$ **4** $\begin{pmatrix} 2 & 3 \\ 6 & 2 \\ 1 & 4 \end{pmatrix}$ **6** $\begin{pmatrix} 6 & 1 & -3 \\ 4 & 1 & 2 \\ 6 & 8 & 1 \end{pmatrix}$

Reminder of matrix multiplication

Remember that you need to multiply the rows of the first matrix by the columns of the second matrix.

Exercise 8b

Find, where possible

a $\begin{pmatrix} 2 & 3 \\ 1 & -3 \\ 4 & -2 \end{pmatrix}\begin{pmatrix} 2 & 3 \\ 1 & 4 \end{pmatrix}$ **b** $(1 \ \ 2)(3 \ \ 4)$

a $\begin{pmatrix} 2 & 3 \\ 1 & -3 \\ 4 & -2 \end{pmatrix}\begin{pmatrix} 2 & 3 \\ 1 & 4 \end{pmatrix} = \begin{pmatrix} 7 & 18 \\ -1 & -9 \\ 6 & 4 \end{pmatrix}$, 7 comes from $(2 \ \ 3)\begin{pmatrix} 2 \\ 1 \end{pmatrix} = 4 + 3$

$ \quad 3 \times 2 \qquad 2 \times 2 \qquad \quad 3 \times 2$

b $(1 \ \ 2)(3 \ \ 4)$ not possible

$ \quad 1 \times 2 \quad 1 \times 2$

Find where possible:

1 $\begin{pmatrix} 3 & 1 \\ 4 & 3 \end{pmatrix}\begin{pmatrix} 1 & 2 \\ 1 & 1 \end{pmatrix}$

5 $\begin{pmatrix} 4 & -1 \\ 2 & 4 \\ 3 & -5 \end{pmatrix}(1 \ 0 \ 1)$

9 $\begin{pmatrix} 4 \\ 3 \\ 2 \end{pmatrix}(1 \ 6 \ 1)$

2 $(1 \ 0 \ 1)\begin{pmatrix} 4 & -1 \\ 2 & 4 \\ 3 & -5 \end{pmatrix}$

6 $\begin{pmatrix} 3 & 5 \\ 4 & -3 \end{pmatrix}\begin{pmatrix} 2 & 1 \\ 4 & 2 \end{pmatrix}$

10 $(1 \ 6 \ 1)\begin{pmatrix} 4 \\ 3 \\ 2 \end{pmatrix}$

3 $\begin{pmatrix} 1 & 2 \\ 3 & 4 \end{pmatrix}\begin{pmatrix} 1 & 6 & -1 \\ 3 & 2 & 1 \end{pmatrix}$

7 $\begin{pmatrix} 1 & 2 \\ 1 & 1 \end{pmatrix}\begin{pmatrix} 3 & 1 \\ 4 & 3 \end{pmatrix}$

11 $\begin{pmatrix} 1 & -1 \\ -1 & 1 \end{pmatrix}\begin{pmatrix} 4 \\ 3 \end{pmatrix}$

4 $\begin{pmatrix} 1 & 6 & -1 \\ 3 & 2 & 1 \end{pmatrix}\begin{pmatrix} 1 & 2 \\ 3 & 4 \end{pmatrix}$

8 $\begin{pmatrix} 4 \\ 3 \\ 2 \end{pmatrix}\begin{pmatrix} 1 \\ 6 \\ 1 \end{pmatrix}$

12 $(2 \ 6)\begin{pmatrix} 5 & -3 \\ 4 & 2 \end{pmatrix}$

The unit matrix and the zero matrix

Exercise 8c

Find:

1 $\begin{pmatrix} 4 & 2 \\ 3 & 4 \end{pmatrix}\begin{pmatrix} 1 & 0 \\ 0 & 1 \end{pmatrix}$

3 $(6 \ 2)\begin{pmatrix} 0 & 0 \\ 0 & 0 \end{pmatrix}$

5 $(3 \ 2)\begin{pmatrix} 1 & 0 \\ 0 & 1 \end{pmatrix}$

2 $\begin{pmatrix} 1 & 0 \\ 0 & 1 \end{pmatrix}\begin{pmatrix} 4 \\ 5 \end{pmatrix}$

4 $\begin{pmatrix} 0 & 0 \\ 0 & 0 \end{pmatrix}\begin{pmatrix} 2 & 3 \\ 4 & 1 \end{pmatrix}$

6 $\begin{pmatrix} 1 & 0 \\ 0 & 1 \end{pmatrix}\begin{pmatrix} 3 & 2 & -1 \\ 4 & 3 & 1 \end{pmatrix}$

The *unit matrix* $\begin{pmatrix} 1 & 0 \\ 0 & 1 \end{pmatrix}$ does not alter the matrix it multiplies or is multiplied by, e.g.:

$$\begin{pmatrix} 1 & 2 \\ 4 & 5 \end{pmatrix}\begin{pmatrix} 1 & 0 \\ 0 & 1 \end{pmatrix} = \begin{pmatrix} 1 & 2 \\ 4 & 5 \end{pmatrix} \text{ and } \begin{pmatrix} 1 & 0 \\ 0 & 1 \end{pmatrix}\begin{pmatrix} 1 & 2 \\ 4 & 5 \end{pmatrix} = \begin{pmatrix} 1 & 2 \\ 4 & 5 \end{pmatrix}$$

This is similar to multiplying by 1 in ordinary number multiplication.
A unit matrix is usually represented by **I**. A unit matrix is also called an *identity matrix*.

The *zero matrix* $\begin{pmatrix} 0 & 0 \\ 0 & 0 \end{pmatrix}$, when multiplied by any compatible matrix, gives the zero matrix:

$$\begin{pmatrix} 1 & 2 \\ 4 & 5 \end{pmatrix}\begin{pmatrix} 0 & 0 \\ 0 & 0 \end{pmatrix} = \begin{pmatrix} 0 & 0 \\ 0 & 0 \end{pmatrix} \text{ and } \begin{pmatrix} 0 & 0 \\ 0 & 0 \end{pmatrix}\begin{pmatrix} 1 & 2 \\ 4 & 5 \end{pmatrix} = \begin{pmatrix} 0 & 0 \\ 0 & 0 \end{pmatrix}$$

This is similar to multiplying by zero in ordinary number multiplication.
A zero matrix is usually represented by **0**.

There are other unit and zero matrices, e.g.

$$\begin{pmatrix} 1 & 0 & 0 \\ 0 & 1 & 0 \\ 0 & 0 & 1 \end{pmatrix} \text{ and } \begin{pmatrix} 0 & 0 & 0 \\ 0 & 0 & 0 \\ 0 & 0 & 0 \end{pmatrix}$$

but we are concerned mainly with 2×2 square matrices in this chapter.

Exercise 8d

Find:

1 $\begin{pmatrix} 1 & 1 \\ 2 & 3 \end{pmatrix}\begin{pmatrix} 3 & -1 \\ -2 & 1 \end{pmatrix}$

5 $\begin{pmatrix} 6 & -3 \\ -3 & 2 \end{pmatrix}\begin{pmatrix} 2 & 3 \\ 3 & 6 \end{pmatrix}$

2 $\begin{pmatrix} 4 & 1 \\ 3 & 2 \end{pmatrix}\begin{pmatrix} 2 & -1 \\ -3 & 4 \end{pmatrix}$

6 $\begin{pmatrix} 2 & 3 \\ 3 & 6 \end{pmatrix}\begin{pmatrix} 6 & -3 \\ -3 & 2 \end{pmatrix}$

3 $\begin{pmatrix} 2 & -1 \\ -3 & 4 \end{pmatrix}\begin{pmatrix} 4 & 1 \\ 3 & 2 \end{pmatrix}$

7 $\begin{pmatrix} 9 & 7 \\ 4 & 3 \end{pmatrix}\begin{pmatrix} 3 & -7 \\ -4 & 9 \end{pmatrix}$

4 $\begin{pmatrix} 9 & 4 \\ 4 & 2 \end{pmatrix}\begin{pmatrix} 2 & -4 \\ -4 & 9 \end{pmatrix}$

8 $\begin{pmatrix} 3 & -2 \\ -4 & 3 \end{pmatrix}\begin{pmatrix} 3 & 2 \\ 4 & 3 \end{pmatrix}$

9 What do you notice about the arrangement of the numbers and signs in each pair of matrices? What do you notice about the answers?

In questions **10** to **19**, find a second matrix **B** such that **B** × **A** is of the form $\begin{pmatrix} k & 0 \\ 0 & k \end{pmatrix}$, i.e. like the answers to questions **1** to **8**.

10 $\mathbf{A} = \begin{pmatrix} 3 & 2 \\ 8 & 6 \end{pmatrix}$

15 $\mathbf{A} = \begin{pmatrix} 5 & 3 \\ 9 & 6 \end{pmatrix}$

11 $\mathbf{A} = \begin{pmatrix} 2 & 1 \\ 2 & 3 \end{pmatrix}$

16 $\mathbf{A} = \begin{pmatrix} 5 & 4 \\ 9 & 8 \end{pmatrix}$

12 $\mathbf{A} = \begin{pmatrix} 6 & -1 \\ -20 & 3 \end{pmatrix}$

17 $\mathbf{A} = \begin{pmatrix} 6 & 2 \\ 14 & 4 \end{pmatrix}$

13 $\mathbf{A} = \begin{pmatrix} 1 & -3 \\ 1 & 5 \end{pmatrix}$

18 $\mathbf{A} = \begin{pmatrix} 1 & 2 \\ 2 & 1 \end{pmatrix}$

14 $\mathbf{A} = \begin{pmatrix} 5 & 7 \\ 3 & 4 \end{pmatrix}$

19 $\mathbf{A} = \begin{pmatrix} 4 & -3 \\ 3 & -2 \end{pmatrix}$

If we are given a 2×2 matrix **A** and we want a matrix **B** such that $\mathbf{A} \times \mathbf{B}$ is of the form $\begin{pmatrix} k & 0 \\ 0 & k \end{pmatrix}$ we

a interchange the numbers in the leading diagonal of **A**

b leave the other two numbers of **A** where they are, but change their signs.

For example, if $\mathbf{A} = \begin{pmatrix} 4 & 2 \\ 5 & 3 \end{pmatrix}$, to get **B**

we change $\begin{pmatrix} 4 & \\ & 3 \end{pmatrix}$ to $\begin{pmatrix} 3 & \\ & 4 \end{pmatrix}$ and $\begin{pmatrix} & 2 \\ 5 & \end{pmatrix}$ to $\begin{pmatrix} & -2 \\ -5 & \end{pmatrix}$

then $\begin{pmatrix} 4 & 2 \\ 5 & 3 \end{pmatrix}\begin{pmatrix} 3 & -2 \\ -5 & 4 \end{pmatrix} = \begin{pmatrix} 2 & 0 \\ 0 & 2 \end{pmatrix}$

This matrix is called a *diagonal matrix*.

The inverse of a matrix

In ordinary numbers we say $\frac{1}{7}$ is the inverse of 7 because $7 \times \frac{1}{7} = 1$

Similarly $\frac{2}{5}$ is the inverse of $\frac{5}{2}$ because $\frac{5}{2} \times \frac{2}{5} = 1$

In the same way $\begin{pmatrix} 3 & 5 \\ 1 & 2 \end{pmatrix} \times \begin{pmatrix} 2 & -5 \\ -1 & 3 \end{pmatrix} = \begin{pmatrix} 1 & 0 \\ 0 & 1 \end{pmatrix}$

so $\begin{pmatrix} 2 & -5 \\ -1 & 3 \end{pmatrix}$ is the inverse of $\begin{pmatrix} 3 & 5 \\ 1 & 2 \end{pmatrix}$

and $\begin{pmatrix} 3 & 5 \\ 1 & 2 \end{pmatrix}$ is the inverse of $\begin{pmatrix} 2 & -5 \\ -1 & 3 \end{pmatrix}$

Exercise 8e

Find the inverse of $\begin{pmatrix} 5 & 3 \\ 3 & 2 \end{pmatrix}$

Try $\begin{pmatrix} 2 & -3 \\ -3 & 5 \end{pmatrix}$ (Interchange the numbers in the leading diagonal and change the signs of the numbers in the other diagonal.)

$$\begin{pmatrix} 5 & 3 \\ 3 & 2 \end{pmatrix}\begin{pmatrix} 2 & -3 \\ -3 & 5 \end{pmatrix} = \begin{pmatrix} 1 & 0 \\ 0 & 1 \end{pmatrix}$$

So $\begin{pmatrix} 2 & -3 \\ -3 & 5 \end{pmatrix}$ is the inverse of $\begin{pmatrix} 5 & 3 \\ 3 & 2 \end{pmatrix}$

Find the inverses of the following matrices:

1 $\begin{pmatrix} 4 & 1 \\ 7 & 2 \end{pmatrix}$ **3** $\begin{pmatrix} 10 & -3 \\ 7 & -2 \end{pmatrix}$ **5** $\begin{pmatrix} 7 & 4 \\ 12 & 7 \end{pmatrix}$

2 $\begin{pmatrix} 11 & 3 \\ 7 & 2 \end{pmatrix}$ **4** $\begin{pmatrix} 3 & 5 \\ 4 & 7 \end{pmatrix}$ **6** $\begin{pmatrix} 1 & 1 \\ 1 & 2 \end{pmatrix}$

Not all inverses can be produced in this way, e.g.

$$\begin{pmatrix} 5 & 4 \\ 3 & 3 \end{pmatrix} \times \begin{pmatrix} 3 & -4 \\ -3 & 5 \end{pmatrix} = \begin{pmatrix} 3 & 0 \\ 0 & 3 \end{pmatrix} \text{ and not } \begin{pmatrix} 1 & 0 \\ 0 & 1 \end{pmatrix}$$

But $\begin{pmatrix} 3 & 0 \\ 0 & 3 \end{pmatrix} = 3\begin{pmatrix} 1 & 0 \\ 0 & 1 \end{pmatrix}$ so our attempt at the inverse is three times too big.

We therefore divide our attempt at the inverse by 3.

The inverse of $\begin{pmatrix} 5 & 4 \\ 3 & 3 \end{pmatrix}$ is $\begin{pmatrix} 1 & -\frac{4}{3} \\ -1 & \frac{5}{3} \end{pmatrix}$

The number 3 is called the *determinant* of $\begin{pmatrix} 5 & 4 \\ 3 & 3 \end{pmatrix}$

In Exercise 8e, questions 1 to 6 the determinant in each case is 1.

Find the inverse of $\begin{pmatrix} 5 & 8 \\ 1 & 2 \end{pmatrix}$

Try $\begin{pmatrix} 2 & -8 \\ -1 & 5 \end{pmatrix}$

$\begin{pmatrix} 5 & 8 \\ 1 & 2 \end{pmatrix}\begin{pmatrix} 2 & -8 \\ -1 & 5 \end{pmatrix} = \begin{pmatrix} 2 & 0 \\ 0 & 2 \end{pmatrix}$

The determinant of $\begin{pmatrix} 5 & 8 \\ 1 & 2 \end{pmatrix}$ is 2,

so the inverse of $\begin{pmatrix} 5 & 8 \\ 1 & 2 \end{pmatrix}$ is $\begin{pmatrix} \frac{2}{2} & \frac{-8}{2} \\ \frac{-1}{2} & \frac{5}{2} \end{pmatrix} = \begin{pmatrix} 1 & -4 \\ -\frac{1}{2} & 2\frac{1}{2} \end{pmatrix}$

Note that this inverse can be written as $\frac{1}{2}\begin{pmatrix} 2 & -8 \\ -1 & 5 \end{pmatrix}$

Find the inverses of the following matrices:

1 $\begin{pmatrix} 6 & 2 \\ 8 & 3 \end{pmatrix}$ **4** $\begin{pmatrix} 1 & -2 \\ 1 & 1 \end{pmatrix}$ **7** $\begin{pmatrix} 4 & -3 \\ -5 & 4 \end{pmatrix}$

2 $\begin{pmatrix} 9 & 2 \\ 3 & 1 \end{pmatrix}$ **5** $\begin{pmatrix} 2 & 0 \\ 0 & 3 \end{pmatrix}$ **8** $\begin{pmatrix} 4 & 3 \\ 5 & 4 \end{pmatrix}$

3 $\begin{pmatrix} 4 & 2 \\ 10 & 6 \end{pmatrix}$ **6** $\begin{pmatrix} 3 & 2 \\ 11 & 8 \end{pmatrix}$ **9** $\begin{pmatrix} 3 & -4 \\ -2 & 4 \end{pmatrix}$

Find the inverses of the following matrices:

10 $\begin{pmatrix} 6 & 2 \\ 7 & 2 \end{pmatrix}$ **13** $\begin{pmatrix} 3 & 2 \\ 4 & 2 \end{pmatrix}$

11 $\begin{pmatrix} 9 & 4 \\ 5 & 2 \end{pmatrix}$ **14** $\begin{pmatrix} 3 & -4 \\ -3 & 3 \end{pmatrix}$

12 $\begin{pmatrix} -4 & 1 \\ 1 & 1 \end{pmatrix}$ **15** $\begin{pmatrix} 5 & 4 \\ 8 & 6 \end{pmatrix}$

Tip The determinants of these matrices are negative numbers.

Sometimes the determinant is zero. In this case, as we cannot divide by zero, we cannot produce an inverse.

16 State whether the given matrix has a zero determinant and hence no inverse:

a $\begin{pmatrix} 6 & 4 \\ 3 & 2 \end{pmatrix}$ **b** $\begin{pmatrix} 11 & 3 \\ 4 & 1 \end{pmatrix}$ **c** $\begin{pmatrix} 8 & 6 \\ 4 & 3 \end{pmatrix}$

Tip Try to find the inverse.
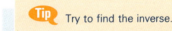

17 Which of the following matrices are their own inverses?

a $\begin{pmatrix} 1 & 0 \\ 0 & -1 \end{pmatrix}$ **b** $\begin{pmatrix} 0 & 1 \\ 1 & 0 \end{pmatrix}$ **c** $\begin{pmatrix} -1 & 0 \\ 0 & -1 \end{pmatrix}$

Find, where possible, the inverses of the following matrices:

18 $\begin{pmatrix} 6 & 5 \\ 5 & 5 \end{pmatrix}$ **20** $\begin{pmatrix} 5 & 0 \\ 0 & 5 \end{pmatrix}$ **22** $\begin{pmatrix} 5 & 3 \\ 9 & 6 \end{pmatrix}$

19 $\begin{pmatrix} 6 & 4 \\ 6 & 4 \end{pmatrix}$ **21** $\begin{pmatrix} 5 & 7 \\ 3 & 4 \end{pmatrix}$ **23** $\begin{pmatrix} 8 & 3 \\ 16 & 6 \end{pmatrix}$

Notation

With ordinary numbers, $\frac{1}{3}$ is the inverse of 3 under multiplication.

$\frac{1}{3}$ can be written as 3^{-1}, so 3^{-1} is the inverse of 3.
In the same way, the inverse of **A** is written $\mathbf{A}^{-1}$.

If $\qquad\qquad \mathbf{A} = \begin{pmatrix} 4 & 7 \\ 1 & 2 \end{pmatrix}$, then $\mathbf{A}^{-1} = \begin{pmatrix} 2 & -7 \\ -1 & 4 \end{pmatrix}$

$$\begin{pmatrix} 4 & 7 \\ 1 & 2 \end{pmatrix}\begin{pmatrix} 2 & -7 \\ -1 & 4 \end{pmatrix} = \begin{pmatrix} 1 & 0 \\ 0 & 1 \end{pmatrix}$$

and $\qquad\qquad \mathbf{AA}^{-1} = \mathbf{I}$

The determinant of **A** is written $|\mathbf{A}|$. In this case $|\mathbf{A}| = 1$

Exercise 8g

The questions in this exercise refer to the following matrices:

$\mathbf{A} = \begin{pmatrix} 5 & 3 \\ 3 & 2 \end{pmatrix}$ $\mathbf{B} = \begin{pmatrix} 2 & 2 \\ 2 & 3 \end{pmatrix}$ $\mathbf{C} = \begin{pmatrix} 6 & 2 \\ 5 & 2 \end{pmatrix}$ $\mathbf{D} = \begin{pmatrix} 7 & 3 \\ 2 & 1 \end{pmatrix}$ $\mathbf{I} = \begin{pmatrix} 1 & 0 \\ 0 & 1 \end{pmatrix}$

Find **a** $|\mathbf{D}|$ **b** $\mathbf{D}^{-1}$ **c** $|\mathbf{C}|$ **d** $\mathbf{C}^{-1}$ **e** $\mathbf{D}^{-1}\mathbf{C}$

a Try $\begin{pmatrix} 1 & -3 \\ -2 & 7 \end{pmatrix}$

$$\begin{pmatrix} 7 & 3 \\ 2 & 1 \end{pmatrix}\begin{pmatrix} 1 & -3 \\ -2 & 7 \end{pmatrix} = \begin{pmatrix} 1 & 0 \\ 0 & 1 \end{pmatrix}$$

$$|\mathbf{D}| = 1$$

b $\mathbf{D}^{-1} = \begin{pmatrix} 1 & -3 \\ -2 & 7 \end{pmatrix}$

c Try $\begin{pmatrix} 2 & -2 \\ -5 & 6 \end{pmatrix}$

$$\begin{pmatrix} 6 & 2 \\ 5 & 2 \end{pmatrix}\begin{pmatrix} 2 & -2 \\ -5 & 6 \end{pmatrix} = \begin{pmatrix} 2 & 0 \\ 0 & 2 \end{pmatrix}$$

$$|\mathbf{C}| = 2$$

d $\mathbf{C}^{-1} = \frac{1}{2}\begin{pmatrix} 2 & -2 \\ -5 & 6 \end{pmatrix} = \begin{pmatrix} 1 & -1 \\ -2\frac{1}{2} & 3 \end{pmatrix}$

e $\mathbf{D}^{-1}\mathbf{C} = \begin{pmatrix} 1 & -3 \\ -2 & 7 \end{pmatrix}\begin{pmatrix} 6 & 2 \\ 5 & 2 \end{pmatrix} = \begin{pmatrix} -9 & -4 \\ 23 & 10 \end{pmatrix}$

Write out the matrices **A**, **B**, **C**, **D** and **I** and hence find:

<u>**1**</u> |**A**| and **A**$^{-1}$ <u>**4**</u> **AB**   <u>**7**</u> **B**$^{-1}$**A**$^{-1}$ <u>**10**</u> (**A**$^{-1}$)^2

<u>**2**</u> |**B**| and **B**$^{-1}$ <u>**5**</u> (**AB**)$^{-1}$ <u>**8**</u> **A**^{2} <u>**11**</u> **C**$^{-1}$**A**

<u>**3**</u> **I**$^{-1}$ <u>**6**</u> **A**$^{-1}$**B**$^{-1}$ <u>**9**</u> (**A**^{2})$^{-1}$ <u>**12**</u> **AC**$^{-1}$

Finding the determinant

The determinant can be found without trying to find the inverse.

> The determinant of the matrix $\begin{pmatrix} a & b \\ c & d \end{pmatrix}$ is $ad - bc$.

We find the product of the entries in the leading diagonal, ad, and then we subtract the product of the entries in the other diagonal, bc.

For example, the determinant of $\begin{pmatrix} 3 & 1 \\ 4 & 2 \end{pmatrix}$ is $(3 \times 2) - (1 \times 4) = 2$

Exercise 8h

Find the determinants of the following matrices:

> **Tip**   Be careful with signs; it is sensible to use brackets, i.e. if
>
> $\mathbf{A} = \begin{pmatrix} a & b \\ c & d \end{pmatrix}$, $|\mathbf{A}| = (a \times d) - (b \times c)$

1 $\begin{pmatrix} 6 & 1 \\ 3 & 2 \end{pmatrix}$ **5** $\begin{pmatrix} 5 & 1 \\ 4 & -2 \end{pmatrix}$

2 $\begin{pmatrix} 4 & 1 \\ 3 & 5 \end{pmatrix}$ **6** $\begin{pmatrix} 6 & -2 \\ 2 & 1 \end{pmatrix}$

3 $\begin{pmatrix} 6 & 2 \\ 6 & 2 \end{pmatrix}$ **7** $\begin{pmatrix} 3 & 4 \\ 1 & 1 \end{pmatrix}$ <u>**9**</u> $\begin{pmatrix} 1 & 4 \\ -1 & 5 \end{pmatrix}$ <u>**11**</u> $\begin{pmatrix} 2 & 3 \\ 1 & 4 \end{pmatrix}$

4 $\begin{pmatrix} 4 & -3 \\ 1 & 4 \end{pmatrix}$ **8** $\begin{pmatrix} 6 & -7 \\ -2 & 1 \end{pmatrix}$ <u>**10**</u> $\begin{pmatrix} -2 & -3 \\ -1 & -4 \end{pmatrix}$ <u>**12**</u> $\begin{pmatrix} 6 & 1 \\ 3 & -1 \end{pmatrix}$

Puzzle

Find the missing number.

8	3
4	6

12	5
10	6

16	3
8	

Solution of simultaneous equations using matrices

A pair of simultaneous equations is sometimes given in matrix form.

Consider the matrix equation

$$\begin{pmatrix} 5 & 2 \\ 3 & 5 \end{pmatrix}\begin{pmatrix} x \\ y \end{pmatrix} = \begin{pmatrix} 1 \\ -7 \end{pmatrix}$$

Multiply the left-hand side

$$\begin{pmatrix} 5x + 2y \\ 3x + 5y \end{pmatrix} = \begin{pmatrix} 1 \\ -7 \end{pmatrix}$$

This represents the pair of equations

$$5x + 2y = 1$$

$$3x + 5y = -7$$

Similarly, a pair of given equations can be written in matrix form.

For example, $\begin{aligned} 4x + 2y &= 7 \\ x + 3y &= 8 \end{aligned}$ gives $\begin{pmatrix} 4x + 2y \\ x + 3y \end{pmatrix} = \begin{pmatrix} 7 \\ 8 \end{pmatrix}$

i.e. $\begin{pmatrix} 4 & 2 \\ 1 & 3 \end{pmatrix}\begin{pmatrix} x \\ y \end{pmatrix} = \begin{pmatrix} 7 \\ 8 \end{pmatrix}$

Exercise 8i   Find the pair of simultaneous equations which are equivalent to the

matrix equation $\begin{pmatrix} 1 & 2 \\ 3 & 1 \end{pmatrix}\begin{pmatrix} x \\ y \end{pmatrix} = \begin{pmatrix} 7 \\ 5 \end{pmatrix}$

$$\begin{pmatrix} 1 & 2 \\ 3 & 1 \end{pmatrix}\begin{pmatrix} x \\ y \end{pmatrix} = \begin{pmatrix} 7 \\ 5 \end{pmatrix}$$

(Multiply the left-hand side)

$$\begin{pmatrix} 1x + 2y \\ 3x + 1y \end{pmatrix} = \begin{pmatrix} 7 \\ 5 \end{pmatrix}$$

∴ the pair of equations is $\begin{aligned} x + 2y &= 7 \\ 3x + y &= 5 \end{aligned}$

Find the pairs of simultaneous equations which are equivalent to the following matrix equations:

1 $\begin{pmatrix} 1 & 2 \\ 3 & 2 \end{pmatrix}\begin{pmatrix} x \\ y \end{pmatrix} = \begin{pmatrix} 3 \\ 5 \end{pmatrix}$

3 $\begin{pmatrix} 9 & 2 \\ 4 & 1 \end{pmatrix}\begin{pmatrix} x \\ y \end{pmatrix} = \begin{pmatrix} 24 \\ 11 \end{pmatrix}$

2 $\begin{pmatrix} 4 & 2 \\ 5 & 3 \end{pmatrix}\begin{pmatrix} x \\ y \end{pmatrix} = \begin{pmatrix} 12 \\ 15 \end{pmatrix}$

4 $\begin{pmatrix} 6 & -1 \\ 2 & 1 \end{pmatrix}\begin{pmatrix} p \\ q \end{pmatrix} = \begin{pmatrix} -8 \\ 0 \end{pmatrix}$

Find the matrix equation which is equivalent to the pair of equations

$2x - 4y = 0$
$9x + y = 19$

$$2x - 4y = 0$$
$$9x + y = 19$$

Strip out the xs and ys to give $\begin{pmatrix} x \\ y \end{pmatrix}$; the numbers left form the 2 × 2 matrix.

$$\begin{pmatrix} 2 & -4 \\ 9 & 1 \end{pmatrix} \begin{pmatrix} x \\ y \end{pmatrix} = \begin{pmatrix} 0 \\ 19 \end{pmatrix}$$

Find the matrix equations which are equivalent to the following pairs of equations:

5 $3x + 2y = 8$
   $x + y = 3$

6 $4x - 3y = 1$
   $2x + y = 3$

7 $4x + 3y = 5$
   $5x + 4y = 6$

8 $3x - 2y = 1$
   $x - y = 0$

9 $7x - 2y = 3$
   $3x + 4y = 11$

10 $5x + y = -8$
   $4x - 3y = -14$

Consider the equation $2x = 6$

If we multiply both sides by $\frac{1}{2}$, we get $\frac{1}{2} \times 2x = \frac{1}{2} \times 6$

i.e. $\qquad\qquad\qquad\qquad\qquad x = 3$

$\frac{1}{2}$ is the reciprocal of 2, i.e. $\frac{1}{2}$ is the inverse of 2 under multiplication.

In the same way, we can simplify a matrix equation if we multiply both sides by the inverse of the first matrix.

Solve the matrix equation $\begin{pmatrix} 3 & 1 \\ 1 & 2 \end{pmatrix} \begin{pmatrix} x \\ y \end{pmatrix} = \begin{pmatrix} 7 \\ 4 \end{pmatrix}$

(Find the inverse first.)

The determinant of $\begin{pmatrix} 3 & 1 \\ 1 & 2 \end{pmatrix} = (3 \times 2) - (1 \times 1) = 5$

∴ the inverse is $\frac{1}{5} \begin{pmatrix} 2 & -1 \\ -1 & 3 \end{pmatrix}$

Multiply both sides of the equation by $\frac{1}{5} \begin{pmatrix} 2 & -1 \\ -1 & 3 \end{pmatrix}$

(The order matters now. *The inverse must go in front on both sides.*)

$$\frac{1}{5} \begin{pmatrix} 2 & -1 \\ -1 & 3 \end{pmatrix} \begin{pmatrix} 3 & 1 \\ 1 & 2 \end{pmatrix} \begin{pmatrix} x \\ y \end{pmatrix} = \frac{1}{5} \begin{pmatrix} 2 & -1 \\ -1 & 3 \end{pmatrix} \begin{pmatrix} 7 \\ 4 \end{pmatrix}$$

$$\begin{pmatrix} 1 & 0 \\ 0 & 1 \end{pmatrix} \begin{pmatrix} x \\ y \end{pmatrix} = \frac{1}{5} \begin{pmatrix} 10 \\ 5 \end{pmatrix}$$

i.e.

$$\begin{pmatrix} x \\ y \end{pmatrix} = \begin{pmatrix} 2 \\ 1 \end{pmatrix}$$

∴ $x = 2$ and $y = 1$

Solve the following matrix equations:

1 $\begin{pmatrix} 4 & 1 \\ 3 & 1 \end{pmatrix} \begin{pmatrix} x \\ y \end{pmatrix} = \begin{pmatrix} 6 \\ 5 \end{pmatrix}$

7 $\begin{pmatrix} 1 & -1 \\ 1 & 3 \end{pmatrix} \begin{pmatrix} x \\ y \end{pmatrix} = \begin{pmatrix} 2 \\ 10 \end{pmatrix}$

2 $\begin{pmatrix} 2 & 1 \\ 5 & 3 \end{pmatrix} \begin{pmatrix} x \\ y \end{pmatrix} = \begin{pmatrix} 7 \\ 19 \end{pmatrix}$

8 $\begin{pmatrix} 5 & 2 \\ 2 & 1 \end{pmatrix} \begin{pmatrix} x \\ y \end{pmatrix} = \begin{pmatrix} 1 \\ 0 \end{pmatrix}$

3 $\begin{pmatrix} 4 & 3 \\ 2 & 2 \end{pmatrix} \begin{pmatrix} x \\ y \end{pmatrix} = \begin{pmatrix} 1 \\ 0 \end{pmatrix}$

9 $\begin{pmatrix} 3 & 2 \\ 3 & 3 \end{pmatrix} \begin{pmatrix} x \\ y \end{pmatrix} = \begin{pmatrix} 16 \\ 18 \end{pmatrix}$

4 $\begin{pmatrix} 3 & -1 \\ 4 & 1 \end{pmatrix} \begin{pmatrix} x \\ y \end{pmatrix} = \begin{pmatrix} 7 \\ 7 \end{pmatrix}$

10 $\begin{pmatrix} 4 & -2 \\ -3 & 1 \end{pmatrix} \begin{pmatrix} p \\ q \end{pmatrix} = \begin{pmatrix} 2 \\ -2 \end{pmatrix}$

5 $\begin{pmatrix} 2 & 1 \\ 1 & 2 \end{pmatrix} \begin{pmatrix} x \\ y \end{pmatrix} = \begin{pmatrix} 6 \\ 3 \end{pmatrix}$

11 $\begin{pmatrix} -5 & 1 \\ 9 & -2 \end{pmatrix} \begin{pmatrix} s \\ t \end{pmatrix} = \begin{pmatrix} 13 \\ -24 \end{pmatrix}$

6 $\begin{pmatrix} 7 & -2 \\ 3 & 4 \end{pmatrix} \begin{pmatrix} x \\ y \end{pmatrix} = \begin{pmatrix} 3 \\ 11 \end{pmatrix}$

Give the following pairs of simultaneous equations in matrix form and hence solve them:

12 $x + y = 2$
$x + 2y = 3$

14 $5x + 4y = 1$
$x + y = 0$

16 $9x + 2y = 11$
$3x + y = 5$

18 $5x + 2y = 16$
$3x - y = 3$

13 $4x - y = 5$
$x + y = 5$

15 $2x + 3y = 15$
$3x + 5y = 23$

17 $2x + 3y = 7$
$3x + 2y = 8$

19 $x + 4y = 11$
$2x + 3y = 7$

 ## Investigation

Try using matrices to solve the equations $\begin{array}{l} x + 2y = 3 \\ x + 2y = 6 \end{array}$

Write a short report on why the method breaks down and how you could predict that it would break down.

What do you think will happen when you try to use matrices to solve

$3x + 2y = 1$
$9x + 6y = 5?$ Explain your answers.

Matrix equations

Exercise 8k

The questions in this exercise refer to the following matrices:

$A = \begin{pmatrix} 4 & 3 \\ 2 & 1 \end{pmatrix}$ $B = \begin{pmatrix} 1 & -2 \\ 0 & 1 \end{pmatrix}$ $C = \begin{pmatrix} 6 & 4 \\ -3 & 1 \end{pmatrix}$

Find **X** if **A** + **X** = **B**

$$\mathbf{A} + \mathbf{X} = \mathbf{B}$$

$$\mathbf{X} = \mathbf{B} - \mathbf{A} \quad \text{(taking } \mathbf{A} \text{ from each side)}$$

$$\mathbf{X} = \begin{pmatrix} 1 & -2 \\ 0 & 1 \end{pmatrix} - \begin{pmatrix} 4 & 3 \\ 2 & 1 \end{pmatrix}$$

$$= \begin{pmatrix} -3 & -5 \\ -2 & 0 \end{pmatrix}$$

Find **X** in the following equations:

1 **B** + **X** = **C**   **3** **X** = **AC**   **5** 2**X** = **A**   **7** **A** + **X** = **B** + **C**

2 **X** − **C** = **B**   **4** **B** − **X** = **C**   **6** $\frac{1}{3}$**X** = **B**   **8** **AX** = **I**

Mixed exercises

Exercise 8l

1 Find $\begin{pmatrix} 2 & 3 & -1 \\ 4 & 1 & 2 \end{pmatrix} + \begin{pmatrix} 3 & 1 & 4 \\ 6 & -9 & 2 \end{pmatrix}$ **4** Find the inverse of $\begin{pmatrix} 4 & 3 \\ 3 & 4 \end{pmatrix}$

2 Find $\frac{1}{2}\begin{pmatrix} 2 & 7 \\ 3 & -1 \end{pmatrix}$ **5** Find $(1 \quad 3 \quad -1)\begin{pmatrix} 1 \\ -3 \\ 1 \end{pmatrix}$

3 Find the determinant of $\begin{pmatrix} 16 & 3 \\ 8 & 3 \end{pmatrix}$ **6** Find $\begin{pmatrix} 1 & 4 \\ 3 & 1 \end{pmatrix}\begin{pmatrix} 2 & -1 \\ 2 & 1 \end{pmatrix}\begin{pmatrix} 1 & 3 \\ 1 & 1 \end{pmatrix}$

Exercise 8m

The questions in this exercise refer to the following matrices:

$$\mathbf{A} = \begin{pmatrix} 2 & 4 \\ 1 & 3 \end{pmatrix} \quad \mathbf{B} = \begin{pmatrix} 3 & -1 \\ -2 & 1 \end{pmatrix} \quad \mathbf{C} = \begin{pmatrix} 2 \\ 3 \end{pmatrix} \quad \mathbf{D} = (4 \quad -2)$$

1 Find **A** + **B**.   **4** Find **BC**.

2 Find the determinant of **A**.   **5** Find **DA**.

3 Find the inverse of **B**.   **6** Find $\frac{3}{4}$**D**.

MATHS IS OUT THERE

Maria Agnesi (1718–1799) was born into a rich Italian family. By the age of nine she was able to speak five languages. As a teenager she argued about mathematics with teachers who visited her home. It is said that the queen sent her a diamond ring, the Pope sent her a letter of praise and a school in Italy was named after her. She was a kind and smart woman. In one of her books however she gave a Latin name to a curve. In a translation the word 'versoria' meaning 'free to move in every direction' was read as 'versiera' meaning 'a witch'. She became known as 'The witch of Agnesi'.

IN THIS CHAPTER...

you have seen that:

- a square matrix has two diagonals; the leading diagonal goes from top left to bottom right

- the unit matrix is $\mathbf{I} = \begin{pmatrix} 1 & 0 \\ 0 & 1 \end{pmatrix}$ and the zero matrix is $\begin{pmatrix} 0 & 0 \\ 0 & 0 \end{pmatrix}$

- the determinant of $\mathbf{A} = \begin{pmatrix} a & b \\ c & d \end{pmatrix}$ is written as $|\mathbf{A}|$ and is equal to $(a \times d) - (b \times c)$

- to find the inverse of the matrix $\mathbf{A} = \begin{pmatrix} a & b \\ c & d \end{pmatrix}$, interchange the numbers in the leading diagonal and change the signs of the numbers in the other diagonal, i.e. $\begin{pmatrix} d & -b \\ -c & a \end{pmatrix}$ then multiply by $\dfrac{1}{|\mathbf{A}|}$

- you can use matrices to solve a pair of simultaneous equations by writing

$$\begin{array}{l} ax + by = e \\ cx + dy = f \end{array} \text{ as } \begin{pmatrix} a & b \\ c & d \end{pmatrix}\begin{pmatrix} x \\ y \end{pmatrix} = \begin{pmatrix} e \\ f \end{pmatrix}$$

9 STRAIGHT LINE GRAPHS

MATHS IS OUT THERE

Although we always tend to think that the shortest distance between two points is a straight line, this is not always the case. The shortest distance between two points on a globe or curved surface is a geodesic.

Straight lines

Straight lines can lie anywhere and at any angle to the x-axis.

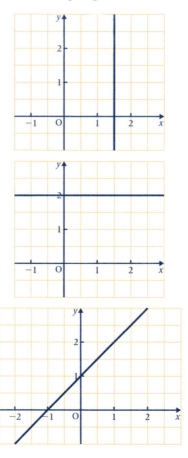

The lines go on for ever in both directions. We draw just part of the line.

Lines parallel to the axes

If we consider any point on the line given in the diagram we find that, no matter what the y coordinate is, the x coordinate is always 2.

We say that the equation of this line is $x = 2$.

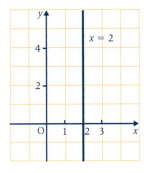

Notice that y is not mentioned because there is no restriction on the value it can take.

In the same way, any point on the line given in this diagram has a y coordinate of 3, while the x coordinate can take any value. The equation of this line is $y = 3$.

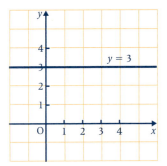

Exercise 9a

Give the equations of the lines in questions 1 to 4:

1

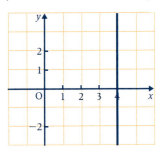

2

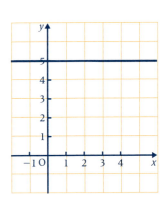

3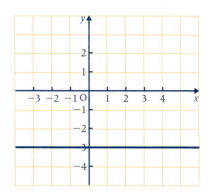

4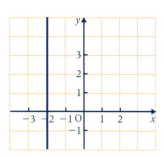

Draw similar diagrams to show the lines with the following equations:

5 $x = 1$ **6** $x = -4$ **7** $y = 4$ **8** $y = -3$

Slant lines

If we take any point on the line shown, say (2, 4) or $(3\frac{1}{2}, 5\frac{1}{2})$ or (−1, 1) or (0, 2), we find that the y coordinate is always 2 units more than the x coordinate.

The equation of the line is therefore $y = x + 2$.

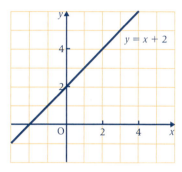

Conversely, if we are given the equation of a line we can draw the line by finding points on it. Two points are enough to draw a straight line but a third is useful as a check. If the three points do not lie in a straight line, at least one point is incorrect. Check all three points.

Suppose that we want to draw the line whose equation is $y = 2x + 1$. Think of this as an instruction for finding the y coordinate to go with a chosen x coordinate.

If $x = 1$, $y = 2 + 1 = 3$ so (1, 3) is a point on the line.
If $x = 3$, $y = 2 \times 3 + 1 = 7$ so (3, 7) is on the line.
If $x = -3$, $y = 2 \times (-3) + 1 = -5$ so (−3, −5) is on the line.

It is simpler to list this information in a table.

x	−3	1	3
y	−5	3	7

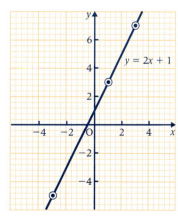

113

Exercise 9b

In all questions in this exercise use 1 cm to 1 unit.

In each of questions **1** to **4**, copy and complete the table. On graph paper, draw x and y axes for the ranges indicated in the brackets. Mark the points and, if they are in a straight line, draw that line:

1 $y = x + 4$ $(-2 \leqslant x \leqslant 4, 0 \leqslant y \leqslant 8)$ **3** $y = 4 - x$ $(-3 \leqslant x \leqslant 3, 0 \leqslant y \leqslant 8)$

x	-2	0	4
y			

x	-3	0	3
y			

2 $y = 2x + 1$ $(-2 \leqslant x \leqslant 3, -4 \leqslant y \leqslant 8)$ **4** $y = 2 - 3x$ $(-1 \leqslant x \leqslant 3, -7 \leqslant y \leqslant 5)$

x	-2	0	3
y			

x	-2	0	4
y			

In each of questions **5** to **8**, make a table, choosing your own values of x within the given range. (Choose one low value, one high value and one in between, such as zero.) Draw x and y axes for the ranges of values indicated. Draw the line.

5 $y = x - 3$ $(-2 \leqslant x \leqslant 5, -5 \leqslant y \leqslant 2)$ **7** $y = 3 - 2x$ $(-2 \leqslant x \leqslant 4, -5 \leqslant y \leqslant 7)$

6 $y = \frac{1}{2}x + 4$ $(-2 \leqslant x \leqslant 4, 0 \leqslant y \leqslant 6)$ **8** $y = 3x - 4$ $(0 \leqslant x \leqslant 6, -5 \leqslant y \leqslant 14)$

Make a table which can be used for drawing the graph of $y = 6 + 2x$, taking values of x in the range $-3 \leqslant x \leqslant 3$. Decide from the table what range of values of y is suitable.

Choose values at each end of the range and one in the middle.

$y = 6 + 2x$

x	-3	0	3
y	0	6	12

The range for y must be $0 \leqslant y \leqslant 12$

For each of questions **9** to **12**, make a table, choosing your own values of x within the given range. Decide on a suitable range of values of y after completing the table.

Draw x and y axes using the given range for the x-axis and the range you have chosen for the y-axis. Draw the line:

9 $y = 3 - x$ $(-3 \leqslant x \leqslant 3)$ **11** $y = 2x + 2$ $(-2 \leqslant x \leqslant 4)$

10 $y = \frac{1}{2}x - 1$ $(-3 \leqslant x \leqslant 3)$ **12** $y = 5 - 3x$ $(-1 \leqslant x \leqslant 3)$

You are given the graph of the
line with equation $y = 1 - \frac{1}{2}x$.
From the graph, find

a the value of y, if x is 0.8

b the value of x, if y is 2.4

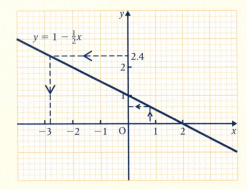

a Find 0.8 on the x-axis (4 small squares from the origin). Go up to the line then
across to the y-axis. Read off the value (3 small squares, which corresponds to 0.6).

From the graph, if $x = 0.8$, $y = 0.6$

b From the graph, if $y = 2.4$, $x = -2.8$

13 You are given the graph of the line with
equation $y = x + 1$.
From the graph, find

a y if $x = \frac{1}{2}$

b x if $y = 1.4$

c x if $y = -0.6$

> **Tip** Make sure you know what
> numbers are represented by the
> intermediate graduations on the axes.

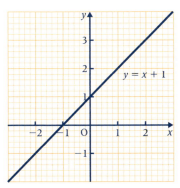

14 You are given the graph of the line with equation $y = 2x - 1$.
From the graph, find

a y if $x = \frac{1}{2}$ **b** x if $y = -2.6$ **c** y if $x = -1.2$

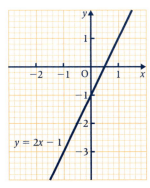

15 You are given the graph of the line with equation $y = -x - 1$. From the graph, find

   a y if $x = 1.6$

   b x if $y = 0.8$

   c x if $y = -2.2$

16 You are given the graph of the line with equation $y = 3 - 3x$. From the graph, find

   a y if $x = -0.2$

   b x if $y = 1.2$

   c x if $y = -0.6$

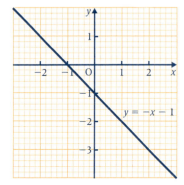

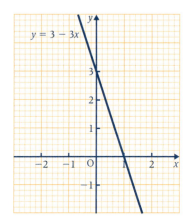

17 Using the graph you drew for question 5, find

   a y if $x = \frac{1}{2}$ **b** x if $y = 1.4$ **c** x if $y = -0.6$

18 Using the graph you drew for question 6, find

   a y if $x = 1.6$ **b** x if $y = 4.6$ **c** x if $y = -1.6$

19 Using the graph you drew for question 7, find

   a y if $x = 2.2$ **b** x if $y = 0.2$ **c** x if $y = -4$

20 Using the graph you drew for question 8, find

   a y if $x = 4.2$ **b** x if $y = 4.4$ **c** x if $y = 5$

Points on a line

Exercise 9c

Do the points (3, 6) and (−2, −10) lie on the line whose equation is $y = 3x - 4$?

$$y = 3x - 4$$

When $x = 3$, the value of y on the line is given by $y = 3 \times 3 - 4 = 5$. But for the point (3, 6), $y = 6$ so the point does not lie on the line.

When $x = -2$, the value of y on the line is given by $y = 3 \times (-2) - 4 = -10$. For the point (−2, −10), $y = -10$ so the point lies on the line.

Find whether the given points lie on the line whose equation is given:

1 $y = 2x - 1$; (5, 9), (1, 2)

4 $y = \frac{1}{2}x + 4$; (4, 5), (3, $5\frac{1}{2}$)

2 $y = 3 + 3x$; (2, 9), (−4, −9)

5 $y = 5 + 4x$; ($\frac{1}{2}$, 7), (−2, −3)

3 $y = 6 - 2x$; (3, 1), (4, −1)

6 $y = 6 - \frac{1}{2}x$; (−2, 5), (−2, 7)

Comparing slopes

Exercise **9d**

1 Draw, on the same pair of axes, the graphs of the lines

   a $y = 2x - 3$ **b** $y = 2x$ **c** $y = 2x - 2$

   for $-3 \leqslant x \leqslant 3$ and $-7 \leqslant y \leqslant 9$ using 1 cm to 1 unit.
   Label each line with its equation.
   What can you say about the lines?
   What do the equations have in common?

2 Draw a pair of axes using $-2 \leqslant x \leqslant 2$ and $-7 \leqslant y \leqslant 9$. On these axes draw the
   three lines with the following equations

   a $y = -3x - 1$ **b** $y = -3x$ **c** $y = -3x + 3$

   What can you say about the lines?
   What do the equations have in common?

3 Draw a pair of axes using $-4 \leqslant x \leqslant 4$, $-7 \leqslant y \leqslant 6$. On these axes draw the
   three lines with the following equations

   a $y = \frac{1}{2}x - 4$ **b** $y = \frac{1}{2}x + 1$ **c** $y = \frac{1}{2}x + 3$

   Comment on the three lines and their equations.

4 Draw x and y axes, marking values from −5 to 5 on each axis. Draw the three
   lines with the equations

   a $y = x$ **b** $y = x - 3$ **c** $y = x + 2$

   What do you notice?

5 Draw x and y axes, using $-3 \leqslant x \leqslant 3$, $-3 \leqslant y \leqslant 5$. Draw the three lines with the
   following equations

   a $y = 2x + 2$ **b** $y = -2x + 1$ **c** $y = 2x$

   Which two lines are parallel?

6 Draw x and y axes, marking values from −5 to 5 on each axis. Draw the lines
   with the equations

   a $y = 4 - x$ **b** $y = -x$ **c** $y = -3 - x$

   What do you notice?

Gradients

Different lines have different gradients or slopes. Some lines point steeply up to the right; they have large *positive* gradients.

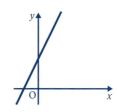

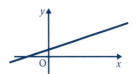

Some have a shallow slope up to the right; they have small *positive* gradients.

Some lines slope the other way; they have *negative* gradients.

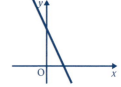

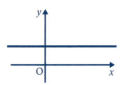

Some have zero gradient and are parallel to the *x*-axis.

Some are parallel to the *y*-axis.

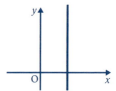

We can see from the questions in the last exercise that the *coefficient* of *x* (2 in question 1, -3 in question 2 and $\frac{1}{2}$ in question 3) has something to do with gradient.

Calculating the gradient of a line

First method

The gradient is found by comparing the amount you move up as you go from A to B, with the amount you move across. In this case the gradient is $\frac{4}{3}$.

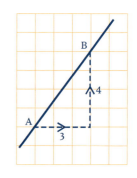

In the second case the distance moved is downwards so we take this distance as negative.

The gradient is $\frac{-3}{5}$, i.e. $-\frac{3}{5}$.

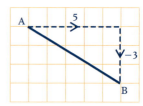

Second method

The distance moved up is given by the difference in the y coordinates and the distance moved across is given by the difference in the x coordinates, so the gradient is

$$\frac{\text{the difference in the } y \text{ coordinates}}{\text{the difference in the } x \text{ coordinates}}$$

If A is the point (3, 5) and B is (2, 7) then the gradient is $\frac{5-7}{3-2} = \frac{-2}{1} = -2$

Notice that the coordinates of B are taken from the coordinates of A for both x and y. We may change the order as long as we change it for both x and y, i.e. the gradient is also $\frac{7-5}{2-3} = -2$

Exercise 9e

Find the gradients of the lines joining the points

a (4, 2) and (6, 7)   **b** (2, 3) and (4, −3)

a *Either:* from the diagram the gradient is $\frac{5}{2}$

(In moving from left to right you go up 5 units (i.e. +5) and go across 2 units to the right, i.e. +2.)

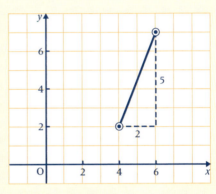

Or: gradient = $\dfrac{\text{difference in the } y \text{ coordinates}}{\text{difference in the } x \text{ coordinates}} = \dfrac{7-2}{6-4} = \dfrac{5}{2}$

b *Either:* from the diagram the gradient is $\frac{-6}{2} = -3$

(In moving from left to right you go down 6 units (−6) and across to the right 2 units, i.e. +2.)

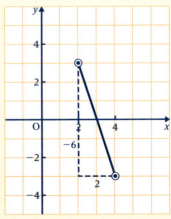

Or: gradient = $\dfrac{-3-3}{4-2} = \dfrac{-6}{2} = -3$

(Notice that we could also say

gradient = $\dfrac{3-(-3)}{2-4} = \dfrac{6}{-2} = -3$)

119

Find the gradients of the lines joining the following pairs of points:

1 (5, 1) and (7, 9)

2 (3, 6) and (5, 2)

3 (3, 4) and (6, 7)

4 (−2, 4) and (2, 1)

5 (1, 2) and (6, −7)

6 (−3, 4) and (−6, 2)

> **Tip** If you understand what is going on and can find the gradient without plotting the points on a grid, do so.

7 Find the gradient of the line joining the points (4, 3) and (7, 3).

8 Which axis is parallel to the line joining the points (4, 3) and (4, 6)? What happens when you try to work out the gradient?

9 If lines are drawn joining the following pairs of points, state which lines have zero gradient and which are parallel to the y-axis:

   a (0, 4) and (0, −2)   **b** (3, 0) and (−10, 0)

   c (−6, 0) and (−2, 0)   **d** (0, 6) and (0, 12)

10 If (2, 1) is a point on a line and its gradient is 3, draw the line and find the coordinates of two other points on it.

Exercise 9f

Find the gradient of the line $y = -2x + 3$

Choose any two points on the line, e.g. (−2, 7) and (2, −1).

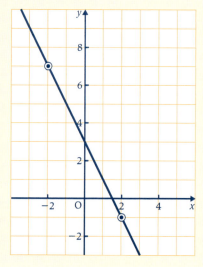

Gradient $= \dfrac{-1-7}{2-(-2)} = \dfrac{-8}{4} = -2$

(This is the same as the coefficient of x in the equation of the line.)

Choose two points on each line and hence find the gradient of the line. In each case compare your answer with the coefficient of x:

1 $y = 2x + 3$ **3** $y = 2x - 1$

2 $y = x + 4$ **4** $y = 3 - 2x$

> **Tip** You may find that drawing a diagram helps.

5 State the gradient of the line $y = 4x + 1$, without calculation if possible.

6 Give the gradients of the lines with equations

   a $y = 4x + 4$ **b** $y = 2 - 3x$ **c** $y = x - 3$ **d** $y = \frac{1}{2}x + 1$

7 Sketch a line with a gradient of

   a 3 **b** -1

8 Sketch a line with a gradient of

   a 4 **b** $\frac{1}{2}$

9 Sketch a line with a gradient of

   a -2 **b** 1

The intercept on the y-axis

Consider the line with equation $y = x + 3$.

When $x = 0$, $y = 3$, so the line crosses the y-axis where $y = 3$.

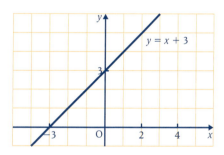

If the equation is $y = 2x - 3$ then, when $x = 0$, $y = -3$, i.e. the line cuts the y-axis where $y = -3$.

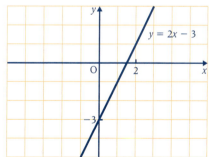

As long as the equation is of the form $y = mx + c$, i.e. like $y = 2x + 3$ or $y = -5x + 1$ then

> the number term c tells us where the line cuts the y-axis and m, the coefficient of x, tells us the gradient.

If the equation is $y = 3x$ then the number term is 0, i.e. the intercept is 0. Therefore, the line goes through the origin.

Exercise 9g   Give the gradient and the intercept on the y-axis of the line with equation $y = x + 5$. Sketch the line.

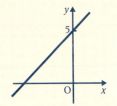

The gradient is the coefficient of x when the equation is written in the form $y = mx + c$, i.e. 1

The intercept on the y-axis is 5

Give the gradients and the intercepts on the y-axis of the lines with the following equations. *Sketch* each line:

1 $y = 2x + 4$ **2** $y = 5x + 3$ **3** $y = 3x - 4$ **4** $y = x - 6$

Give the gradient and the intercept on the y-axis of the line with equation $y = 4 - 2x$. Sketch the line.

Rewrite the equation in the form $y = mx + c$,
i.e. as $y = -2x + 4$

The gradient is -2 so the lines slope back to the left.
The intercept on the y-axis is 4.

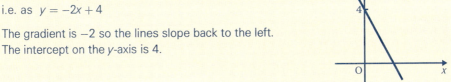

Give the gradients and the intercepts on the y-axis of the lines with the following equations. *Sketch* each line:

5 $y = 3 - 2x$ **10** $y = 3x - 7$

6 $y = -4x + 2$ **11** $y = 7 - 3x$

7 $y = 2 + 5x$ **12** $y = \frac{1}{3}x + 7$

8 $y = \frac{1}{2}x - 1$ **13** $y = 9 - 0.4x$

9 $y = -\frac{1}{3}x + 4$ **14** $y = 4 + 5x$

Give the gradient and the intercept on the y-axis of the line with equation $2y = 3x - 1$. Sketch the line.

$2y = 3x - 1$ (Write this equation in the form $y = mx + c$)

Divide both sides by 2 $y = \frac{3}{2}x - \frac{1}{2}$

The gradient is $\frac{3}{2}$ so the line slopes up to the right

The intercept on the y-axis is $-\frac{1}{2}$

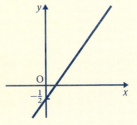

Give the gradients and the intercepts on the y-axis of the lines with the following equations. *Sketch* each line:

15 $2y = 4x + 5$ **16** $3y = x - 6$ **17** $5y = 5 + 2x$ **18** $4y = 8 - 3x$

A line has a gradient of -2 and the intercept on the y-axis is 4. Find the equation of the line.

Comparing with the equation $y = mx + c$, $m = -2$ and $c = 4$.

The equation is $y = -2x + 4$

Write down the equations of the lines with the given gradients and intercepts on the y-axis (y-intercepts):

	Gradient	y-intercept
19	2	7
20	3	1
21	1	3
22	2	-5

	Gradient	y-intercept
23	$\frac{1}{2}$	6
24	-2	1
25	1	-2
26	$-\frac{1}{2}$	4

Puzzle

A farmer grows cabbages on a square plot. He says he has 151 more cabbages this year than last year, when he also had a square plot. How many cabbages did he raise last year?

Parallel lines

Exercise 9h

Which of the lines with the following equations are parallel:

$y = 2x + 3$, $y = 4 - 2x$, $y = 4 + 2x$, $2y = x + 1$, $y = x + 3$?

Written in the form $y = mx + c$ the equations of these lines are
$y = 2x + 3$, $y = -2x + 4$, $y = 2x + 4$, $y = \frac{1}{2}x + \frac{1}{2}$ and $y = x + 3$.

The gradients of the lines are 2, -2, 2, $\frac{1}{2}$ and 1, so the first and third lines are parallel.

In questions **1** to **4**, state which of the lines with the given equations are parallel.

1 $y = 3x + 1$, $\quad y = \frac{1}{3}x - 4$, $\quad y = x + 1$,
$\quad y = 4 - 3x$, $\quad y = 5 + 3x$, $\quad y = 3x - 4$

2 $y = 2 - x$, $\quad y = x + 2$, $\quad y = 4 - x$,
$\quad 2y = 3 - 2x$, $\quad y = -x + 1$, $\quad y = -x$

> **Tip** Write each equation in the form $y = mx + c$ first.

3 $3y = x$, $y = \frac{1}{3}x + 2$, $y = \frac{1}{3} + x$,
$\quad y = \frac{1}{3} + \frac{1}{3}x$, $y = \frac{1}{3}x - 4$

4 $y = \frac{1}{2}x + 2$, $y = 2 - \frac{1}{2}x$, $y = -x - 4$, $y = \frac{1}{2}x - 1$, $2y = 3 - x$

5 What is the gradient of the line with equation $y = 2x + 1$? Give the equation of the line that is parallel to the first line and which cuts the y-axis at the point $(0, 3)$.

6 What is the gradient of the line with equation $y = 6 - 3x$?
If a parallel line goes through the point $(0, 1)$, what is its equation?

7 Give the equation of the line through the origin that is parallel to the line with equation $y = 4x + 2$.

8 Give the equations of any three lines that are parallel to the line with equation $y = 4 - x$.

9 Give the equations of the lines through the point $(0, 4)$ that are parallel to the lines with equations

   a $y = 4x + 1$ **b** $y = 6 - 3x$ **c** $y = \frac{1}{2}x + 1$

10 Give the equations of the lines, parallel to the line with equation $y = \frac{1}{3}x + 1$, that pass through the points

   a $(0, 6)$ **b** $(0, 0)$ **c** $(0, -3)$

11 Give the equations of the lines with gradient 2 which pass through the points

   a $(0, 2)$ **b** $(0, 10)$ **c** $(0, -4)$

12 Which two of the lines with the following equations are parallel?
$y = 3 + 2x$, $y = 3 - 2x$, $y = 2x - 3$

13 Find the gradients and the intercepts on the y-axis of the lines with equations $y = 4 - 3x$ and $y = 4x - 3$. Give the equation of the line that is parallel to the first line and cuts the y-axis at the same point as the second line.

14 A line of gradient -4 passes through the origin.

   a Give its equation

   b Give the equation of the line that is parallel to the first line and that passes through the point $(0, -7)$

Different forms of the equation of a straight line

The terms in the equation of a straight line can only be x terms, y terms or number terms.

An equation containing terms like x^2, y^2, $\frac{1}{x}$, $\frac{1}{x^2}$ is not the equation of a straight line.

Sometimes the equation of a straight line is not given exactly in the form $y = mx + c$.

It could be $2x + y = 6$ or $\frac{x}{4} + \frac{y}{2} = 1$

An easy way to draw a line when its equation is in one of these forms is to start by finding the points where it cuts each axis.

Exercise 9i

Draw on graph paper the line with equation

$$3x - 4y = 12 \quad (\text{use } 0 \leqslant x \leqslant 5)$$

Find the gradient of the line.

$$3x - 4y = 12$$

When $x = 0$, $-4y = 12$, i.e. $y = -3$

When $y = 0$, $3x = 12$, i.e. $x = 4$

So the points $(0, -3)$ and $(4, 0)$ lie on the line.

(Draw the line using these two points only. Choose a point on the line, e.g. $(2, -1\frac{1}{2})$, to check.)

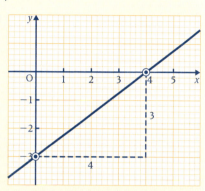

Check: when $x = 2$, $y = -1\frac{1}{2}$,

$3x - 4y = 6 - 4 \times (-1\frac{1}{2}) = 6 + 6 = 12$, which is correct.

From the graph, the gradient $= \frac{3}{4}$, (lines that slope up to the right have a positive gradient.)

Draw on graph paper the lines with the following equations. Use 1 cm to 1 unit. Find the gradient of each line:

1 $3x + 5y = 15$ $0 \leqslant x \leqslant 6$

2 $2x + 6y = 12$ $0 \leqslant x \leqslant 6$

3 $x - 4y = 8$ $0 \leqslant x \leqslant 8$

4 $x + y = 6$ $0 \leqslant x \leqslant 7$

5 $2x + y = 5$ $-2 \leqslant x \leqslant 4$

6 $x + 3y = 5$ $0 \leqslant x \leqslant 5$

7 $x - 3y = 6$ $0 \leqslant x \leqslant 6$

8 $2x - y = 3$ $-2 \leqslant x \leqslant 2$

9 On the same axes $(-6 \leqslant x \leqslant 6$ and $-6 \leqslant y \leqslant 6)$ draw the lines $x + y = 1$, $x + y = 4$, $x + y = 6$, $x + y = -4$. Find their gradients.

Exercise 9j

Draw on graph paper the line with equation

$$\frac{x}{3} + \frac{y}{2} = 1 \quad (-1 \leqslant x \leqslant 4)$$

Find its gradient.

When $x = 0$, $\frac{y}{2} = 1$, so $y = 2$

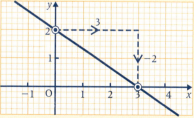

When $y = 0$, $\frac{x}{3} = 1$, so $x = 3$

Check: from the graph, when $x = 1\frac{1}{2}$, $y = 1$

Therefore $\frac{x}{3} + \frac{y}{2} = \frac{1\frac{1}{2}}{3} + \frac{1}{2} = \frac{1}{2} + \frac{1}{2} = 1$ which is correct.

The gradient is $-\frac{2}{3}$, (lines that slope up to the left have a negative gradient).

Draw on graph paper the lines whose equations are given below. Find the gradient of each line:

1 $\dfrac{x}{4} + \dfrac{y}{3} = 1$ **3** $\dfrac{x}{4} - \dfrac{y}{2} = 1$ **5** $\dfrac{x}{1} - \dfrac{y}{2} = 1$

2 $\dfrac{x}{5} + \dfrac{y}{3} = 1$ **4** $\dfrac{x}{3} + \dfrac{y}{6} = 1$ **6** $\dfrac{y}{3} - \dfrac{x}{4} = 1$

7 Without drawing a diagram, state where the lines with the following equations cut the axes

  a $\dfrac{x}{2} + \dfrac{y}{4} = 1$ **b** $\dfrac{x}{12} - \dfrac{y}{9} = 1$

8 Form the equations of the lines which cut the axes at

  a (0, 5) and (6, 0)   **b** (0, −3) and (4, 0)

9 Sketch the line with equation $\dfrac{x}{6} + \dfrac{y}{2} = 1$ and find its gradient.

Getting information from the equation of a line

From the last exercise, we can see that if the equation of a line is in the form

$$\frac{x}{a} + \frac{y}{b} = 1 \quad \text{(i.e. like questions 1 to 6),}$$

then the line cuts the x-axis at $x = a$

and the y-axis at $y = b$.

Then if we sketch the line we can work out the gradient.

If the equation is in the form $ax + by = c$, like those in Exercise 9i, we need to rearrange the equation so that it is in the form $y = mx + c$. Then the gradient and the intercept on the y-axis can be seen.

Exercise 9k   Find the gradient and the intercept on the y-axis of the line with equation $2x + 3y = 6$.

Change the equation to the form $y = mx + c$.

Take $2x$ from both sides $3y = 6 - 2x$

Divide both sides by 3 $y = 2 - \frac{2}{3}x$

i.e. $y = -\frac{2}{3}x + 2$

The gradient is $-\frac{2}{3}$ and the intercept on the y-axis is 2

Find the gradient and the intercept on the y-axis of each of the following lines:

1 $3x + 5y = 15$ **3** $x - 4y = 8$ **5** $y - 3x = 6$

2 $2x + 6y = 12$ **4** $x - 3y = 6$ **6** $x + 3y = 6$

Find the gradient and the intercept on the y-axis of the line with equation $\dfrac{x}{2} + \dfrac{y}{5} = 1$

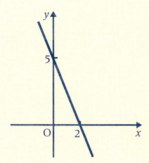

Either

If $y = 0$ then $x = 2$ and if $x = 0$ then $y = 5$ so this line cuts the axes at (2, 0) and (0, 5).

Hence the gradient is $-\dfrac{5}{2}$ and the intercept on the y-axis is 5

Or	$\dfrac{x}{2} + \dfrac{y}{5} = 1$
Multiply both sides by 5	$\dfrac{5x}{2} + y = 5$
Take $\dfrac{5x}{2}$ from both sides	$y = 5 - \dfrac{5x}{2}$
i.e.	$y = -\dfrac{5}{2}x + 5$

Then the gradient is $-\dfrac{5}{2}$ and the intercept is 5

Find the gradient and the intercept on the y-axis of each of the following lines:

7   $\dfrac{x}{4} + \dfrac{y}{3} = 1$

8   $\dfrac{x}{5} + \dfrac{y}{3} = 1$

9   $\dfrac{x}{4} - \dfrac{y}{2} = 1$

10 $\dfrac{x}{2} + \dfrac{y}{6} = 1$

11 $\dfrac{x}{3} + \dfrac{y}{4} = 1$

12 $\dfrac{x}{3} - \dfrac{y}{4} = 1$

13 $y = 4x + 2$

14 $x + y = 4$

15 $\dfrac{x}{2} + \dfrac{y}{4} = 1$

16 $2x + 5y = 15$

17 $y = 5 - \dfrac{1}{2}x$

18 $2y = 4x + 5$

19 $\dfrac{x}{2} - \dfrac{y}{4} = 1$

20 $x + y = -3$

21 $3x + 4y = 12$

The equation of a line through two given points

Find the gradient and the intercept on the *y*-axis of the line through the points (4, 2) and (0, 4). Hence give the equation of the line.

(A sketch only is needed.)

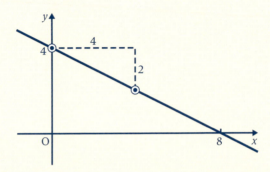

The equation of any line is $y = mx + c$, so to find the equation of this line we need to find the value m, i.e. the gradient, and c, i.e. where it cuts the *y*-axis.

$$\text{Gradient} = \frac{2-4}{4-0} = \frac{-2}{4} = -\frac{1}{2}$$

When $x = 0$, $y = 4$ so the *y*-intercept $= 4$

∴ the equation is $y = -\frac{1}{2}x + 4$

(Multiplying by 2 and rearranging, allows this equation to be written in the form $2y = 8 - x$ or $2y + x = 8$.)

Find the gradient and the intercept on the *y*-axis of the line through the given points. Hence give the equation of the line:

1 (0, 4) and (3, 0)

2 (0, 7) and (2, 3)

3 (5, 4) and (0, 1)

4 (0, 2) and (3, −2)

5 (0, −4) and (2, 3)

6 (−6, −3) and (0, −1)

7 (6, 2) and (0, 1)

8 (5, 1) and (0, −3)

9 (0, −4) and (3, 1)

10 (−5, 0) and (0, −5)

11 (0, 12) and (−6, 0)

12 (−6, 1) and (0, 6)

13 A, B and C are the points (0, 4), (5, 6) and (3, −1). Find the equations of lines AB and AC.

Find the gradient and the equation of the line through (6, 2) and (4, 8).

The gradient is

$$\frac{2-8}{6-4} = \frac{-6}{2} = -3$$

(We do not know the intercept on the y-axis.)

Let the equation be $y = -3x + c$, i.e. $m = -3$ in the equation $y = mx - c$.

When $x = 6$, $y = 2$ so

$$2 = -18 + c$$

∴

$$c = 20$$

∴ the equation is $y = -3x + 20$

(Check: when $x = 4$, the equation gives $y = -12 + 20$, i.e. $y = 8$, which is correct.)

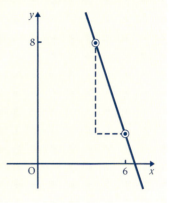

In questions **14** to **34**, find the gradient and the equation of the line through the given pair of points:

14 (4, 1) and (7, 10)

15 (1, 4) and (2, 1)

16 (3, 7) and (5, 12)

17 (−2, 3) and (−1, 5)

18 (4, −1) and (3, −6)

19 (5, −2) and (−4, 7)

20 (−6, 7) and (1, 0)

21 (4, −3) and (2, −7)

22 (−9, −3) and (6, 0)

Questions **23** to **34** can be done by using either the method of questions **1** to **6** or the method and ideas of Exercise 9j:

23 (4, 0) and (0, 5)

24 (3, 0) and (0, 2)

25 (3, 0) and (0, −2)

26 (2, 0) and (0, 6)

27 (4, 2) and (6, 8)

28 (0, 4) and (3, 1)

29 (2, 1) and (0, −6)

30 (−1, 4) and (5, −2)

31 (0, 5) and (−2, 0)

32 (−6, 3) and (5, 5)

33 (−1, −2) and (−5, −6)

34 (3, 2) and (7, 1)

Problems

Exercise **9m**

1 Find the equation of the line through the point (6, 2), which is parallel to the line $y = 3x - 1$.

> **Tip** Remember that parallel lines have the same gradient.

2 On the same pair of axes, for $-8 \leqslant x \leqslant 8$ and $-12 \leqslant y \leqslant 4$, using 1 cm to 1 unit, draw the four lines $y = 2x - 4$, $y = 2x + 6$, $x + 2y + 10 = 0$ and $2y + x = 0$. What type of quadrilateral is formed by the four lines?

3 On the same pair of axes, for $-4 \leqslant x \leqslant 10$ and $-6 \leqslant y \leqslant 10$ using 1 cm to 1 unit, draw the four lines $y = 3x + 4$, $y = 3x - 6$, $y = -3x$ and $y = 10 - 3x$. Name the type of quadrilateral formed by the lines.

4 Find the point where the lines $y = 2x + 2$ and $y = 4 - 2x$ meet.

5 Find the equation of the line with gradient -2 that passes through the midpoint of the line joining the points (3, 2) and (7, 4). (The midpoint can be found either from a drawing on squared paper or from a rough sketch.)

6 Find the equation of the line that is parallel to the line joining $(-1, 4)$ and (3, 2) and which passes through the point (4, 0).

7 On the same pair of axes, for $-2 \leqslant x \leqslant 5$ and $-5 \leqslant y \leqslant 2$, using 1 cm to 1 unit, draw the four lines $4y = x + 1$, $4y = x - 16$, $4x + y - 13 = 0$ and $4x + y + 4 = 0$. Name the type of quadrilateral formed by the lines.

Mixed exercises

Exercise 9n

1 What is the gradient of the line with equation $y = 2x + 1$?

2 At what point does the line with equation $y = 4 - 2x$ cut the y-axis?

3 At what point does the line with equation $\dfrac{x}{4} - \dfrac{y}{3} = 1$ cut the x-axis?

4 On the line with equation $y = 6 - 2x$, if $x = -3$, what is y?

5 What is the equation of the line whose gradient is 5 and which passes through the origin?

6 At what point does the line with equation $2x + 3y = 24$ cut the x-axis?

7 Does the point (2, 4) lie on the line with equation $y = 6x - 8$?

8 Find the gradient of the line joining the points (6, 4) and (1, 1).

Exercise 9p

1 What is the gradient of the line with equation $y = 4 - 3x$?

2 Does the point $(6, -1)$ lie on the line $3x + 11y = 8$?

3 What is the equation of the line that passes through the origin and has a gradient of -4?

4 At what point does the line with equation $y = 4 - x$ cut the y-axis?

5 At what points does the line with equation $x + y = 6$ cut the two axes?

6 Find the gradient of the line through the points (3, 1) and $(5, -2)$.

7 Give the equation of the line that is parallel to the line with equation $y = 4 + \frac{1}{2}x$ and which passes through the origin.

8 At what points does the line $\frac{x}{2} + \frac{y}{3} = 1$ cut the two axes?

Puzzle

A hunter met two shepherds. One shepherd had three small loaves while the other shepherd had five similar loaves. All the loaves were the same size. They decided to divide the eight loaves equally between them. The hunter thanked the shepherds and paid them eight dinars.

How should the shepherds divide the money?

Did you know that there is something special about the integers 1, 2 and 3? Their sum is equal to their product. There is only one other set of three integers that have this property. Can you find them?

IN THIS CHAPTER...

you have seen that:

- the equation $x = h$ gives a straight line parallel to the y-axis

- the equation $y = k$ gives a straight line parallel to the x-axis

- if you know one coordinate of a point on a line you can use the equation of the line to find the other coordinate

- the gradient of the straight line joining two points is
  $$\frac{\text{the difference in the } y \text{ coordinates}}{\text{the difference in the } x \text{ coordinates}}$$

- a line that slopes up to the right has a positive gradient while a line that slopes the other way has a negative gradient; the bigger the gradient the steeper the slope

- any equation that can be arranged in the form $y = mx + c$ represents a straight line, with gradient m, that crosses the y-axis at the point $(0, c)$. c is called the y-intercept.

- For example, the equation $2x + 3y = 8$ can be rewritten as $y = -\frac{2}{3}x + \frac{8}{3}$ and so represents a straight line with gradient $\frac{-2}{3}$ and y-intercept $\frac{8}{3}$

- lines that have the same gradient are parallel.

10 GRAPHS

AT THE END OF THIS CHAPTER...

you should be able to:

1 Draw a graph when corresponding values of two variables are given.

2 Construct a table of values when a formula connecting two variables is given.

3 Use a graph to find the value of one quantity, given the corresponding value of another.

4 Use a graph to find the least or greatest value of a quadratic function.

MATHS IS OUT THERE

Teacher: Why are you working on the floor?

Student: Because you said I couldn't use tables.

BEFORE YOU START

you need to know:
✓ how to draw and read graphs
✓ how to find squares and square roots of numbers
✓ how to substitute values into a formula

KEY WORDS

area, capacity, compound interest, data, quadratic graphs, volume

Graphs from tables

Graphs are used to give a visual representation of information about two, related, varying quantities.

You can see examples of graphs in many newspapers and periodicals. The best of them give a clear visual impression of the way in which the related quantities vary and allow us to get more information from the graph. The worst graphs are misleading, often because the axes are incorrectly labelled or because the scales are distorted.

Before we can draw a graph we have to know how the quantities vary. Sometimes we are given this information in a table, while at other times we are told that the two quantities are connected by a formula. The following example shows how to draw a clear, informative graph.

The Agricultural Department collected data for a particular type of tree. The data is given in the following table:

Age of tree in years (A)	0.8	1.4	2.2	3.5	5.4	6.7	7.8
Girth of tree in cm (G)	15.8	20.9	26.5	33.2	41.2	45.9	49.5

To draw a graph to represent the data we must:

a   draw axes that intersect at right angles in the bottom left-hand corner of the graph paper

b   state the scale used on each axis

c   name the axes
(We put time along the horizontal axis and the distance around the tree, or *girth*, along the vertical axis.)

d   plot carefully the points
representing the data
in the table
(We shall use a dot for
each point.)

e   draw a smooth curve to
pass through the points

f   give the graph a title.

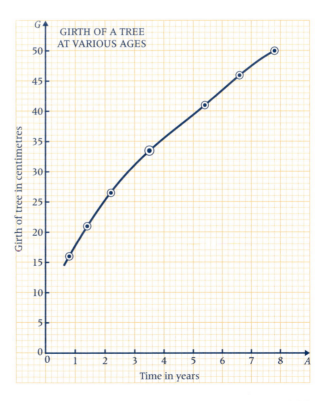

GIRTH OF A TREE
AT VARIOUS AGES

Girth of tree in centimetres

Time in years

We can use this graph to find the girth of the tree at any given age, or the age of the tree when it has a particular girth, provided that we keep within the range of values on our axes.

From the graph:

a   when the tree is four years old its girth is 35.5 cm

b   when the girth is 40 cm the tree is just over five years old.

Exercise 10a

1   The table shows the weight (W) in tonnes which a chain of diameter (D) centimetres will support before breaking.

D	1.2	1.8	2.2	2.6	2.8	3	3.4	3.8
W	72	162	242	338	392	450	578	722

Draw a graph for values of D from 0 to 4. Take 4 cm as 1 unit for D and 1 cm as 50 units for W.
Use your graph to find

    a   the greatest weight that a chain of diameter 2 cm will support

    b   the smallest diameter of chain that will support a load of 500 tonnes.

2   When a sum of money is invested, the interest added to the original value gives an increased value called the amount. The table shows the amount (V) of $100 when it has been invested for T years at 10% compound interest.

T	0	1	2	3	4	5	6	7	8	9
V	100	110	121	133	146	161	177	195	214	236

Draw the graph connecting T and V. Take 2 cm as 1 unit for T and 4 cm as 50 units for V. From your graph find

    a   the amount of $100 after $5\frac{1}{2}$ years

    b   the time, in years, in which $100 will double in value.

3   The table shows the connection between two quantities X and Y as X takes different values from 1 to 8

X	1	1.5	2	3	4	5	6	8
Y	14	10	8	6	5	4.4	4	3.5

Draw a graph connecting X and Y taking 2 cm as 1 unit for X and 1 cm as 1 unit for Y. Start at 0 for each scale. From your graph find

    a   the value of Y when X is 7.2

    b   the value of X when Y is 7.2

4   The table gives values for the surface area, A cm^2, of a closed rectangular box of fixed volume, which has square ends of side x cm.

x (cm)	1	1.5	2	3	4	5	6
A (cm^2)	110	76.5	62	54	59	71.6	90

Draw a graph connecting A and x taking 4 cm to represent 1 unit on the x-axis and 1 cm to represent 5 units on the A-axis. Let 40 be the lowest value on the A-axis. Use your graph to find

a   the value of x that gives the lowest value of A

b   the value of A when x is 1.8

5   The table shows the capacity (C), in litres, for jugs that are mathematically similar, but of different heights (H cm).

Height of jug in cm (H)	4	5.25	7.5	10.25	12.75	15.5	18.25
Capacity of jug in litres (C)	0.05	0.11	0.33	0.84	1.67	2.90	4.74

Draw a graph to represent this data using 4 cm to represent 5 units on the H-axis and 1 unit on the C-axis. Use your graph to find

a   the height of a similar jug with a capacity of 3.5 litres

b   the capacity of a similar jug that is 14 cm high.

Constructing a table from a formula

In the exercise we have just completed, all the data was given in tables. However, in many questions we are given a formula and have to construct our own table for given values of the variables.

Consider a car that starts from rest and travels a distance D metres in T seconds. Suppose that the formula connecting D and T for the first 10 seconds of the journey is

$$D = 50\sqrt{T}$$

and that we wish to draw a graph to show this relationship. Before we can draw the graph, we must work our some corresponding values of D and T.

If we take the whole number values of T from 0 to 10 we get:

when			
$T = 1$	$D = 50\sqrt{1}$	$= 50 \times 1$	$= 50$
$T = 2$	$D = 50\sqrt{2}$	$= 50 \times 1.414$	$= 71$
$T = 3$	$D = 50\sqrt{3}$	$= 50 \times 1.732$	$= 87$
$T = 4$	$D = 50\sqrt{4}$	$= 50 \times 2$	$= 100$
$T = 5$	$D = 50\sqrt{5}$	$= 50 \times 2.236$	$= 112$
$T = 6$	$D = 50\sqrt{6}$	$= 50 \times 2.449$	$= 122$
$T = 7$	$D = 50\sqrt{7}$	$= 50 \times 2.646$	$= 132$
$T = 8$	$D = 50\sqrt{8}$	$= 50 \times 2.828$	$= 141$
$T = 9$	$D = 50\sqrt{9}$	$= 50 \times 3$	$= 150$
$T = 10$	$D = 50\sqrt{10}$	$= 50 \times 3.162$	$= 158$

Each value of D is calculated correct to the nearest whole number.

These values of T in seconds and D in metres can be set out in table form:

T	0	1	2	3	4	5	6	7	8	9	10
D	0	50	71	87	100	112	122	132	141	150	158

We plot time (T) along the horizontal axis and distance (D) along the vertical axis.

When all the points have been plotted, draw a smooth curve to pass through them. The resulting graph is shown in the diagram. From this graph we can find D for any given value of T from 0 to 10, e.g. the distance travelled in 4.4 s is 105 m. Similarly, the time to travel 146 m is 8.5 s.

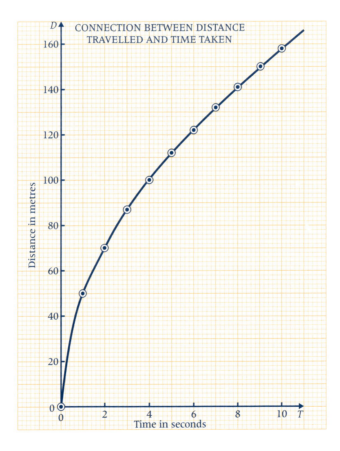

CONNECTION BETWEEN DISTANCE TRAVELLED AND TIME TAKEN

Distance in metres

Time in seconds

Exercise 10b

1 A stone is dropped from the top of a cliff. After t seconds it has fallen s metres, where $s = 5t^2$. Construct a table to show the relation between s and t, for values of t from 0 to 6, at half-unit intervals.
Use these values to draw a graph, and use your graph to find

  a how far the stone has fallen in 3.4 s   b how long the stone takes to fall 100 m.

2   Two quantities P and Q are connected by the formula

$$P = \frac{36}{Q}$$

Construct a table to show the relation between P and Q for values of Q from 1 to 8. Plot these points on a graph and from this graph find

a   the value of P when Q is 7.5   b   the value of Q when P is 4.8

3   Draw the graph of $y = \frac{6}{x}$ for values of x from 1 to 6 inclusive, plotting points at

whole number values of x and at $x = 1\frac{1}{2}$. Take 2 cm as 1 unit on each axis. Use your graph to find

a   the value of x when $y = 3.6$ b   the value of y when $x = 5.5$

4   Taking 2 cm as the unit on both axes, draw the graph of $y = x + \frac{7}{x}$ for values of

x between 1 and 7 at unit intervals. Use your graph to find the lowest value of y and the corresponding value of x.

5   Draw the graph of $xy = 8$ for values of x from -8 to -1 and from 1 to 8. (We cannot take $x = 0$ since there is no value of y which enables the product xy to be 8.) Let each axis range from -8 to 8, and take 1 cm as 1 unit on both axes. Use your graph to find

a   the value of y when x is 2.4   b   the value of x when y is -5.6

 ## Puzzle

'As I was going to St. Ives, I met a man with seven wives. Every wife had seven sacks, every sack had seven cats, every cat had seven kittens. Kittens, cats, sacks, wives, how many were going to St. Ives?'

Points to remember when drawing graphs of curves

1   Do not take too few points. About ten are usually necessary.

2   To decide where you need to draw the y-axis, look at the range of x-values.

3   To decide where to draw the x-axis, look at the range of y-values.

4   In some questions you will be given most of the y-values but you may have to calculate a few more for yourself. In this case always plot first those points that you were given and, from these, get an idea of the shape of the curve. Then you can plot the points you calculated and see if they fit on to the curve you have in mind. If they do not, go back and check your calculations.

5   When you draw a smooth curve to pass through the points, always turn the page into a position where your wrist is on the inside of the curve.

Quadratic graphs

The most important family of curves we consider give what are called *quadratic graphs*. The simplest of these is the graph of $y = x^2$.

We will draw the graph of $y = x^2$ for values of x ranging from -3 to $+3$ at half-unit intervals. The corresponding values for x and y are given in the table below:

x	-3	-2.5	-2	-1.5	-1	-0.5	0	0.5	1	1.5	2	2.5	3
$y (=x^2)$	9	6.25	4	2.25	1	0.25	0	0.25	1	2.25	4	6.25	9

A suitable scale for you to take is 2 cm to represent 1 unit on both axes but we have taken 1 cm to represent 1 unit on both axes.

We draw the y-axis vertically in the centre of the page, since the x-values range from -3 to $+3$, i.e. they are symmetrical about O.

The x-axis is drawn along the bottom of the page since all the y-values are positive.

The table gives us thirteen points. In drawing any quadratic graph, aim for at least ten points. It is especially important to have plenty of points where the graph is changing direction most quickly. For quadratic graphs this is about the lowest (or highest) point.

From the graph we can find the value of y that corresponds to any value of x within the range -3 to $+3$. For any value of y between 0 and 9 we can find the two corresponding values of x.

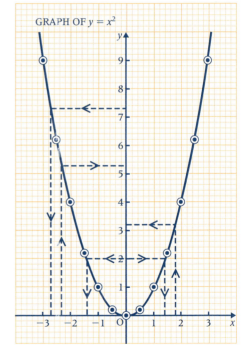

GRAPH OF $y = x^2$

For example:

a if $x = 1.8$, $y = 3.2$

b if $x = -2.3$, $y = 5.3$

c if $y = 2$, $x = 1.4$ or -1.4

   (There are two values of x for which $y = 2$)

Exercise 10c

1 Draw on the same axes the graphs of $y = x^2$, $y = 2x^2$ and $y = 3x^2$, taking half-unit intervals for x in the range -2 to $+2$. Take 4 cm as the unit on the x-axis and 2 cm as the unit on the y-axis for values of y from 0 to 12.
What can you deduce about the graph of $y = ax^2$ for any positive value of a?

2 Draw the graph of $y = x^2 + 3$ for values of x in the range -3 to 3. Take 4 cm as the unit for x and 1 cm as the unit for y.

a Use your graph to find the values of x when $y = 6$.

b Are there any values of x for which y has the value 0?

3 Draw the graph of $y = -x^2 + 4$ for values of x in the range -3 to 3. Take 2 cm as the unit for both x and y. Use your graph to find the value of x when

a $y = 0$ **b** $y = 3$.

Is the graph upside down compared with those you drew in questions 1 and 2?

4 Draw on the same axes the graphs of $y = x^2$, $y = x^2 + 4$ and $y = x^2 - 4$. Use values of x from -3 to $+3$ at unit intervals, taking 2 cm as 1 unit on the x-axis and 1 cm as 1 unit on the y-axis. Let the scale on the y-axis range from -5 to $+14$. What can you say about the shapes of the three graphs? What can you deduce about the graph of $y = x^2 + C$ for different positive or negative values of C?

5 Complete the following table which gives values of $x(x - 3)$ for values of x in the range -1 to 4 at half-unit intervals.

x	-1	-0.5	0	0.5	1	1.5	2	2.5	3	3.5	4
$x - 3$	-4			-2.5		-1.5					1
$x(x - 3)$	4			-1.25		-2.25					4

Hence draw the graph of $y = x(x - 3)$ within the given range taking 2 cm as the unit on both axes. Use your graph to write down

a the values of x where the graph crosses the x-axis

b the values of x when $x(x - 3) = 3$ (i.e. when $y = 3$).

6 Draw the graph of $y = x(2x - 3)$ for values of x in the range -2 to 3 taking values of x at half-unit intervals. Use a scale of 2 cm for 1 unit on the x-axis and 1 cm for 1 unit on the y-axis. Use your graph to find

a the values of x where the graph crosses the x-axis

b the lowest value of $x(2x - 3)$, i.e. the lowest value of y and the corresponding value of x.

7 Draw the graph of $y = 2x(2 + x)$ for values of x in the range -5 to 3. Take values of x at unit intervals, with extra values where you think they are needed. Let the scale on your y-axis range from -4 to $+32$. Let 1 cm represent 2 units. Use your graph to find

a the smallest value of $2x(2 + x)$ and the value of x for which it occurs

b the value of $2x(2 + x)$ when $x = -3.5$

c the values of x when $2x(2 + x) = 0$

Draw the graph of $y = x^2 + x - 6$ for whole number values of x from -4 to $+4$. Take 1 cm as 1 unit on the x-axis and 1 cm as 2 units on the y-axis. Use your graph to find

a the lowest value of $x^2 + x - 6$ and the corresponding value of x

b the values of x when $x^2 + x - 6$ is 4.

x	−4	−3	−2	−1	0	1	2	3	4
x^2	16	9	4	1	0	1	4	9	16
x	−4	−3	−2	−1	0	1	2	3	4
−6	−6	−6	−6	−6	−6	−6	−6	−6	−6
$x^2 + x - 6$	6	0	−4	−6	−6	−4	0	6	14

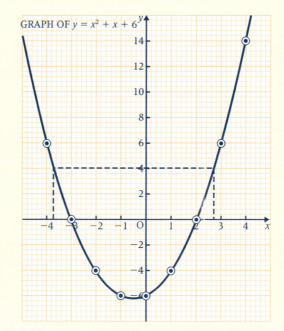

GRAPH OF $y = x^2 + x + 6$

a From the graph, the lowest value of $x^2 + x - 6$ is $-6\frac{1}{4}$. This occurs when $x = -\frac{1}{2}$.

b The values of x when $x^2 + x - 6$ is 4 are -3.70 and 2.70

8 Draw the graph of $y = x^2 - 2x - 3$ for whole number values of x in the range -3 to 5. Take 2 cm as 1 unit for x and 1 cm as 1 unit for y. Use your graph to find

a the lowest value of $x^2 - 2x - 3$ and the corresponding value of x

b the values of x when $x^2 - 2x - 3$ has a value of   i 1   ii 8

9 Draw the graph of $y = 6 + x - x^2$ for whole number values of x from -3 to 4. Take 2 cm as 1 unit on both axes. Use your graph to find

a the highest value of $6 + x - x^2$ and the corresponding value of x

b the values of x when $6 + x - x^2$ has a value of   i −2   ii 4

Investigation

Curve Stitching

Interesting curves can be produced by joining points that are equally spaced on straight lines.

This example uses two straight lines drawn at right angles.

The coloured lines form what is called the envelope of the curve.

a Draw two lines, approximately 10 cm long that bisect each other, as shown in the diagram. Mark equally spaced points on the lines about 5 mm apart. Use a ruler or compasses (which should give a more accurate result if used carefully) to mark the points. Use a ruler and a coloured line to join the points as shown in the diagram, then continue the pattern to complete the curve. Repeat the pattern on the other half of the diagram to give two curves.

b Investigate with two lines drawn at different angles.

c Now investigate with more than two lines.

Apollonius of Perga (250–175 BCE) was an astronomer. He gained immortality from his work with double cones which he sliced to get different shaped curves – called conic sections.

You have drawn examples of one of these in this chapter – the quadratic curve (also called a parabola). The diagram shows this together with two of the other curves Apollonius investigated.

IN THIS CHAPTER...

you have seen that:

- when you have a formula connecting two quantities you can draw a graph that shows the relationship. To do this work out corresponding values of the quantities and then plot the points on a set of axes.

- the simplest relationship that gives a quadratic graph is $y = x^2$

- a quadratic graph comes from any relationship of the form $y = ax^2 + bx + c$, where a, b and c are numbers as long as a is not equal to zero.

REVIEW TEST 1
CHAPTERS 1–10

In questions **1** to **10**, choose the letter for the correct answer.

1 What is the value of $\left(\frac{4}{5}\right)^{-2}$?

    A $\frac{25}{16}$ **B** $\frac{16}{25}$ **C** $-\frac{8}{10}$ **D** $-\frac{4}{10}$

2 The solution of $-3x > 6$, is

    A $x > 2$ **B** $x > -2$ **C** $x < 2$ **D** $x < -2$

3 $n = 0.15$ correct to 2 d.p.

    A $0.1 \leqslant n \leqslant 0.2$ **C** $0.145 \leqslant n < 0.155$

    B $0.145 \leqslant n \leqslant 0.155$ **D** $0.14 \leqslant n < 0.16$

4 $3(a^2 b^{-1})^3$ is equal to

    A $27a^6 b^{-3}$ **B** $3a^6 b^{-3}$ **C** $3a^5 b^2$ **D** $27a^5 b^2$

5 Given $\mathbf{X} = \begin{pmatrix} 2 & 3 \\ -4 & 5 \end{pmatrix}$, then $2\mathbf{X} =$

    A $\begin{pmatrix} 4 & 3 \\ -8 & 5 \end{pmatrix}$ **B** $\begin{pmatrix} 2 & 6 \\ -4 & 10 \end{pmatrix}$ **C** $\begin{pmatrix} 4 & 6 \\ -8 & 10 \end{pmatrix}$ **D** $\begin{pmatrix} 4 & 6 \\ -4 & 5 \end{pmatrix}$

6 The determinant of $\begin{pmatrix} 2 & 5 \\ 7 & -3 \end{pmatrix} =$

    A -41 **B** -29 **C** 29 **D** 41

7 Which of the points lie on the line $y = 2x + 3$?

    i $(-2, 1)$ **ii** $(1, 5)$ **iii** $(-1, 1)$

    A i and ii only   **B** i and iii only   **C** ii and iii only   **D** i, ii and iii

8 What is the gradient of $3x + 4y = 9$?

    A -3 **B** $-\frac{3}{4}$ **C** $\frac{3}{4}$ **D** 3

9 To make a skeleton model of a cube, 48 cm of wire was used. What is the length of one side?

    A 4 cm   **B** 6 cm   **C** 8 cm   **D** 12 cm

10 Which of the following is the solution to the pair of equations $3x + y = 5$ and $x + y = 1$?

    A $(2, -1)$ **B** $(0, 1)$ **C** $(1, 2)$ **D** $(1, 0)$

11 a If $3x - 7y = 8$ and $4x + 5y = 8$, find the ratio of $x : y$.

   b Solve the equation $\dfrac{(x+1)}{2} + \dfrac{(x-2)}{3} = \dfrac{(x+4)}{4}$

12 a A commission agent received a commission of \$980 on a sale for \$28 000. At what rate per cent was the commission paid to the agent?

   b A car is bought for \$50 000 and depreciates at 10% each year. What is its value at the end of two years?

13 a Without using a set-square or a protractor, construct a triangle ABC in which $AB = 8\,cm$, $\hat{A} = 45°$, $\hat{B} = 60°$. Measure the length of BC.

   b

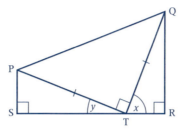

In the figure, $PT = TQ$ and $\hat{S} = \hat{R} = P\hat{T}Q = 90°$.
If the angle marked x measures 60°, what is the measure of y?
Name two congruent triangles, giving reasons for your answer.

14 a If $x = 9$, $y = 3$, $p = \frac{1}{2}$, $q = 2$, find the value of $x^p + q^y$

   b If x is an integer, find the values of x which satisfy $x - 5 < 3x + 2 < 7$

15 a Write the equations $2x + 3y = 7$ and $3x - y = 5$ in matrix form and hence solve them.

   b Find the equation of the line through the point $(5, -2)$ which is parallel to $y = 4x + 1$

11 AREAS

AT THE END OF THIS CHAPTER...

you should be able to:

1   Calculate the areas of triangles, rectangles and parallelograms.

2   Calculate the area of a polygon by subdividing it into triangles, rectangles or parallelograms.

3   Calculate the area of a trapezium given the lengths of its parallel sides and height.

4   Identify triangles or parallelograms which have equal area.

5   State the ratio of the areas of two triangles which have equal heights (bases) but different bases (heights).

6   Solve problems using the fact that triangles with equal heights (bases) have areas proportional to their bases (heights).

MATHS IS OUT THERE

The word trapezium comes from the Greek word *trapeza* meaning 'table'. Since the sixteenth century its meaning has been 'a quadrilateral with one pair of opposite sides parallel' but originally it applied to any quadrilateral that was not a parallelogram.

BEFORE YOU START

you need to know:
✓ how to work with decimals
✓ the meaning of symmetry
✓ the properties of special quadrilaterals

KEY WORDS

parallelogram, perpendicular height, ratio, rectangle, rhombus, slant height, symmetry, trapezium, triangle

Areas of familiar shapes

Rectangle  $A = lb$

Parallelogram 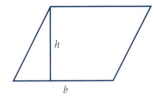 $A = bh$

Triangle 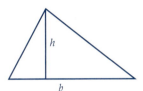 $A = \frac{1}{2}bh$

Remember that when we talk about the height of a figure we mean the *perpendicular height*, not the slant height.

Remember also that both the lengths we use must be measured in the *same* unit.

Exercise 11a

Find the areas of the following figures:

1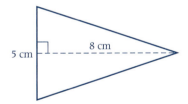

5 cm   8 cm

3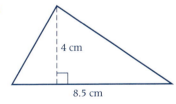

4 cm

8.5 cm

2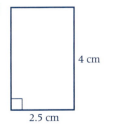

4 cm

2.5 cm

4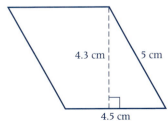

4.3 cm   5 cm

4.5 cm

In questions **5** to **9** use squared paper and draw axes for x and y in the ranges $-6 \leqslant x \leqslant 6$, $-6 \leqslant y \leqslant 6$ using 1 square to 1 unit. Draw the figure and find its area in square units:

5 Triangle ABC with A(0, 6), B(6, 6) and C(5, 2).

6 Parallelogram ABCD with A(0, 1), B(0, 6), C(6, 4) and D(6, −1).

7 Rectangle ABCD with A(-4, 2), B(0, 2) and C(0, -1).

8 Square ABCD with A(0, 0), B(0, 4) and C(4, 4).

9 Triangle ABC with A(-5, -4), B(2, -4), C(-2, 3).

For each of the following figures, find the missing measurement. Draw a diagram in each case:

	Figure	Base	Height	Area
10	Triangle	8 cm		16 cm^2
11	Rectangle	3 cm	15 mm	
12	Parallelogram	4 cm		20 cm^2
13	Square	5 m		
14	Triangle	70 mm		14 cm^2

In questions **15** to **18** give answers correct to three significant figures where necessary.

15

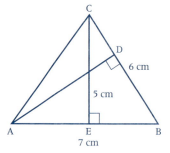

> **Tip** Find the area of △ABC in terms of AD. Put this value equal to the value you found in part **a**.

In △ABC, AB $=$ 7 cm, CB $=$ 6 cm and CE $=$ 5 cm. Find

a the area of △ABC   **b** the length of AD.

16

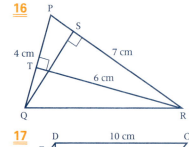

In △PQR, PQ $=$ 4 cm, PR $=$ 7 cm and RT $=$ 6 cm. Find

a the area of △PQR   **b** the length of QS.

17

In parallelogram ABCD, DC $=$ 10 cm, BC $=$ 6 cm and DE $=$ 4 cm. Find

a the area of ABCD   **b** the length of BF.

18

In △ABC, AC $=$ 4 cm, BC $=$ 7 cm and BE $=$ 3.5 cm. Find

a the area of △ABC   **b** the length of AD.

Areas of compound shapes

ABCD is a rhombus. AC = 8 cm and BD = 12 cm.

Find the area of ABCD.

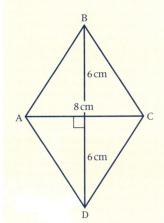

Split the rhombus into two triangles.

(The diagonals of a rhombus bisect each other at right angles.)

Area $\triangle$ABC $= \frac{1}{2}$ base $\times$ height

$\quad = \frac{1}{2} \times 8 \times 6$ cm^2

$\quad = 24$ cm^2

Area $\triangle$ACD = area $\triangle$ABC

(AC is an axis of symmetry)

$\therefore$ total area = 48 cm^2

Find the area of each of the following shapes.

Tip Draw a diagram for each question and mark in all the measurements. Then mark in any other facts that you know.

1

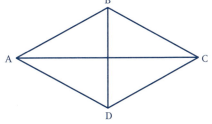

ABCD is a rhombus.
AC = 15 cm and BD = 8 cm

2

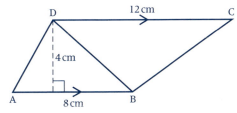

3

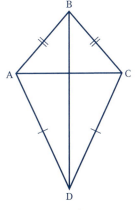

ABCD is a kite.
AC = 6 cm and BD = 10 cm

4

B ——— 8 cm ——— C

5 cm

A ——— 10 cm ——— D

In questions **5** and **6**, find the area of the shaded figure (find the area of the complete figure, then subtract the areas of the unshaded parts).

5

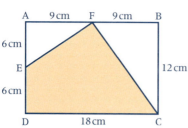

ABCD is a rectangle.

6

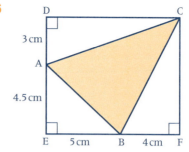

Puzzle

How many different rectangles can you find in this shape?

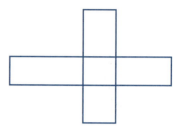

Area of a trapezium

In the last exercise we found the areas of several trapeziums. A trapezium is a shape that occurs often enough to justify finding a formula for its area.

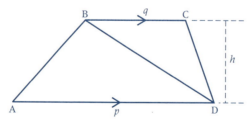

$$\text{Area of } \triangle ABD = \tfrac{1}{2} \text{ base} \times \text{height} = \tfrac{1}{2}p \times h$$

$$\text{Area of } \triangle BCD = \tfrac{1}{2} \text{ base} \times \text{height} = \tfrac{1}{2}q \times h$$

The heights of both triangles are the same, as each is the distance between the parallel sides of the trapezium.

∴   total area of $ABCD = \tfrac{1}{2}ph + \tfrac{1}{2}qh = \tfrac{1}{2}(p+q) \times h$

i.e.   the area of a trapezium is equal to

$\tfrac{1}{2}$(sum of parallel sides) × (distance between them)

Exercise 11c Find the area of the trapezium in the diagram.

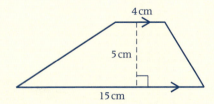

$$\text{Area} = \tfrac{1}{2}(\text{sum of parallel sides}) \times (\text{distance between them})$$
$$= \tfrac{1}{2}(4 + 15) \times 5 \text{ cm}^2$$
$$= \tfrac{1}{2} \times 19 \times 5 \text{ cm}^2$$
$$= 47.5 \text{ cm}^2$$

Find the area of each of the following trapeziums:

1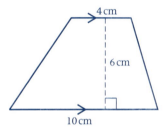

3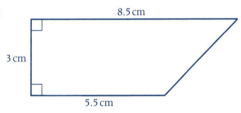

2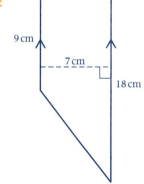

4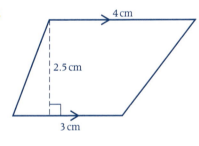

For questions **5** to **10** use squared paper and draw axes for x and y using ranges $-6 \leqslant x \leqslant 6$ and $-6 \leqslant y \leqslant 6$ and a scale of one square to 1 unit. Plot the points and join them up in alphabetical order. Find, in square units, the area of the resulting shape:

5 A(6, 1), B(4, −3), C(−2, −3), D(−3, 1)

6 A(4, 4), B(−2, 2), C(−2, −2), D(4, −3)

7 A(3, 5), B(−4, 4), C(−4, −2), D(3, −5)

8 A(1, 0), B(5, 0), C(5, 3), D(3, 5), E(1, 3)

9 A(6, −4), B(6, 1), C(2, 5), D(−5, 3), E(−5, −4)

10 A(2, 0), B(6, 4), C(−4, 4), D(−4, −2), E(5, −2)

Shapes that have equal areas

*Exercise **11d***

1

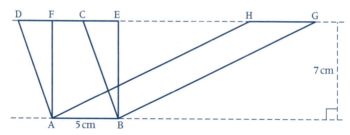

There are three parallelograms in the diagram. Write down the area of each one.

2

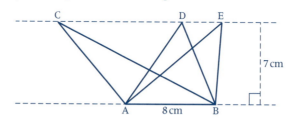

Write down the area of each of the three triangles ABC, ABD and ABE.

In questions **3** and **4**, take the side of a square as the unit of length.

3

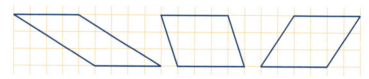

What is the length of the base and the height of each of these parallelograms? What can you conclude about the areas of the three parallelograms?

4

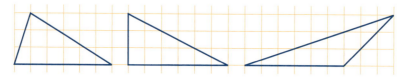

What is the length of the base and the height of each of these triangles? What can you conclude about the areas of the three triangles?

5 Using squared paper draw x and y axes for $-8 \leqslant x \leqslant 8$ and $-4 \leqslant y \leqslant 12$. Draw parallelogram ABCD where A($2, 2$), B($2, 7$), C($7, 4$), D($7, -1$). Using AB as the base in each case, draw three other parallelograms whose areas are equal to the area of ABCD.

6 Use the same set of axes as you used for question 5. Draw the triangle LMN where L($-6, 1$), M($-1, 1$), N($-5, 5$). Using LM as the base draw two other triangles whose areas are equal to the area of triangle LMN.

7 Draw again the diagram that you used for question 6. Draw a triangle which is equal in area to triangle LMN and which has LN for one of its sides.

8 There are four triangles in the diagram all with the same base AB. Find, in ascending order, the ratio of their heights. Find, in ascending order, the ratio of their areas. Comment on your results.

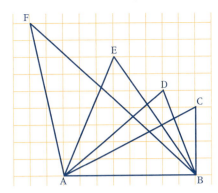

9 Using squared paper draw x and y axes for $0 \leqslant x \leqslant 10$ and $0 \leqslant y \leqslant 12$. Plot the points A(2, 1), B(10, 1) and C(8, 5). Draw a triangle ABD whose area is twice that of triangle ABC. Give the y coordinate of D.

10 Using the same set of axes as in question 9, draw a triangle ABE whose area is half the area of △ABC. Give the y coordinate of E.

11 The triangles XAB, YBD, ZBC and ZCE all have equal heights of 5 units.

  a Find the ratio of their bases.

  b Find the ratio of their areas.

Comment on your results.

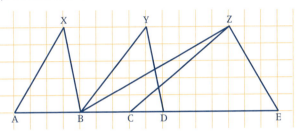

From the last exercise we see that:

  parallelograms with equal bases and equal heights have the same area.

  triangles with equal bases and equal heights have the same area.

  triangles on equal bases but with different heights have areas proportional to (i.e. in the same ratio as) their heights.

  triangles with equal heights but different bases have areas proportional to (i.e. in the same ratio as) their bases.

Exercise 11e

ABCD is a trapezium. Show that the shaded triangles have the same area. Area △ADB = area △ACB

(Same base AB and same height, as DC ∥ AB)

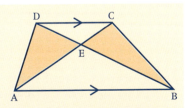

Area △AEB is common to both △ADB and △ACB. Removing it from each triangle in turn leaves the shaded areas.

∴   area △AED = area △BEC

1

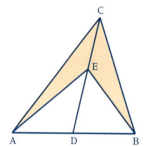

Tip Triangles with the same base and height are equal in area.

D is the midpoint of AB. Show that the shaded triangles have the same area.

2

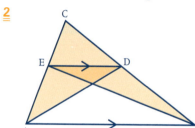

ED is parallel to AB. Show that area △ACD = area △BCE.

4

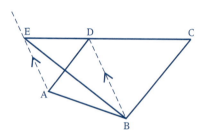

AE is parallel to BD. Show that the area of △BCE is equal to the area of the quadrilateral ABCD.

3

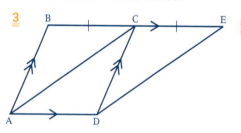

BC = CE, AB is parallel to DC and AD is parallel to BE. Show that area △ADC = area △DCE.

5

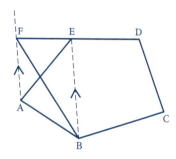

AF is parallel to BE. Show that the area of the pentagon ABCDE is equal to the area of the quadrilateral BCDF.

ABCD is a trapezium. E is the midpoint of DC and F is the midpoint of AB. Show that EF divides the area of ABCD into two equal parts.

6

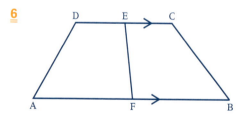

ABCD is a rectangle.
The height of △ABE (i.e. the distance of E above AB) is two thirds of the height, AD, of the rectangle.
The area of ABCD is 12 cm².
Find the area of △AEB.

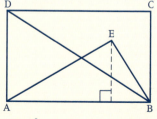

Area of △ADB = 6 cm² (half area of ABCD)

△s ABD, ABE are on the same base AB, so their areas are in the same ratio as their heights.

$$\therefore \quad \frac{\text{area } \triangle ABE}{\text{area } \triangle ADB} = \frac{2}{3}$$

$$\frac{\text{area } \triangle ABE}{6\,\text{cm}^2} = \frac{2}{3} \quad \text{(Multiply both sides by 6 cm}^2\text{)}$$

$$\text{area } \triangle ABE = \frac{2}{3} \times \overset{2}{6}\,\text{cm}^2$$

$$\therefore \quad \text{area } \triangle ABE = 4\,\text{cm}^2$$

7

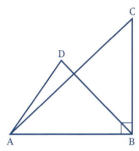

The area of △ABD is $\frac{3}{5}$ of the area of △ABC. BC = 20 cm. Find the height of D above AB.

8

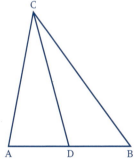

D is the midpoint of AB. Find the ratio of the area of △ABC to the area of △ADC.

9

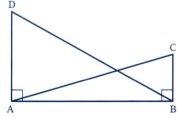

The area of △ABC = $\frac{7}{12}$ of the area of △ADB. AD = 24 cm. Find the length of BC.

10

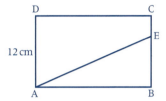

In the rectangle ABCD, E is a point on BC such that area of △ABE is $\frac{1}{3}$ of the area of the rectangle ABCD.

Find the length of BE.

11

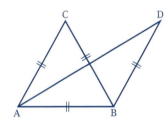

AB = AC = CB = BD.
Area △ABC = area △ABD.
Find the size of AD̂B.

12

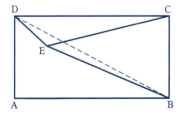

E is a point inside the rectangle ABCD such that the distance of E from BC is $\frac{3}{4}$ of the distance of A from BC and the distance of E from DC is $\frac{1}{3}$ of the distance of A from DC. The area of the rectangle ABCD is 72 cm².

Find the areas of △BEC and △DEC.

In questions **13** and **14**, before beginning the construction draw a rough sketch and on it mark any extra lines that you will need:

13 Construct a parallelogram ABCD with AB = 4 cm, BC = 6 cm and $A\hat{B}C = 60°$. Construct a parallelogram ABEF that is equal in area to ABCD such that BE = 7 cm, and such that E and D are on opposite sides of BC. Measure $A\hat{B}E$.

14 Construct △ABC with AB = 12 cm, $\hat{A} = 30°$ and $\hat{B} = 30°$. On AC as base, construct a triangle ADC that is equal in area to △ABC, such that $C\hat{A}D = 90°$. Measure AD.

15 In the △ABC, D is the midpoint of BC. E is a point on AD so that AE = $\frac{1}{4}$AD. If BC = 16 cm and the area of △ABC = 96 cm², find the area of △DEC.

Puzzle

The sketch shows seventeen identical sticks laid out to form six equal squares.

Remove six sticks to leave two perfect squares.

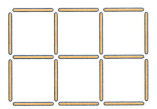

Did you know that if one rectangle measuring *l* cm by *b* cm overlaps another larger rectangle measuring *L* cm by *B* cm, the difference between the two shaded areas, *P* and *Q*, is always the same? This is quite easy to prove. Try it.

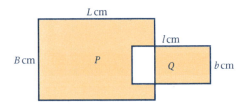

IN THIS CHAPTER...

you have seen that:

- the area of a trapezium is equal to $\frac{1}{2}$(sum of the parallel sides) × (distance between them)

- parallelograms with equal bases and the same heights have the same area

- triangles with equal bases and equal heights have the same area

- triangles on equal bases but with different heights have areas proportional to their heights

- triangles with equal heights but different bases have areas proportional to their bases.

12 AVERAGES

To mathematicians', solutions are answers, but to chemists, solutions are all mixed up.

BEFORE YOU START

you need to know:
- ✓ units of time and distance
- ✓ how to work with fractions and decimals

KEY WORDS arithmetic average, average speed, centilitre, mean

Arithmetic average

The arithmetic average or mean of a set of numbers is their sum divided by the number of numbers in the set.

For example, the mean of the six numbers 5, 10, 15, 21, 25 and 32 is

$$\frac{5 + 10 + 15 + 21 + 25 + 32}{6} = \frac{108}{6} = 18$$

Exercise 12a

Find the arithmetic average or mean of the following sets of numbers:

1 40, 48, 44, 63, 35

2 6, 36, 21, 59, 47, 80, 34, 14, 63

3 9.3, 26.5, 14.2, 18.7, 32.4, 18.2, 20.7

4 47.4, 40.6, 53.8, 41.1, 29.9, 35.6

5 In five frames of snooker, Ray scored 76, 12, 43, 83 and 41.

   a How many did he score altogether?

   b What was his average score per frame?

6 Janet's marks in six consecutive geography examinations were 72, 54, 58, 73, 47 and 62.

   a Find the total marks she scored.

   b What was her average mark?

7 The heights of five girls were 147 cm, 151 cm, 149 cm, 160 cm and 158 cm. Find the total of the heights of the five girls, and hence find their average height.

8 Jennifer's average mark for nine examination papers was 62. How many marks did she score altogether?

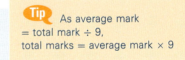

Tip As average mark = total mark ÷ 9, total marks = average mark × 9

9 During a certain week in June, the midday temperatures in degrees Celsius at a holiday resort were 30°, 29°, 22°, 31°, 32°, 26° and 33°. What was the average midday temperature for the week?

I bought six apples at $1.00 each and nine apples at 75 c each. Find the average price of the apples I bought.

To find the average price, you need to find the total cost of the 6 plus 9, i.e. 15, apples.

6 apples at $1.00 each cost   6 × 100 c = $6.00

9 apples at 75 c each cost   9 × 75 c = $6.75

$$\therefore \quad 15 \text{ apples cost} \qquad \$6.00 + \$6.75 = \$12.75$$

$$\text{Average price of an apple} = \frac{\text{total cost}}{\text{total number}}$$

$$= \frac{\$12.75}{15}$$

$$= 85\,c$$

10 I bought four white loaves at 98 c each and three brown loaves at $1.12 each. Find the average price I paid for a loaf of bread.

11 Andrew buys seven books at $6.20 each and three books at $4.70 each. What is the average amount that Andrew pays for a book?

12 Betty buys eight bottles of Cola each containing 70 cl and four bottles of Cola each containing 1 litre. Find, in centilitres, the average amount in the bottles she buys.

13 A shipping company owns three ships of average gross tonnage 18 400 t, eight ships of average gross tonnage 11 500 t and 12 ships of average gross tonnage 4600 t. Calculate the average gross tonnage of the company's ships.

The average height of the ten boys in a class is 160 cm and the average height of the 15 girls is 154 cm. Find the average height of the class. As a result of a new boy joining the class, the average height of the class increases by 0.1 cm. How tall is he?

To find the average height of the class, first you need to find the total heights of the boys plus the total heights of the girls.

$$\text{Total height of the 10 boys} = 160 \times 10\,\text{cm} = 1600\,\text{cm}$$

$$\text{Total height of the 15 girls} = 154 \times 15\,\text{cm} = 2310\,\text{cm}$$

$$\therefore \quad \text{Total height of the 25 pupils in the class} = 3910\,\text{cm}$$

$$\text{Average height of the class} = \frac{3910}{25}\,\text{cm}$$

$$= 156.4\,\text{cm}$$

When a new boy joins, the average height increases by 0.1 cm to 156.5 cm.

There are now 26 pupils in the class so the total height of the 26 pupils in the class

$$= 156.5 \times 26\,\text{cm}$$

$$= 4069\,\text{cm}$$

But, total height of the original 25 pupils = 3910 cm

Therefore, the height of the boy who joins is

$$(4069 - 3910)\,\text{cm} = 159\,\text{cm}$$

14 The average daily 'takings' in the corner shop from Monday to Friday of a certain week was $120.50, while the average daily 'takings' from Monday to Saturday of the same week was $132.40.

   a How much was taken over the five days Monday to Friday?

   b How much was taken over the six days Monday to Saturday?

   c How much was taken on the Saturday?

15 The average number of birthday cards sold by a newsagent from Monday to Friday during a certain week was 26, while the average number sold from Monday to Saturday of the same week was 32.

   a How many cards were sold over the five days Monday to Friday?

   b How many cards were sold over the six days Monday to Saturday?

   c How many birthday cards were sold on the Saturday?

16 In the first year of Stanley School, the average size of forms 1E, 1L, 1M and 1W is 30; the average size of forms 1P, 1D and 1B is 23; and the average size of 1L and 1K is 18.

   a How many pupils are there in the first year?

   b What is the average size of a first year form?

17 The average age of a family of five is 28 years 11 months. If the average age of the parents is 46 years 2 months, find the average age of the children. If the father's age is 48 years 7 months, how old is the mother?

Susan's average mark for her first five examinations was 54. What must she score in her sixth examination to raise her average mark to 60?

You need to find the total marks scored in the five examinations and then the total marks that must be scored in the six examinations to give an average mark of 60.

Total marks scored in her first five examinations is

$$54 \times 5 = 270$$

Total marks needed in the six examinations is

$$60 \times 6 = 360$$

∴ the mark needed in sixth examination must be $360 - 270 = 90$

18 In a boat race the average weight of the eight oarsmen was 63.2 kg and the average weight of the crew was 60.9 kg. How heavy was the cox?

19 The average weight of the eleven players in a school football team was 55.6 kg, while the average weight of the team plus a travelling reserve was 56.5 kg. How heavy was the reserve?

20 In twelve completed innings, a batsman's average score was 47.5. After a further innings his average fell to 44. How many runs did he score in his thirteenth innings?

Average speeds

If a car travels 180 km in three hours, we say that the average distance it travels in each hour is $\dfrac{180}{3}$ km $= 60$ km

i.e. its average speed is 60 km/h.

The general result was given in Book 2

$$\text{average speed} = \frac{\text{total distance travelled}}{\text{total time taken}}$$

or $$S = \frac{D}{T}$$

It follows that $$D = ST$$

i.e. $$\text{distance} = \text{average speed} \times \text{time}$$

and $$T = \frac{D}{S}$$

i.e. $$\text{time} = \frac{\text{distance}}{\text{average speed}}$$

Exercise 12b   Find the average speed, in km/h, for a journey of 24 kilometres in 36 minutes.

To find a speed in km/h, the distance must be in kilometres and the time must be in hours.

The distance travelled is 24 km.

The time taken is 36 minutes $= \frac{36}{60}$ h $= \frac{3}{5}$ h

$$\text{Average speed} = \frac{\text{distance}}{\text{time}}$$

$$= \frac{24\,\text{km}}{\frac{3}{5}\,\text{h}}$$

$$= 24 \times \frac{5}{3}\ \text{km/h}$$

i.e. $$\text{average speed} = 40\,\text{km/h}$$

Find the average speed, in km/h, for a journey of:

1 180 km in 3 hours

2 75 km in $2\frac{1}{2}$ hours

3 12 km in 20 minutes

4 99 km in 66 minutes

5 27 km in $1\frac{1}{2}$ hours

6 175 km in 40 minutes

Find the average speed, in m.p.h., for a journey of:

7 200 miles in 4 hours

8 12 miles in $1\frac{1}{2}$ hours

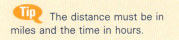
Tip The distance must be in miles and the time in hours.

9 30 miles in 40 minutes

10 90 miles in 100 minutes

How far will a car travel in 45 minutes at an average speed of 56 m.p.h.?

Average speed is 56 m.p.h.

Time taken is 45 minutes $= \frac{45}{60}$ h $= \frac{3}{4}$ h

$$\text{Distance travelled} = \text{average speed} \times \text{time}$$

$$= 56 \times \tfrac{3}{4} \text{ miles}$$

$$= 42 \text{ miles}$$

How far will a car travel:

11 In 4 hours at an average speed of 65 m.p.h.?

12 In $2\frac{1}{2}$ hours at an average speed of 32 m.p.h.?

Tip Units must be consistent, i.e.
m.p.h. → miles and hours,
km/h → kilometres and hours.

13 In 50 minutes at an average speed of 186 km/h?

14 In 15 minutes at an average speed of 324 m.p.h.?

How long will it take to travel 77 km at 140 km/h?

$$\text{Time taken} = \frac{\text{distance travelled}}{\text{average speed}}$$

$$= \frac{77}{140} \text{ hours}$$

$$= 0.55 \text{ hours}$$

$$= 0.55 \times 60 \text{ minutes}$$

$$= 33 \text{ minutes}$$

How long will it take to travel:

15 96 km at an average speed of 32 km/h?

16 135 miles at an average speed of 27 m.p.h.?

17 4 km at an average speed of 12 km/h?

18 27 miles at an average speed of 36 m.p.h.?

Puzzle

John and Sam are on their skateboards 200 m apart. They start moving towards each other as fast as they can; John manages an average of 6 metres per second and Sam manages an average of 4 metres per second. As they start, Sam's dog decides to run between them; he runs from Sam to John, back to Sam, back to John and so on until John and Sam collide. The dog runs at 8 metres per second (he is a big dog!). How far does the dog run?

Problems

Exercise 12c

Adrian lives on a farm. To go to school he walks one kilometre to the bus stop at an average speed of 6 km/h and immediately gets on the bus. He spends 20 minutes on the bus, which is travelling at an average speed of 39 km/h, before he arrives at school.

a How far does he travel on the bus?

b How long, in hours, does he take to go from home to school?

c What is his average speed for the whole journey?

(We must work in the same units throughout and we will use kilometres and hours.)

a Time spent on the bus is 20 minutes $= \frac{20}{60}$ h $= \frac{1}{3}$ h

Distance travelled on bus $=$ average speed $\times$ time

$$= 39 \times \tfrac{1}{3} \text{ km}$$

$$= 13 \text{ km}$$

b Time spent walking $= \dfrac{\text{distance walked}}{\text{average walking speed}}$

$$= \frac{1 \text{ km}}{6 \text{ km/h}}$$

$$= \tfrac{1}{6} \text{ h}$$

Total journey time $=$ time walking $+$ time on bus

$$= \left(\tfrac{1}{6} + \tfrac{1}{3} \right) \text{ h}$$

$$= \tfrac{1}{2} \text{ h}$$

c Average speed for the journey $= \dfrac{\text{total distance}}{\text{total time}}$

$$= \frac{(1 + 13) \text{ km}}{\tfrac{1}{2} \text{ h}}$$

$$= 14 \times \tfrac{2}{1} \text{ km/h}$$

$$= 28 \text{ km/h}$$

> **Tip** When solving problems it is particularly important to remember that when calculating the average speed for a distance you should never add or subtract average speeds for parts of the total distance. Always consider total distances and total times.

1 A cyclist travels the 20 kilometres from St Peter to St Michael at an average speed of 30 km/h and immediately continues the journey to St Andrew, which is a further 40 kilometres away, at an average speed of 20 km/h. Find the average speed for the whole journey.

2 A motorist travels for one hour at an average speed of 80 km/h and then for two hours at an average speed of 110 km/h. Find his average speed for the whole journey.

3 For the first 50 km of its journey a train travels at an average speed of 100 km/h, for the next 50 km at an average speed of 200 km/h and for the last 40 km at an average speed of 160 km/h. Find its average speed for the whole journey.

4 Jean jogs $\frac{2}{3}$ km at 12 km/h and then walks $1\frac{1}{3}$ km at 8 km/h. Find her average speed for the whole journey.

In New York a train is scheduled to make a 120 km journey at an average speed of 150 km/h. It leaves three minutes late and its average speed is increased so that it arrives on time. What is the increased average speed?

From the information given we can work out how long the journey should take. We then know how long the journey must take for the train to get there on time.

Time taken if train leaves on time $= \dfrac{\text{distance travelled}}{\text{average speed}}$

$$= \frac{120}{150} \text{ hours}$$

$$= \frac{4}{5} \times 60 \text{ minutes}$$

$$= 48 \text{ minutes}$$

If the train leaves three minutes late it must complete the journey in 45 minutes, i.e. $\frac{45}{60}$ hours $= \frac{3}{4}$ hour

$$\text{Average speed} = \frac{\text{distance travelled}}{\text{time taken}}$$

$$= \frac{120 \text{ km}}{\frac{3}{4} \text{ hours}}$$

$$= 120 \times \frac{4}{3} \text{ km/h}$$

$$= 160 \text{ km/h}$$

5 A coach is scheduled to make a 60 km journey at an average speed of 80 km/h. If it leaves five minutes late, what must its average speed be increased to, so that it arrives on time?

6 The distance from Georgetown to New Amsterdam is 110 km. A bus is due to leave Georgetown at 0800 hr and to arrive in New Amsterdam at 1026 hr. If it leaves Georgetown six minutes late, find its average speed if it is to arrive on time.

7 I normally drive the 20 km to school at an average speed of 60 km/h. If I am five minutes late leaving home, calculate my average speed if I am to arrive on time.

8 A continental holiday maker is motoring at a steady 20 m/s oblivious to the world. When he is 500 m from an unmanned railway crossing an express train, 100 m long, is travelling towards the crossing at a steady 45 m/s, and is 1 km from it. Which of the following happens?

   a The car crosses safely before the train arrives.

   b There is an unfortunate accident.

   c The train passes through the crossing before the car reaches it.

MATHS IS OUT THERE

Q: Why is 6 not happy being next to 7?

A: 7 8 9

What do you call two students who cheat on mathematics tests?

Co-efficients!

How would you wake up twenty mathematics students living on a farm?

By roster.

First exponent: Will you give me a loan?

Second exponent: If it's within my power.

IN THIS CHAPTER...

you have seen that:

- the arithmetic average value of a set of values is the total of the values divided by the number of values

- if you know the arithmetic average and the number of values, you can find the total of the set of values by multiplying the average by the number of values

- to find the average speed you need the total distance covered and the total time taken

- the relationship between average speed, S, time, T, and distance, D, is given by $D = S \times T$. It follows that if you know two of these quantities, you can find the third.

ALGEBRAIC PRODUCTS

AT THE END OF THIS CHAPTER...

you should be able to:

1   Calculate the product of expressions in two brackets each of which contains two terms.

2   Square an expression of the form $(ax + b)$.

The branch of mathematics called algebra gets its name from the Arabic word *al-jebr* which means 'the pulling together of broken parts'. In the seventeenth century it was used to describe the surgical treatment of fractures!

BEFORE YOU START

you need to know:
✓ how to work with directed numbers
✓ how to multiply out expressions such as $3(5x - 2)$ and $3a(3b + 2c)$

KEY WORDS   collecting like terms, product

Brackets

Remember that $\quad 5(x+1) = 5x + 5$

and that $\qquad 4x(y+z) = 4xy + 4xz$

Expand:

1 $2(x+1)$	**7** $5(1-b)$	**13** $5x(3y+z)$
2 $3(x-1)$	**8** $2(3a-1)$	**14** $4y(4x+3z)$
3 $4(x+3)$	**9** $4(2+3b)$	**15** $2n(3p-5q)$
4 $5(a+4)$	**10** $5a(b-c)$	**16** $8r(2t-s)$
5 $3(b+7)$	**11** $4a(b-2c)$	**17** $3a(b-5c)$
6 $3(1-a)$	**12** $3a(2a+b)$	**18** $4x(3y+2z)$

The product of two brackets

Frequently, we wish to find the product of two brackets, each of which contains two terms, e.g. $(a+b)(c+d)$. The meaning of this product is that each term in the first bracket has to be multiplied by each term in the second bracket.

Always multiply the brackets together in the following order:

1. the first terms in the brackets
2. the outside terms
3. the inside terms
4. the second terms in the brackets.

Thus
$$(a+b)(c+d) = ac + ad + bc + bd$$

Expand $(x+2y)(2y-z)$

$$(x+2y)(2y-z) = 2xy - xz + 4y^2 - 2yz$$

Multiply the first term in the first bracket by each term in the second, then the second term in the first bracket by each term in the second. Try to keep the same order when you multiply two brackets. You are less likely to leave a term out.

Expand:

1 $(a + b)(c + d)$	**7** $(x + y)(y + z)$	**13** $(6u - 5v)(w - 5r)$
2 $(p + q)(s + t)$	**8** $(2a + b)(3c + d)$	**14** $(3a + 4b)(2c - 3d)$
3 $(2a + b)(c + 2d)$	**9** $(5x + 4y)(z + 2)$	**15** $(3x + 2y)(3z + 2)$
4 $(5x + 2y)(z + 3)$	**10** $(3x - 2y)(5 - z)$	**16** $(3p - q)(4r - 3s)$
5 $(x + y)(z - 4)$	**11** $(p + q)(2s - 3t)$	**17** $(3a - 4b)(3c + 4d)$
6 $(a - b)(c + d)$	**12** $(a - 2b)(c - d)$	**18** $(7x - 2y)(3 - 2z)$

We get a slightly simpler form when we find the product of two brackets such as $(x + 2)$ and $(x + 3)$,

i.e. using the order we chose earlier

$$(x + 2)(x + 3) = x^2 + 3x + 2x + 6$$

$$= x^2 + 5x + 6 \quad \text{(since } 2x \text{ and } 3x \text{ are like terms)}$$

i.e. $\qquad (x + 2)(x + 3) = x^2 + 5x + 6$

Exercise 13c

Expand:

1 $(x + 3)(x + 4)$	**4** $(x + 5)(x + 2)$	**7** $(b + 2)(b + 7)$
2 $(x + 2)(x + 4)$	**5** $(x + 8)(x + 3)$	**8** $(c + 4)(c + 6)$
3 $(x + 1)(x + 6)$	**6** $(a + 4)(a + 5)$	**9** $(p + 3)(p + 12)$

Expand $(x - 4)(x - 6)$

$$(x - 4)(x - 6) = x^2 - 6x - 4x + 24$$

Collect the like terms: $\qquad = x^2 - 10x + 24$

Expand:

10 $(x-2)(x-3)$

11 $(x-5)(x-7)$

12 $(a-2)(a-8)$

13 $(x-10)(x-3)$

14 $(b-5)(b-5)$

15 $(x-3)(x-4)$

16 $(x-4)(x-8)$

17 $(b-4)(b-2)$

Tip Remember to keep to the same order when you multiply out the brackets.

18 $(a-4)(a-4)$

Expand $(x+3)(x-6)$

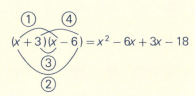

$$(x+3)(x-6) = x^2 - 6x + 3x - 18$$

Collect the like terms: $\qquad = x^2 - 3x - 18$

Expand:

19 $(x+3)(x-2)$

20 $(x-4)(x+5)$

21 $(x-7)(x+4)$

22 $(a+3)(a-10)$

23 $(p+5)(p-5)$

24 $(x+7)(x-2)$

25 $(x-5)(x+6)$

26 $(x+10)(x-1)$

27 $(b-8)(b-7)$

 ## Investigation

Four 5s can be written as $5 \times 5 \div (5 \div 5)$. The answer is 25.

Investigate other ways of writing four 5s, together with any of the signs $+$, $-$, $\times$ or $\div$ to give an answer of 25.

Finding the pattern

You may have noticed in the previous exercise, that when you expanded the brackets and simplified the answers, there was a definite pattern,

e.g.
$$(x+5)(x+9) = x^2 + 9x + 5x + 45$$
$$= x^2 + 14x + 45$$

We could have written it

$$(x+5)(x+9) = x^2 + (9+5)x + (5) \times (9)$$
$$= x^2 + 14x + 45$$

Similarly
$$(x+4)(x-7) = x^2 + (-7+4)x + (4) \times (-7)$$
$$= x^2 - 3x - 28$$

and
$$(x - 3)(x - 8) = x^2 + (-8 - 3)x + (-3) \times (-8)$$
$$= x^2 - 11x + 24$$

In each case there is a pattern:
the *product* of the two numbers in the brackets gives the number term in the expansion,
while *collecting* them gives the number of xs.

Exercise 13d

Use the pattern given above to expand the following products:

1 $(x + 4)(x + 5)$	**9** $(a + 2)(a - 5)$	**17** $(x - 5)(x - 1)$
2 $(a + 2)(a + 5)$	**10** $(y - 6)(y + 3)$	**18** $(b + 9)(b + 7)$
3 $(x - 4)(x - 5)$	**11** $(z + 4)(z - 10)$	**19** $(a + 4)(a - 4)$
4 $(a - 2)(a - 5)$	**12** $(p + 5)(p - 8)$	**20** $(r - 14)(r + 2)$
5 $(x + 8)(x + 6)$	**13** $(a - 10)(a + 7)$	**21** $(p + 12)(p + 2)$
6 $(a + 10)(a + 7)$	**14** $(y + 10)(y - 2)$	**22** $(t + 5)(t - 12)$
7 $(x - 8)(x - 6)$	**15** $(z - 12)(z + 1)$	**23** $(c - 5)(c + 8)$
8 $(a - 10)(a - 7)$	**16** $(p + 2)(p - 13)$	**24** $(x + 5)(x - 5)$

The pattern is similar when the brackets are slightly more complicated.

Exercise 13e

Expand the product $(2x + 3)(x + 2)$

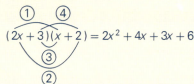

$$(2x + 3)(x + 2) = 2x^2 + 4x + 3x + 6$$

Collect like terms: $\qquad = 2x^2 + 7x + 6$

Expand the following products:

1 $(2x + 1)(x + 1)$	**5** $(3x + 2)(x + 1)$
2 $(x + 2)(5x + 2)$	**6** $(x + 3)(3x + 2)$
3 $(5x + 2)(x + 3)$	**7** $(4x + 3)(x + 1)$
4 $(3x + 4)(x + 5)$	**8** $(7x + 2)(x + 3)$

Tip Remember to stick to the same order when you multiply out the brackets.

Expand the product $(3x - 2)(2x + 5)$

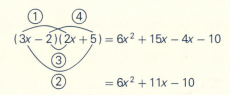

$$(3x - 2)(2x + 5) = 6x^2 + 15x - 4x - 10$$

$$= 6x^2 + 11x - 10$$

Expand:

9 $(3x+2)(2x+3)$	**13** $(5x+3)(2x+5)$	**17** $(2a+3)(2a-3)$	**21** $(4x+3)(4x-3)$
10 $(4x-3)(3x-4)$	**14** $(7x-2)(3x-2)$	**18** $(3b-7)(3b+7)$	**22** $(5y-2)(5y+2)$
11 $(5x+6)(2x-3)$	**15** $(3x-2)(4x+1)$	**19** $(7y-5)(7y+5)$	**23** $(3x-1)(3x+1)$
12 $(7a-3)(3a-7)$	**16** $(3b+5)(2b-5)$	**20** $(5a+4)(4a-3)$	**24** $(4x-7)(4x+5)$

Expand $(3x-2)(5-2x)$

$$(3x-2)(5-2x) = 15x - 6x^2 - 10 + 4x \qquad ((-2x) \times (-2) = +4x)$$
$$= 19x - 6x^2 - 10$$
$$= -6x^2 + 19x - 10$$

Expand:

25 $(2x+1)(1+3x)$	**28** $(5a-2)(3-7a)$	**31** $(5x+2)(4+3x)$	**34** $(3-p)(4+p)$
26 $(5x+2)(2-x)$	**29** $(3x+2)(4-x)$	**32** $(7x+4)(3-2x)$	**35** $(x-5)(2+x)$
27 $(6x-1)(3-x)$	**30** $(4x-5)(3+x)$	**33** $(4x-3)(3-5x)$	**36** $(4x-3)(3+x)$

Important products

Three very important products are:

$$(x+a)^2 = (x+a)(x+a)$$
$$= x^2 + xa + ax + a^2$$
$$= x^2 + 2ax + a^2 \quad \text{(since } xa \text{ is the same as } ax\text{)}$$

i.e. $\qquad (x+a)^2 = x^2 + 2ax + a^2$

so $\qquad (x+3)^2 = x^2 + 6x + 9$

$\qquad (x-a)^2 = (x-a)(x-a)$
$$= x^2 - xa - ax + a^2$$

i.e. $\qquad (x-a)^2 = x^2 - 2ax + a^2$

so $\qquad (x-4)^2 = x^2 - 8x + 16$

$\qquad (x+a)(x-a) = x^2 - xa + ax - a^2$
$$= x^2 - a^2$$

i.e. $\qquad (x+a)(x-a) = x^2 - a^2$

and $\qquad (x-a)(x+a) = x^2 - a^2$

so $\qquad (x+5)(x-5) = x^2 - 25$

and $\qquad (x-3)(x+3) = x^2 - 9$

You should learn these three results thoroughly, for they will appear time and time again. Given the left-hand side you should know the right-hand side and vice versa.

Did you know that Euclid of Alexandria – a mathematician more closely associated with geometry than algebra – was the first person to expand $(a + b)^2$ as $a^2 + 2ab + b^2$?

Exercise **13f**

Expand $(x + 5)^2$

Comparing with $(x + a)^2 = x^2 + 2ax + a^2$, $a = 5$.

$$(x + 5)^2 = x^2 + 10x + 25$$

Expand:

1 $(x + 1)^2$	5 $(t + 10)^2$	9 $(x + y)^2$	13 $(p + q)^2$
2 $(x + 2)^2$	6 $(x + 12)^2$	10 $(y + z)^2$	14 $(a + b)^2$
3 $(a + 3)^2$	7 $(x + 8)^2$	11 $(c + d)^2$	15 $(e + f)^2$
4 $(b + 4)^2$	8 $(p + 7)^2$	12 $(m + n)^2$	16 $(u + v)^2$

Expand $(2x + 3)^2$

$$(2x + 3)^2 = (2x)^2 + 2(2x)(3) + (3)^2$$

i.e.
$$(2x + 3)^2 = 4x^2 + 12x + 9$$

Expand:

17 $(2x + 1)^2$	19 $(5x + 2)^2$	21 $(3a + 1)^2$	23 $(3a + 4)^2$
18 $(4b + 1)^2$	20 $(6c + 1)^2$	22 $(2x + 5)^2$	24 $(4y + 3)^2$

Expand $(2x + 3y)^2$

$$(2x + 3y)^2 = (2x)^2 + 2(2x)(3y) + (3y)^2$$

i.e.
$$(2x + 3y)^2 = 4x^2 + 12xy + 9y^2$$

Expand:

25 $(x + 2y)^2$	27 $(2x + 5y)^2$	29 $(3a + b)^2$	31 $(7x + 2y)^2$
26 $(3x + y)^2$	28 $(3a + 2b)^2$	30 $(p + 4q)^2$	32 $(3s + 4t)^2$

Expand $(x - 5)^2$

$$(x - 5)^2 = x^2 - 10x + 25$$

Expand:

33 $(x - 2)^2$	35 $(a - 10)^2$	37 $(x - 3)^2$	39 $(a - b)^2$
34 $(x - 6)^2$	36 $(x - y)^2$	38 $(x - 7)^2$	40 $(u - v)^2$

Expand $(2x - 7)^2$

$$(2x - 7)^2 = (2x)^2 + 2(2x)(-7) + (-7)^2$$

i.e.

$$(2x - 7)^2 = 4x^2 - 28x + 49$$

Expand:

41	$(3x - 1)^2$	43	$(10a - 9)^2$	45	$(2a - 1)^2$	47	$(7b - 2)^2$
42	$(5z - 1)^2$	44	$(4x - 3)^2$	46	$(4y - 1)^2$	48	$(5x - 3)^2$

Expand $(7a - 4b)^2$

$$(7a - 4b)^2 = (7a)^2 + 2(7a)(-4b) + (-4b)^2$$

i.e.

$$(7a - 4b)^2 = 49a^2 - 56ab + 16b^2$$

Expand:

49	$(2y - x)^2$	51	$(3m - 2n)^2$	53	$(a - 3b)^2$	55	$(5a - 2b)^2$
50	$(5x - y)^2$	52	$(7x - 3y)^2$	54	$(m - 8n)^2$	56	$(3p - 5q)^2$

The difference between two squares

Exercise 13g   Expand   a $(a + 2)(a - 2)$ b $(2x + 3)(2x - 3)$

a $(a + 2)(a - 2) = a^2 - 4$

b $(2x + 3)(2x - 3) = 4x^2 - 9$

Expand:

1	$(x + 4)(x - 4)$	6	$(a - 7)(a + 7)$	11	$(7a + 2)(7a - 2)$		
2	$(b + 6)(b - 6)$	7	$(q + 10)(q - 10)$	12	$(5a - 4)(5a + 4)$		
3	$(c - 3)(c + 3)$	8	$(x - 8)(x + 8)$	13	$(5x + 1)(5x - 1)$		
4	$(x + 12)(x - 12)$	9	$(2x - 1)(2x + 1)$	14	$(2a - 3)(2a + 3)$		
5	$(x + 5)(x - 5)$	10	$(3x + 1)(3x - 1)$	15	$(10m - 1)(10m + 1)$		

Expand $(3x + 2y)(3x - 2y)$

$$(3x + 2y)(3x - 2y) = (3x)^2 - (2y)^2$$
$$= 9x^2 - 4y^2$$

Expand:

16	$(3x + 4y)(3x - 4y)$	19	$(7y + 3z)(7y - 3z)$	22	$(1 + 3x)(1 - 3x)$		
17	$(2a - 5b)(2a + 5b)$	20	$(10a - 9b)(10a + 9b)$	23	$(3 - 5x)(3 + 5x)$		
18	$(1 - 2a)(1 + 2a)$	21	$(5a - 4b)(5a + 4b)$	24	$(5m + 8n)(5m - 8n)$		

The results from this exercise are very important when written the other way around,

i.e. $a^2 - b^2 = (a + b)(a - b)$

We refer to this as 'factorising the difference between two squares' and we will deal with it in detail in the next chapter.

Harder expansions

Exercise 13h Simplify $(x + 2)(x + 5) + 2x(x + 7)$

Work out the brackets first.

$$(x + 2)(x + 5) + 2x(x + 7) = x^2 + 5x + 2x + 10 + 2x^2 + 14x$$

Collect like terms.

$$= 3x^2 + 21x + 10$$

Simplify:

1 $(x + 3)(x + 4) + x(x + 2)$

2 $x(x + 6) + (x + 1)(x + 2)$

3 $(x + 4)(x + 5) + 6(x + 2)$

4 $(a - 6)(a - 5) + 2(a + 3)$

5 $(a - 5)(2a + 3) - 3(a - 4)$

6 $(x + 3)(x + 5) + 5(x + 2)$

7 $(x - 3)(x + 4) - 3(x + 3)$

> **Tip** Expand the brackets first then collect like terms.

8 $(x + 7)(x - 5) - 4(x - 3)$

9 $(2x + 1)(3x - 4) + (2x + 3)(5x - 2)$

10 $(5x - 2)(3x + 5) - (3x + 5)(x + 2)$

Expand $(xy - z)^2$

$$(xy - z)^2 = (xy)^2 - 2(xy)(z) + z^2$$

i.e.

$$(xy - z)^2 = x^2 y^2 - 2xyz + z^2$$

Expand:

11 $(xy - 3)^2$

12 $(5 - yz)^2$

13 $(xy + 4)^2$

14 $(3pq + 8)^2$

15 $(a - bc)^2$

16 $(ab - 2)^2$

17 $(6 - pq)^2$

18 $(mn + 3)^2$

19 $(uv - 2w)^2$

Summary

The following is a summary of the most important types of examples considered in this chapter that will be required in future work.

1 $2(3x + 4) = 6x + 8$

2 $(x + 2)(x + 3) = x^2 + 5x + 6$

3 $(x - 2)(x - 3) = x^2 - 5x + 6$

4 $(x - 2)(x + 3) = x^2 + x - 6$

5 $(2x + 1)(3x + 2) = 6x^2 + 7x + 2$

6 $(2x - 1)(3x - 2) = 6x^2 - 7x + 2$

7 $(2x + 1)(3x - 2) = 6x^2 - x - 2$

Note that **a** if the signs in the brackets are the same, i.e. both + or both −, then the number term is +
(numbers **2**, **3**, **5** and **6**)

whereas **b** if the signs in the brackets are different, i.e. one + and one −, then the number term is −

(numbers **4** and **7**)

c the middle term is given by collecting the product of the outside terms in the brackets and the product of the inside terms in the brackets,

i.e. in **2** the middle term is $3x + 2x$ or $5x$

in **3** the middle term is $-3x - 2x$ or $-5x$

in **4** the middle term is $3x - 2x$ or x

in **5** the middle term is $4x + 3x$ or $7x$

in **6** the middle term is $-4x - 3x$ or $-7x$

in **7** the middle term is $-4x + 3x$ or $-x$.

Most important of all we must remember the general expansions:

$$(x + a)^2 = x^2 + 2ax + a^2$$
$$(x - a)^2 = x^2 - 2ax + a^2$$
$$(x + a)(x - a) = x^2 - a^2$$

Mixed exercises

Exercise 13i

Expand:

1 $5(x + 2)$
2 $8p(3q - 2r)$
3 $(3a + b)(2a - 5b)$
4 $(4x + 1)(3x - 5)$

5 $(x + 6)(x + 10)$
6 $(x - 8)(x - 12)$
7 $(4y + 3)(4y - 7)$

8 $(4y - 9)(4y + 9)$
9 $(5x + 2)^2$
10 $(2a - 7b)^2$

Exercise 13j

Expand:

1 $4(2 - 5x)$
2 $8a(2 - 3a)$
3 $(4a + 3)(3a - 11)$
4 $(x + 11)(x - 9)$

5 $(2x + 5)(1 - 10x)$
6 $(y + 2z)^2$
7 $(6y - z)(6y + 5z)$

8 $(4a + 1)^2$
9 $(5a - 7)^2$
10 $(6z - 13y)^2$

Exercise *13k*

Expand:

1 $3(2 - a)$

2 $4a(2b + c)$

3 $(5a + 2b)(2c + 5d)$

4 $(x - 7)(x - 12)$

5 $(a + 7)(a + 9)$

6 $(a + 4)(a - 5)$

7 $(3x + 1)(2x + 3)$

8 $(5x - 2)(5x + 2)$

9 $(3x - 7)^2$

10 $(5x + 2y)(5x - 2y)$

Puzzle

A gambler visits three casinos. At the first he doubles his money and spends $30 on food. At the second he trebles his money and spends $54 on a round of drinks. At the third casino he increases his money fourfold and spends $72 on presents. If he now has $48, how much did he have to start with?

Did you ever find the meaning of TRISKAIDEKAPHOBIA? We have come to the end of Chapter 13. If the number 13 did not cause you any fear, then you do not suffer from triskaidekaphobia, which means 'the fear of the number 13'.

'Thirteen' is usually associated with awkwardness and bad luck. For this reason many buildings do not have a 13th floor.

IN THIS CHAPTER...

you have seen that:

- two brackets are multiplied together by multiplying each term in one bracket by every term in the other bracket, e.g.

$(3a + 2b)(2c - 3d) = 6ac - 9ad + 4bc - 6bd$

(Always try to do the multiplying in the same order.)

- three important products which you should commit to memory are:

$(x + a)^2 = x^2 + 2ax + a^2$

$(x - a)^2 = x^2 - 2ax + a^2$

$(x + a)(x - a) = x^2 - a^2$ and
$(x - a)(x + a) = x^2 - a^2$

- multiplying brackets of the form $(x + a)(x + b)$ gives $x^2 + (a + b)x + ab$

AT THE END OF THIS CHAPTER...

you should be able to factorize expressions of the form

1   $ax + ab$

2   $ax^2 - bx$

3   $ax^2 + bx + c$

4   $a^2x^2 - b^2y^2$

5   $ax + ay + bx + by$.

Did you know that the way of writing decimals is not the same throughout the world? In most English speaking counties, one point five is written as 1.5, but in much of mainland Europe it is written 1,5.

BEFORE YOU START

you need to know:
- ✓ how to expand algebraic expressions containing brackets
- ✓ how to collect like terms
- ✓ how to work with negative numbers

KEY WORDS

coefficient, common factor, difference between two squares, expression, factor, factorising, perfect squares, product, quadratic

Finding factors

In a previous chapter we removed brackets and expanded expressions. Frequently we need to be able to do the reverse, i.e. to find the factors of an expression. This is called *factorising*.

Common factors

In the expression $7a + 14b$ we could write the first term as $7 \times a$ and the second term as $7 \times 2b$,

i.e. $\qquad 7a + 14b = 7 \times a + 7 \times 2b$

The 7 is a common factor.

However we already know that $7(a + 2b) = 7 \times a + 7 \times 2b$

$\therefore \qquad 7a + 14b = 7 \times a + 7 \times 2b = 7(a + 2b)$

Exercise 14a   Factorise $3x - 12$

3 is a factor of $3x$ and of 12,

so $\qquad\qquad\qquad\qquad 3x - 12 = 3(x - 4)$

Factorise:

1 $4x + 4$	**4** $5a - 10b$	**7** $12a + 4$
2 $12x - 3$	**5** $3t - 9$	**8** $2a + 4b$
3 $6a + 2$	**6** $10a - 5$	**9** $14x - 7$

Factorise $x^2 - 7x$

$x^2 - 7x = x \times x - 7 \times x \qquad$ so x is a common factor.

$\qquad\quad = x(x - 7)$

Factorise:

10 $x^2 + 2x$	**13** $2x^2 + x$	**16** $x^2 - 4x$
11 $x^2 - 7x$	**14** $4t - 2t^2$	**17** $b^2 + 4b$
12 $a^2 + 6a$	**15** $x^2 + 5x$	**18** $4a^2 - a$

Factorise $9ab + 12bc$

$$9ab + 12bc = 3b \times 3a + 3b \times 4c$$
$$= 3b(3a + 4c)$$

You may not at first see that both 3 and b are common factors. If you spot only 3 you would have $9ab + 12bc = 3(3ab + 4bc)$, then a check inside the bracket shows that b is also a common factor.

Factorise:

19 $2x^2 - 6x$

20 $2z^3 + 4z$

21 $25a^2 - 5a$

22 $12x^2 + 16x$

23 $5ab - 10bc$

24 $3y^2 + 27y$

25 $2a^2 - 12a$

26 $6p^2 + 2p$

27 $9y^2 - 6y$

> **Tip** Always check the terms inside the bracket to make sure that you have not missed any common factors.

Factorise $ab + 2bc + bd$

$$ab + 2bc + bd = b(a + 2c + d)$$

Factorise:

28 $2x^2 + 4x + 6$

29 $10a^2 - 5a + 20$

30 $ab + 4bc - 3bd$

31 $8x - 4y + 12z$

32 $9ab - 6ac - 3ad$

33 $3x^2 - 6x + 9$

34 $4a^2 + 8a - 4$

35 $5xy + 4xz + 3x$

36 $5ab + 10bc + 5bd$

37 $2xy - 4yz + 8yw$

> **Tip** Check all three terms.

Factorise $8x^3 - 4x^2$

$$8x^3 - 4x^2 = 4x^2(2x - 1)$$

You may do this in stages:

$$8x^3 - 4x^2 = 4(2x^3 - x^2) \quad \text{(take out 4)}$$
$$= 4x(2x^2 - x) \quad \text{(take out } x\text{)}$$
$$= 4x^2(2x - 1) \quad \text{(take out another } x\text{)}$$

Factorise:

38 $x^3 + x^2$

39 $x^2 - x^3$

40 $20a^2 - 5a^3$

41 $12x^3 - 16x^2$

42 $4x^4 + 12x^2$

43 $a^2 + a^3$

44 $b^3 - b^2$

45 $4x^3 - 2x^2$

46 $27a^2 - 18a^3$

47 $10x^2 - 15x^4$

Factorise:

48 $12x + 8$

49 $8x^2 + 12x$

50 $9x^2 - 6x + 12$

51 $5x^3 - 10x$

52 $8pq + 4qr$

53 $x^2 - 8x$

54 $12 + 9y^2$

55 $12xy + 16xz + 8x$

56 $4x^3 + 6x$

> **Tip** Check inside the bracket for any missed common factors.

---

57 $\frac{1}{2}ah + \frac{1}{2}bh$

58 $mg - ma$

59 $\frac{1}{2}mv^2 + \frac{1}{2}mu^2$

60 $P + \dfrac{PRT}{100}$

61 $2\pi r^2 + \pi rh$

62 $\pi R^2 + \pi r^2$

63 $2gh_1 - 2gh_2$

64 $\frac{1}{2}mv^2 - mgh$

65 $\frac{4}{3}\pi r^3 - \frac{1}{3}\pi r^2 h$

66 $3\pi r^2 + 2\pi rh$

67 $\frac{1}{2}mu^2 + \frac{1}{2}mu$

68 $\frac{1}{2}bc - \frac{1}{4}ca$

Factorising by grouping

The expression $ax + ay + bx + by$ can be factorised by grouping the terms in pairs. If we group the first two terms followed by the remaining two terms, i.e. $(ax + ay)$ followed by $(bx + by)$, and factorise each group, we have $a(x + y)$ and $b(x + y)$

Therefore $\quad ax + ay + bx + by = a(x + y) + b(x + y)$

We can think of this as $aB + bB$, where B stands for $(x + y)$.

We have reduced the original four terms to two terms and these two terms now have the bracket B as a common factor.

Therefore $\quad ax + ay + bx + by = a(x + y) + b(x + y)$

$$= aB + bB$$

$$= B(a + b)$$

$$= (x + y)(a + b)$$

If, on the other hand, we pair the first and third terms followed by the remaining terms we have

$$ax + ay + bx + by = x(a + b) + y(a + b)$$

$$= (a + b)(x + y)$$

This shows that, while it is often possible to pair the terms with a common factor in more than one way, the result is the same. Always check your factors by multiplying out.

Exercise 14b   Factorise $xy + 2x + 3y + 6$

$$xy + 2x + 3y + 6 = xy + 2x \quad \text{(common factor } x\text{)}$$

$$+ 3y + 6 \quad \text{(common factor 3)}$$

$$= x(y + 2) + 3(y + 2)$$

$$\text{(common factor } y + 2\text{)}$$

$$= (y + 2)(x + 3)$$

Alternatively

$$xy + 2x + 3y + 6 = y(x + 3) + 2(x + 3)$$

$$= (x + 3)(y + 2)$$

Check your answer by expanding the brackets.

Factorise the following expressions by grouping:

1 $xy + 3x + 3y + 9$

2 $a + ab + 2b + 2b^2$

3 $a^2 + ab + ac + bc$

4 $xy - 3y + 2x - 6$

5 $xz + z + xy + y$

6 $xy + 4x + 2y + 8$

7 $ac + 4a + bc + 4b$

8 $xy - 2x + 4y - 8$

9 $pr + ps + qr + qs$

10 $xy - 3y + 4x - 12$

11 $xy - 5x + 2y - 10$

12 $pr - ps + qr - qs$

13 $ab - 3a + 2b - 6$

14 $pr - qr + ps - qs$

15 $2p + pq + 4q + 8$

16 $6 + 2b + 3a + ab$

Factorise $2x - 2xy - y + y^2$

$$2x - 2xy - y + y^2 = 2x(1 - y) - y(1 - y)$$

Check: $2x(1 - y) = 2x - 2xy$ and $-y(1 - y) = -y + y^2$

$\therefore \quad 2x - 2xy - y + y^2 \qquad\qquad = (1 - y)(2x - y)$

Check: $(1 - y)(2x - y) = 2x - y - 2xy + y^2$

Factorise:

17 $pr - ps - qr + qs$

18 $9a - 3b - 3ab + b^2$

19 $2a - b - 2ab + b^2$

20 $a^2 + 2ab - 2a - 4b$

21 $6 - 3x - 2y + xy$

22 $4a^2 - ab - 8a + 2b$

23 $6a^2 - 9a - 2ab + 3b$

24 $2m - 3n - 2mn + 3n^2$

25 $t^2 + tr + st + sr$

26 $x^2 - x + xy - y$

27 $4a - 4a^2 + 2b - 2ab$

28 $x - xy + y - y^2$

29 $4a + 6b - 6a^2 - 9ab$

30 $2a^2 + 2ab + bc + ac$

31 $4x - 4xy + 2y - 2y^2$

32 $xy + xz + y^2 + yz$

> **Tip**   Be careful with the signs. Check at each stage by mentally expanding the brackets.

Factorise $ac - ad + bd - bc$

$$ac - ad + bd - bc = a(c - d) + b(d - c)$$

$d - c = -1(c - d)$ so $b(d - c) = -b(c - d)$

$\therefore \quad ac - ad + bd - bc \qquad\qquad = a(c - d) - b(c - d)$

$$= (c - d)(a - b)$$

Factorise:

33 $5x - xy + 2y - 10$

34 $ab - 3a - 12 + 4b$

35 $xy - xz - 3z + 3y$

36 $2p - pq + 4q - 8$

37 $6 - 2b + ab - 3a$

38 $3a - ab + 4b - 12$

Factorise $a^2 - ab + a - b$

$$a^2 - ab + a - b = a(a - b) + 1(a - b)$$

$$= (a - b)(a + 1)$$

Alternatively $a^2 - ab + a - b = a^2 + a - ab - b = a(a + 1) - b(a + 1)$

Factorise:

39 $m^2 + mn + m + n$	**44** $a^2 - ab + a - b$	**49** $3x^2 + xy - 3x - y$
40 $a^2 - ab + a - b$	**45** $x^2 + xy - x - y$	**50** $2p^2 + 4pq - p - 2q$
41 $2p^2 - 4pq + p - 2q$	**46** $2a^2 + ab - 2a - b$	**51** $3a + b - 3a^2 - ab$
42 $x - xy + 1 - y$	**47** $5x^2 + 10xy - x - 2y$	**52** $2x + y - 2xz - yz$
43 $a^2 + ab + a + b$	**48** $mn - m - n + 1$	

Factorising quadratic expressions

The type of expression we are most likely to want to factorise is one of the form $ax^2 + bx + c$ where a, b and c are numbers. $ax^2 + bx + c$ is called a quadratic expression.

To factorise such an expression we look for two brackets whose product is the original expression.

When we expanded $(x + 2)(x + 4)$ we had

$$(x + 2)(x + 4) = x^2 + 6x + 8$$

If we write $x^2 + 6x + 8 = (x + 2)(x + 4)$ we say we have factorised $x^2 + 6x + 8$,

i.e. just as $\quad$ 10 is 2×5 so $x^2 + 6x + 8$ is $(x + 2) \times (x + 4)$.

To factorise an expression of the form $x^2 + 7x + 10$, i.e. where all the terms are positive, we remind ourselves of the patterns we observed in the previous chapter and summarised on pages 172–173.

We found when expanding brackets that:

a if the sign in each bracket is $+$ then the number term in the expansion is $+$

b the x^2 term comes from $x \times x$

c the number term in the expansion comes from multiplying the numbers in the brackets together

d the middle term, or x term in the expansion, comes from collecting the product of the outside terms in the brackets and the product of the inside terms in the brackets.

Using these ideas in reverse order

$$x^2 + 7x + 10 = (x + \quad)(x + \quad)$$
$$= (x + 2)(x + 5)$$

(choosing two numbers whose product is 10 and whose sum is 7).

Exercise **14c** $\quad$ Factorise $x^2 + 8x + 15$

$$x^2 + 8x + 15 = (x + 3)(x + 5) \quad \text{or} \quad (x + 5)(x + 3)$$

(The product of 3 and 5 is 15, and their sum is 8.)

181

Factorise:

1 $x^2 + 3x + 2$	6 $x^2 + 8x + 7$	11 $x^2 + 8x + 16$	16 $x^2 + 6x + 9$
2 $x^2 + 6x + 5$	7 $x^2 + 8x + 12$	12 $x^2 + 15x + 36$	17 $x^2 + 20x + 36$
3 $x^2 + 7x + 12$	8 $x^2 + 13x + 12$	13 $x^2 + 19x + 18$	18 $x^2 + 9x + 18$
4 $x^2 + 8x + 15$	9 $x^2 + 16x + 15$	14 $x^2 + 22x + 40$	19 $x^2 + 11x + 30$
5 $x^2 + 21x + 20$	10 $x^2 + 12x + 20$	15 $x^2 + 9x + 8$	20 $x^2 + 14x + 40$

To factorise an expression of the form $x^2 - 6x + 8$ remember the pattern:

a   the numbers in the brackets must multiply to give $+8$, i.e. they must have the same sign. Since the middle term in the expression is $-$ they must both be $-$

b   the x^2 term comes from $x \times x$

c   the middle term, or x term, comes from collecting the product of the outside terms and the product of the inside terms.

Thus $x^2 - 6x + 8 = (x - 2)(x - 4)$

Since $(-2) \times (-4) = +8$

and $x \times (-4) + (-2) \times x = -4x - 2x = -6x$

Exercise 14d   Factorise $x^2 - 7x + 12$

$$x^2 - 7x + 12 = (x - 3)(x - 4)$$

(The product of -3 and -4 is $+12$.

The outside product is $-4x$ and the inside product is $-3x$.

Collecting these gives $-7x$.)

Factorise:

1 $x^2 - 9x + 8$	4 $x^2 - 11x + 28$	7 $x^2 - 16x + 15$
2 $x^2 - 7x + 12$	5 $x^2 - 13x + 42$	8 $x^2 - 6x + 9$
3 $x^2 - 17x + 30$	6 $x^2 - 5x + 6$	9 $x^2 - 18x + 32$

Similarly $x^2 + x - 12 = (x + 4)(x - 3)$

If the number term in the expansion is negative the signs in the brackets are different.

Thus $(+4) \times (-3) = -12$

Working as before, the product of the outside terms is $-3x$

and the product of the inside terms is $+4x$

Therefore the total is $+x$.

Similarly $x^2 + 2x - 15 = (x + 5)(x - 3)$

or $x^2 + 2x - 15 = (x - 3)(x + 5)$

Exercise 14e

Factorise:

1 $x^2 - x - 6$

2 $x^2 + x - 20$

3 $x^2 - x - 12$

4 $x^2 + 3x - 28$

5 $x^2 + 2x - 15$

6 $x^2 - 2x - 24$

7 $x^2 + 6x - 27$

8 $x^2 - 9x - 22$

9 $x^2 - 2x - 35$

Most of the values in the previous three exercises have been easy to spot. Should you have difficulty, set out all possible pairs of numbers, as shown below, until you find the pair that gives the original expression when you multiply back.

Factorise

a $x^2 - 11x + 24$

(Because the number term is + the two numbers in the brackets must have the same sign.)

Possible numbers		Sum
−1	−24	−25
−2	−12	−14
−3	−8	−11

$\therefore \qquad x^2 - 11x + 24 = (x - 3)(x - 8)$

b $\qquad x^2 + 5x - 24$

(Because the number term is − the two numbers in the brackets have different signs)

Possible numbers		Sum
−1	+24	+23
−2	+12	+10
−3	+8	+5

$\therefore \qquad x^2 + 5x - 24 = (x - 3)(x + 8)$

Remember that a + before the number term means that the signs in the brackets are the same, whereas a − before the number term means that they are different.

Exercise 14f

Factorise:

1 $x^2 + 9x + 14$

2 $x^2 - 10x + 21$

3 $x^2 + 5x - 14$

4 $x^2 + x - 30$

5 $x^2 + 9x + 8$

6 $x^2 - 10x + 25$

7 $x^2 + 8x - 9$

8 $x^2 - 15x + 26$

9 $x^2 + x - 56$

10 $x^2 + 32x + 60$

11 $x^2 - 6x - 27$

12 $x^2 + 16x - 80$

13 $x^2 + 14x + 13$

14 $x^2 + 12x - 28$

15 $x^2 + 2x - 80$

16 $x^2 - 11x + 30$

17 $x^2 + 8x - 48$

18 $x^2 + 18x + 72$

19 $x^2 + 17x + 52$

20 $x^2 - 12x - 28$

21 $x^2 + 11x + 24$

22 $x^2 - 11x - 42$

23 $x^2 - 18x + 32$

24 $x^2 + 7x - 60$

Sometimes the terms need rearranging before we try to factorise.

Exercise 14g

Factorise:

1 $8 + x^2 + 9x$

2 $9 + x^2 - 6x$

3 $11x + 28 + x^2$

4 $20 + x^2 - x$

5 $9 + x^2 + 6x$

6 $8 + x^2 - 9x$

7 $17x + 30 + x^2$

8 $27 - 6x + x^2$

9 $x^2 + 22 + 13x$

10 $x^2 - 11x - 26$

11 $7 + x^2 - 8x$

12 $x + x^2 - 42$

13 $x^2 - 5x - 24$

14 $14 + x^2 - 9x$

15 $28x + 27 + x^2$

16 $2x - 63 + x^2$

> **Tip** Rearrange so that these expressions are in the form (x^2 term) then (x term) lastly (constant).

Factorise:

17 $x^2 + 10x + 25$

18 $x^2 - 10x + 25$

19 $x^2 + 4x + 4$

20 $x^2 - 14x + 49$

21 $x^2 + 12x + 36$

22 $x^2 - 12x + 36$

23 $x^2 - 4x + 4$

24 $x^2 + 16x + 64$

> **Tip** The expressions are perfect squares.

Exercise 14h

Factorise $6 - 5x - x^2$

When the x^2 term is negative do not rearrange. Treat it as the last term.

$$6 - 5x - x^2 = (6 + x)(1 - x)$$

Factorise:

1 $2 - x - x^2$

2 $6 + x - x^2$

3 $4 - 3x - x^2$

4 $8 + 2x - x^2$

5 $6 - x - x^2$

6 $2 + x - x^2$

7 $8 - 2x - x^2$

8 $5 - 4x - x^2$

9 $10 - 3x - x^2$

10 $12 + 4x - x^2$

11 $5 + 4x - x^2$

12 $14 - 5x - x^2$

13 $6 + 5x - x^2$

14 $20 - x - x^2$

15 $15 - 2x - x^2$

16 $12 + x - x^2$

The difference between two squares

In the last chapter, one of the expansions we listed was

$$(x + a)(x - a) = x^2 - a^2$$

If we reverse this we have

$$x^2 - a^2 = (x + a)(x - a)$$
$$\text{or} \quad x^2 - a^2 = (x - a)(x + a)$$

(the order of multiplication of two brackets makes no difference to the result).

This result is known as *factorising the difference between two squares* and is *very important*.

When factorising do not confuse $x^2 - 4$ with $x^2 - 4x$.

$$x^2 - 4 = (x + 2)(x - 2)$$

whereas $\quad x^2 - 4x = x(x - 4) \quad$ ($4x$ is *not* a perfect square)

Exercise 14i

Factorise $x^2 - 9$

$$x^2 - 9 = x^2 - 3^2$$
$$= (x + 3)(x - 3) \text{ or } (x - 3)(x + 3)$$

Factorise:

1 $x^2 - 25$	**3** $x^2 - 100$	**5** $x^2 - 64$	**7** $x^2 - 36$
2 $x^2 - 4$	**4** $x^2 - 1$	**6** $x^2 - 16$	**8** $x^2 - 81$

Factorise $4 - x^2$

$$4 - x^2 = 2^2 - x^2$$
$$= (2 + x)(2 - x) \text{ or } (2 - x)(2 + x)$$

Factorise:

9 $9 - x^2$	**12** $a^2 - b^2$	**15** $25 - x^2$
10 $36 - x^2$	**13** $9y^2 - z^2$	**16** $81 - x^2$
11 $100 - x^2$	**14** $16 - x^2$	**17** $x^2 - y^2$

We began this chapter by considering common factors. A little revision is now necessary followed by expressions of the form $ax^2 + bx + c$ where a is a common factor.

Exercise 14j

Factorise $12x - 6$

6 is a common factor so $12x - 6 = 6(2x - 1)$

Factorise:

1 $3x + 12$	**4** $14x + 21$	**7** $9x^2 - 18x$
2 $25x^2 + 10x$	**5** $4x^2 + 2$	**8** $20x + 12$
3 $12x^2 - 8$	**6** $21x - 7$	**9** $4x - 14$

Factorise $3x^2 + 9x + 6$

$$3x^2 + 9x + 6 = 3(x^2 + 3x + 2) \quad \text{(Take out the common factor)}$$
$$= 3(x + 1)(x + 2) \quad \text{(Factorise the quadratic.)}$$

Factorise:

10 $2x^2 + 14x + 24$ **13** $4x^2 - 4x - 48$ **16** $4x^2 - 24x + 36$

11 $3x^2 - 27x + 24$ **14** $5x^2 + 40x + 35$ **17** $5x^2 - 5x - 30$

12 $7x^2 + 14x + 7$ **15** $3x^2 + 24x + 36$ **18** $2x^2 - 18x - 44$

In the previous exercises we considered the factors of the expression $ax^2 + bx + c$ when either a was 1 or a was a common factor. We must now consider other values of a which mean that the two brackets do not both start with x.

When the coefficient of x^2 was 1 as in $x^2 + 7x + 10$, we chose two numbers whose product was 10 and whose sum was 7.

It can be shown that this condition is also required when the coefficient of x^2 is not 1. In order to factorise $ax^2 + bx + c$ we find two numbers whose product is ac and sum is b, for example, to factorise $6x^2 + 19x + 10$ we find two numbers whose product is 60, i.e. (6×10) and sum equals 19. These are 15 and 4. We then write

$$6x^2 + 19x + 10 = 6x^2 + 15x + 4x + 10$$
$$= (6x^2 + 15x) + (4x + 10)$$
$$= 3x(2x + 5) + 2(2x + 5)$$
$$= (2x + 5)(3x + 2)$$

Exercise **14k**

Factorise:

1 $2x^2 + 3x + 1$ **6** $3x^2 - 8x + 4$ **11** $2x^2 - 3x - 2$ **16** $7x^2 - 19x - 6$

2 $3x^2 - 5x + 2$ **7** $2x^2 + 9x + 4$ **12** $3x^2 + x - 4$ **17** $6x^2 - 7x - 10$

3 $4x^2 + 7x + 3$ **8** $5x^2 - 17x + 6$ **13** $5x^2 - 13x - 6$ **18** $5x^2 - 19x + 12$

4 $2x^2 - 7x + 3$ **9** $2x^2 + 11x + 12$ **14** $4x^2 + 5x - 6$ **19** $3x^2 - 11x - 20$

5 $3x^2 + 13x + 4$ **10** $7x^2 - 29x + 4$ **15** $3x^2 + 10x - 8$ **20** $4x^2 + 17x - 15$

After a little practice, you should be able to find the factors without going into too much detail.

Exercise **14l**

Factorise:

1 $6x^2 + 7x + 2$ **6** $6x^2 - 11x + 3$ **11** $8x^2 - 10x - 3$ **16** $6a^2 - a - 15$

2 $6x^2 + 19x + 15$ **7** $9x^2 - 18x + 8$ **12** $15x^2 - x - 2$ **17** $6t^2 - t - 2$

3 $15x^2 + 11x + 2$ **8** $16x^2 - 10x + 1$ **13** $21x^2 + 2x - 8$ **18** $9b^2 - 12b + 4$

4 $12x^2 + 28x + 15$ **9** $15x^2 - 44x + 21$ **14** $80x^2 - 6x - 9$ **19** $5x^2 - 7xy - 6y^2$

5 $35x^2 + 24x + 4$ **10** $20x^2 - 23x + 6$ **15** $24x^2 + 17x - 20$ **20** $4x^2 - 11x + 6$

Factorise $4x^2 - 9$

$$4x^2 - 9 = (2x)^2 - 3^2$$
$$= (2x + 3)(2x - 3)$$

Factorise:

1 $4x^2 - 25$ **3** $36a^2 - 1$ **5** $9x^2 - 25$

2 $9x^2 - 4$ **4** $16a^2 - b^2$ **6** $4a^2 - 1$

Factorise $4x^2 - 9y^2$

$$4x^2 - 9y^2 = (2x)^2 - (3y)^2$$
$$= (2x + 3y)(2x - 3y)$$

Factorise:

7 $16a^2 - 9b^2$ **9** $100x^2 - 49y^2$ **11** $4x^2 - 49y^2$ **13** $9a^2 - 4b^2$

8 $25s^2 - 9t^2$ **10** $9y^2 - 16z^2$ **12** $81x^2 - 100y^2$ **14** $64p^2 - 81q^2$

Factorise $2 - 18a^2$

$$2 - 18a^2 = 2(1 - 9a^2)$$
$$= 2(1^2 - (3a)^2)$$
$$= 2(1 + 3a)(1 - 3a)$$

Factorise:

15 $3a^2 - 27b^2$ **18** $45x^2 - 20$ **21** $\frac{1}{2}a^2 - 2b^2$

16 $18t^2 - 50s^2$ **19** $5a^2 - 20$ **22** $\frac{a^2}{4} - \frac{b^2}{9}$

17 $27x^2 - 3y^2$ **20** $45 - 5b^2$ **23** $27x^2 - \frac{1}{3}y^2$

Calculations using factorising

Find $1.7^2 + 0.3 \times 1.7$

$$1.7^2 + 0.3 \times 1.7 = 1.7(1.7 + 0.3)$$
$$= 1.7 \times 2$$
$$= 3.4$$

Find, without using a calculator:

1 $2.5^2 + 0.5 \times 2.5$ **4** $8.76^2 - 4.76 \times 8.76$ **7** $4.3^2 - 1.3 \times 4.3$

2 $1.3 \times 3.7 + 3.7^2$ **5** $5.2^2 + 0.8 \times 5.2$

3 $5.9^2 - 2.9 \times 5.9$ **6** $2.6 \times 3.4 + 3.4^2$

Find $100^2 - 98^2$

$$100^2 - 98^2 = (100 + 98)(100 - 98)$$

$$= 198 \times 2 = 396$$

8 $55^2 - 45^2$ **10** $7.82^2 - 2.82^2$ **12** $10.2^2 - 9.8^2$ **14** $8.79^2 - 1.21^2$

9 $20.6^2 - 9.4^2$ **11** $2.667^2 - 1.333^2$ **13** $13.5^2 - 6.5^2$ **15** $0.763^2 - 0.237^2$

Investigation

Using the digits 3 and 6 it is possible to make two two-digit numbers, namely 36 and 63.

The difference between the squares of these two numbers is

$$63^2 - 36^2 = 2673$$

$$= 99 \times 27$$

$$= 99 \times 9 \times 3$$

$$= 99 \times \text{(the sum of the original digits)}$$
$$\times \text{(the difference between the original digits)}$$

Investigate whether or not this is true for other pairs of digits.

If you cannot find a pair of digits for which the above result is not true, then start again letting the two digits be x and y. Write the two numbers in terms of x and y. (They are not xy and yx.)

Finally in this chapter we take out a common factor that does not leave the first term as x^2; and factorise expressions that do not begin with the x^2 term.

Exercise 14p   Factorise $8x^2 + 28x + 12$

$$8x^2 + 28x + 12 = 4(2x^2 + 7x + 3)$$

$$= 4(2x + 1)(x + 3)$$

Factorise:

1 $15x^2 + 25x + 10$ **5** $8x^2 + 34x - 30$ **9** $6x^2 + 26x + 8$ **13** $12x^2 + 14x + 4$

2 $4x^2 - 6x - 4$ **6** $8x^2 + 14x + 6$ **10** $15x^2 + 50x - 40$ **14** $100x^2 - 115x + 30$

3 $6x^2 + 9x + 3$ **7** $25x^2 - 65x - 30$ **11** $18x^2 - 36x + 16$ **15** $24x^2 - 4x - 8$

4 $18x^2 - 21x - 30$ **8** $9x^2 + 3x - 12$ **12** $48x^2 - 30x + 3$ **16** $21x^2 + 70x - 56$

Factorise $12 + 7x - 10x^2$

Leave $-10x^2$ at the end and treat it like the number term.

$$12 + 7x - 10x^2 = (3 - 2x)(4 + 5x)$$

Factorise:

17 $4 - 5x - 6x^2$	**20** $24 - 16x + 2x^2$	**23** $12 - 11x - x^2$
18 $12 + 7x - 12x^2$	**21** $16 - 20x - 6x^2$	**24** $8 + 24x + 18x^2$
19 $21 + 25x - 4x^2$	**22** $9 + 8x - x^2$	**25** $45 - 30x + 5x^2$

Mixed questions

Some quadratic expressions such as $x^2 + 9$ and $x^2 + 3x + 1$ will not factorise. The next exercise in this chapter includes some expressions that will not factorise.

Exercise 14q

Factorise where possible:

1 $x^2 + 13x + 40$	**13** $x^2 + 8x + 12$	**25** $6x^2 + 5x - 4$	**37** $x^2 + 13x - 68$
2 $6x^2 + 5x + 1$	**14** $8x^2 - 2x - 1$	**26** $30x^2 + 35x + 10$	**38** $2x^4 - x^3 + 4x - 2$
3 $x^2 - 36$	**15** $x^2 - 49$	**27** $x^2 + 11x + 18$	**39** $6x^2 - 9x - 6$
4 $x^2 + 4$	**16** $x^2 - 7x + 2$	**28** $x^2 - 10x + 24$	**40** $p^3 + p^2 + p + 1$
5 $x^2 - 8x + 12$	**17** $6x^2 - 11x - 10$	**29** $4x^2 - 16y^2$	**41** $(a + b)^2 - c^2$
6 $2x^2 + 7x - 15$	**18** $x^2 + 13x + 42$	**30** $x^2 - 11x - 10$	**42** $116x^2 - 25x - 1$
7 $x^2 + 6x - 7$	**19** $4x^2 - 9y^2$	**31** $12x^2 - 22x - 20$	**43** $a^2 + 23a + 112$
8 $5x^2 + 3x - 2$	**20** $15x^2 - 22x + 8$	**32** $x^2 + 13x - 30$	**44** $x^4 - y^2 - 2y - 1$
9 $x^2 - 11x + 24$	**21** $6x^2 - 5x - 6$	**33** $28 - 12x - x^2$	**45** $3a^2 + 56 - 31a$
10 $3x^2 + 11x + 6$	**22** $x^2 + 11x - 26$	**34** $a^2 - 16a + 63$	**46** $2x^2 - 8x - 154$
11 $x^2 + 14x - 15$	**23** $30x^2 - 2x - 4$	**35** $6 - 16x + 8x^2$	**47** $4x^2 - (y - z)^2$
12 $12x^2 - 7x + 1$	**24** $28 + 3x - x^2$	**36** $1 + 2x + 4x^2 + 8x^3$	**48** $a^2b^2 - ab - 342$

The quadratic expression $ax^2 + bx + c$ will factorise only if we can find two numbers whose product is ac and whose sum is b .

Mixed exercises

Exercise 14r

1 Expand

   a $7(a + 3)$ **b** $3(x - 2y)$

2 Expand

   a $(x + 4)(x + 10)$ **b** $(2x - 3)(3x - 5)$

3 Expand

   a $(5 + x)^2$ **b** $(5 - x)^2$ **c** $(5 + x)(5 - x)$

4 Factorise

   a $10a + 20$ **b** $15p^2 - 10p$

5 Factorise

   a $a^3 + a^2 + a + 1$ **b** $2km - kn + 2lm - ln$

6 Factorise

   a $x^2 + 6x - 27$ **b** $5x^2 - 42x + 49$ **c** $a^2 - \dfrac{b^2}{4}$

7 Factorise

   a $10x^2 - 11x - 6$ **b** $100a^2 - 81b^2$

Exercise 14s

1 Expand

   a $5a(a + 3)$ **b** $4x(3x - 2y)$

2 Expand

   a $(y - 4)(y - 5)$ **b** $(3x - 4y)(5x + 2y)$

3 Expand

   a $(2p + 3q)^2$ **b** $(2p - 3q)^2$ **c** $(2p + 3q)(2p - 3q)$

4 Factorise

   a $8z^3 - 4z^2$ **b** $5xy - 20yz$

5 Factorise

   a $2m + 3mn + 3n + 2$ **b** $ac - 2ad + 2bc - 4bd$

6 Factorise

   a $x^2 - 6x - 27$ **b** $4x^2 + 27x - 7$ **c** $4m^2 - 81n^2$

7 Factorise

   a $15x^2 - 54x + 27$ **b** $15 + 25x - 20x^2$

Exercise 14t

1 Expand   **a** $4(a+7)$ **b** $3x(2x-3y)$

2 Expand   **a** $(x+3)(x+9)$ **b** $(5x-2)(3x+1)$

3 Expand   **a** $(5x+2)^2$ **b** $(5x-2)^2$ **c** $(5x+2)(5x-2)$

4 Factorise   **a** $12z^2-6z$ **b** $8xy-12yz$

5 Factorise   **a** z^3+2z^2+z+2 **b** $3ac+bc+6a+2b$

6 Factorise   **a** $x^2-2x-24$ **b** $4a^2+4a-15$ **c** $9m^2-\dfrac{n^2}{9}$

7 Factorise   **a** $15x^2+x-6$ **b** $6+x-15x^2$

8 If n is a three-digit number whose hundreds digit is a, tens digit is b and units digit is c, then $n = 100a + 10b + c$.

    a Prove that if $a+b=7$ and $b+c=7$ then n is divisible by 7.

    b Prove that if $a+b=13$ and $b+c=13$ then n is divisible by 13.

MATHS IS OUT THERE

Did you know that Descartes used the symbol ⊃⊂ to mean 'is equal to'?

Find out what other symbols for 'is equal to' have been used in the past.

IN THIS CHAPTER...

you have seen that:

- when two or more terms in an algebraic expression have a common factor, you can write that expression as a product of the common factor and a bracket containing the terms without that factor

- an expression of the form $ax^2 + bx + c$ is called quadratic

- some, but not all, quadratic expressions can be written as the product of two brackets

- the difference of two squares can be factorised: for example
  $x^2 - y^2 = (x-y)(x+y)$ or $(x+y)(x-y)$

- some quadratic expressions have a common factor; this should be taken out before trying to factorise the quadratic

- sometimes an expression with four terms can be factorised by grouping the terms in pairs and and taking out a common factor of each pair.

AT THE END OF THIS CHAPTER...

you should be able to:

1   Solve quadratic equations of the
    form $(ax + b)(cx + d) = 0$.

2   Solve quadratic equations by factorisation.

3   Solve equations of the form $ax^3 + bx^2 + cx = 0$.

4   Form quadratic equations to solve given problems.

Did you know that the Babylonians found an approximate value for the

square root of a number using the formula $\sqrt{a^2 + b} \approx a + \dfrac{b}{2a}$?

For example $\sqrt{53} = \sqrt{49 + 4} \approx 7 + \frac{4}{14} = 7.285\ldots$ Using a calculator
$\sqrt{53} = 7.280\ldots$ so the estimate is quite good.

Use the formula to find the square root of 20, 42 , 93 and 131. Check the
accuracy of your result using a calculator.

BEFORE YOU START

you need to know:
✓ how to work with fractions
✓ how to work with directed numbers
✓ that if a number is multiplied by zero the answer is zero

KEY WORDS

factorisation, linear equation, quadratic equation, root

Multiplication by zero

Find the value of $(x - 3)(x - 7)$ if

    a $\ x = 8$ **b** $\ x = 7$ **c** $\ x = 3$

    a If $x = 8$, $\ (x - 3)(x - 7) = (8 - 3)(8 - 7)$

$$= (5)(1)$$
$$= 5$$

    b If $x = 7$, $\ (x - 3)(x - 7) = (4)(0)$

$$= 0$$

    (If any quantity is multiplied by 0 the answer is 0)

    c If $x = 3$, $\ (x - 3)(x - 7) = (0)(-4)$

$$= 0$$

1 Find the value of $(x - 4)(x - 2)$ if

    a $\ x = 6$ **b** $\ x = 4$ **c** $\ x = 2$

2 Find the value of $(x - 5)(x - 9)$ if

    a $\ x = 5$ **b** $\ x = 10$ **c** $\ x = 9$

3 Find the value of $(x - 7)(x - 1)$ if

    a $\ x = 1$ **b** $\ x = 8$ **c** $\ x = 7$

4 Find the value of $(x - 4)(x - 6)$ if

    a $\ x = 4$ **b** $\ x = 6$ **c** $\ x = 3$

Find the value of $(x - 2)(x + 4)$ if

    a $\ x = 2$ **b** $\ x = 4$ **c** $\ x = -4$

    a If $x = 2$ $(x - 2)(x + 4) = (0)(6)$

$$= 0$$

    b If $x = 4$ $(x - 2)(x + 4) = (2)(8)$

$$= 16$$

    c If $x = -4$ $(x - 2)(x + 4) = (-6)(0)$

$$= 0$$

5 Find the value of $(x - 3)(x + 5)$ if

    a $\ x = 6$ **b** $\ x = 3$ **c** $\ x = -5$

6 Find the value of $(x - 4)(x + 6)$ if

    a $\ x = 0$ **b** $\ x = -6$ **c** $\ x = 4$

7 Find the value of $(x - 7)(x + 2)$ if

    a $x = -7$ **b** $x = -2$ **c** $x = 7$

The results of this exercise show that if the product of two factors is 0, then either one or both of these factors must be 0.

In general we can say

$$\text{if} \qquad A \times B = 0$$
$$\text{then either} \quad A = 0 \qquad \text{or/and} \quad B = 0$$

Exercise 15b

In questions **1** to **12** find, if possible, the value or values of A. Note that if $A \times 0 = 0$ then A can have any value.

1 $A \times 6 = 0$ **4** $A \times 0 = 0$ **7** $A \times 10 = 0$ **10** $A \times 3 = 21$

2 $A \times 7 = 0$ **5** $3 \times A = 12$ **8** $A \times 9 = 18$ **11** $0 \times A = 0$

3 $A \times 4 = 0$ **6** $8 \times A = 8$ **9** $A \times 20 = 0$ **12** $4 \times A = 0$

13 If $AB = 0$ find   **15** If $AB = 0$ find

    a A if $B = 2$ **b** B if $A = 10$ **a** A if $B = 10$ **b** B if $A = 3$

14 If $AB = 0$ find   **16** If $AB = 0$ find

    a A if $B = 5$ **b** B if $A = 5$ **a** B if $A = 6$ **b** A if $B = 0$

> Find a and b if $a(b - 3) = 0$
>
> Either $a = 0$ or/and $b - 3 = 0$
>
> i.e., either $a = 0$ or/and $b = 3$

Find a and b if:

17 $a(b - 1) = 0$ **20** $(a - 3)b = 0$ **23** $a(b - 10) = 0$

18 $a(b - 5) = 0$ **21** $(a - 9)b = 0$ **24** $(a - 1)b = 0$

19 $a(b - 2) = 0$ **22** $a(b - 4) = 0$ **25** $(a - 7)b = 0$

Quadratic equations

Previously we have considered equations such as $x - 1 = 0$ and $3x + 2 = 0$. These are examples of *linear equations*. The first equation is true only for $x = 1$ and the second only for $x = -\frac{2}{3}$.

If, however, we consider the equation

$$(x - 1)(x - 2) = 0$$

we find that it is true either when $x - 1 = 0$ or when $x - 2 = 0$,

i.e. either when $x = 1$ or when $x = 2$.

There are, therefore, two values of x that satisfy the equation $(x - 1)(x - 2) = 0$

Expanding the left-hand side gives

$$x^2 - 3x + 2 = 0$$

Equations like this, which contain an x^2 term, are called *quadratic equations*.

When we are given a quadratic equation we can often factorise the left-hand side into two linear factors,

e.g. $\qquad x^2 - 5x + 4 = 0$

gives $\qquad (x - 4)(x - 1) = 0$

It is this technique that concerns us in the present chapter.

Exercise **15c**

What values of x satisfy the equation $x(x - 9) = 0$?

$$x(x - 9) = 0$$

Either $\qquad\qquad x = 0 \quad$ or $\quad x - 9 = 0$

i.e., either $\qquad\qquad x = 0 \quad$ or $\quad x = 9$

What values of x satisfy the following equations?

1 $\ x(x - 3) = 0$ **4** $\ x(x + 4) = 0$ **7** $\ x(x - 10) = 0$

2 $\ x(x - 5) = 0$ **5** $\ (x + 5)x = 0$ **8** $\ (x - 7)x = 0$

3 $\ (x - 3)x = 0$ **6** $\ x(x - 6) = 0$ **9** $\ x(x + 7) = 0$

What values of x satisfy the equation $(x - 3)(x + 5) = 0$?

$$(x - 3)(x + 5) = 0$$

Either $\qquad\qquad x - 3 = 0 \quad$ or $\quad x + 5 = 0$

i.e., either $\qquad\qquad x = 3 \quad$ or $\quad x = -5$

What values of x satisfy the following equations?

10 $\ (x - 1)(x - 2) = 0$ **16** $\ (x - 3)(x + 5) = 0$ **22** $\ (x + a)(x + b) = 0$

11 $\ (x - 5)(x - 9) = 0$ **17** $\ (x + 7)(x - 2) = 0$ **23** $\ (x - 4)(x + 1) = 0$

12 $\ (x - 10)(x - 7) = 0$ **18** $\ (x + 2)(x + 3) = 0$ **24** $\ (x + 9)(x - 8) = 0$

13 $\ (x - 4)(x - 7) = 0$ **19** $\ (x + 4)(x + 9) = 0$ **25** $\ (x + 6)(x + 7) = 0$

14 $\ (x - 6)(x - 1) = 0$ **20** $\ (x + 1)(x + 8) = 0$ **26** $\ (x + 10)(x + 11) = 0$

15 $\ (x - 8)(x + 11) = 0$ **21** $\ (x - p)(x - q) = 0$ **27** $\ (x - a)(x - b) = 0$

Exercise **15d**

Solve the following equations:

1 $(2x - 5)(x - 1) = 0$

2 $(x - 4)(3x - 2) = 0$

3 $(5x - 4)(4x - 3) = 0$

4 $x(4x - 5) = 0$

5 $x(10x - 3) = 0$

6 $(5x + 2)(x - 7) = 0$

7 $(6x + 5)(3x - 2) = 0$

8 $(8x - 3)(2x + 5) = 0$

9 $(7x - 8)(4x + 15) = 0$

10 $(4x + 3)(2x + 3) = 0$

11 $(3x - 7)(x - 2) = 0$

12 $(3x - 5)(2x - 1) = 0$

> **Tip** Either $x - 1 = 0$ (so $x = 1$) or $2x - 5 = 0$ which can be solved to find x.

13 $x(3x - 1) = 0$

14 $x(7x - 3) = 0$

15 $(2x + 3)(x - 3) = 0$

16 $(4x + 3)(2x - 5) = 0$

17 $(10x + 9)(5x - 4) = 0$

18 $(3x - 2)(4x + 9) = 0$

19 $(5x - 12)(2x + 7) = 0$

20 $(5x + 8)(4x + 3) = 0$

Solution by factorisation

The previous two exercises suggest that if the left-hand side of a quadratic equation can be expressed as two factors of the form $ax + b$, we can use these factors to solve the equation.

Exercise **15e**

Solve the equation $x^2 - 10x + 9 = 0$

If $\qquad\qquad x^2 - 10x + 9 = 0$

then $\qquad\qquad (x - 1)(x - 9) = 0$

∴ either $\qquad\qquad x - 1 = 0$ or $x - 9 = 0$

i.e. $\qquad\qquad x = 1$ or 9

Solve the equations:

1 $x^2 - 3x + 2 = 0$

2 $x^2 - 8x + 7 = 0$

3 $x^2 - 5x + 6 = 0$

4 $x^2 - 7x + 10 = 0$

5 $x^2 - 7x + 12 = 0$

6 $x^2 - 6x + 5 = 0$

7 $x^2 - 12x + 11 = 0$

8 $x^2 - 6x + 8 = 0$

9 $x^2 - 8x + 12 = 0$

10 $x^2 - 13x + 12 = 0$

> **Tip** Start by factorising the left hand side.

Solve the equations:

11 $x^2 + 6x - 7 = 0$

12 $x^2 - 2x - 8 = 0$

13 $x^2 + x - 12 = 0$

14 $x^2 - 2x - 15 = 0$

15 $x^2 + 7x - 18 = 0$

16 $x^2 - 12x - 13 = 0$

17 $x^2 + x - 6 = 0$

18 $x^2 - 4x - 12 = 0$

19 $x^2 + x - 20 = 0$

20 $x^2 - 5x - 24 = 0$

Tip These equations have one negative solution.

Solve the equations:

21 $x^2 + 3x + 2 = 0$

22 $x^2 + 8x + 7 = 0$

23 $x^2 + 8x + 15 = 0$

24 $x^2 + 8x + 12 = 0$

25 $x^2 + 11x + 18 = 0$

26 $x^2 + 7x + 6 = 0$

27 $x^2 + 7x + 10 = 0$

28 $x^2 + 14x + 13 = 0$

29 $x^2 + 16x + 15 = 0$

30 $x^2 + 9x + 18 = 0$

Tip These equations have two negative solutions.

Solve the equation $x^2 - 49 = 0$

$$x^2 - 49 = 0$$

$$(x + 7)(x - 7) = 0$$

Either $\quad\quad\quad x + 7 = 0 \quad$ or $\quad x - 7 = 0$

i.e. $\quad\quad\quad\quad\quad x = -7 \quad$ or $\quad 7$

Solve the equations:

31 $x^2 - 1 = 0$

32 $x^2 - 9 = 0$

33 $x^2 - 16 = 0$

34 $x^2 - 81 = 0$

35 $x^2 - 169 = 0$

36 $x^2 - 4 = 0$

37 $x^2 - 25 = 0$

38 $x^2 - 100 = 0$

39 $x^2 - 144 = 0$

40 $x^2 - 36 = 0$

Tip The difference between two squares can always be factorised.

The equations we have solved by factorising have all been examples of the equation $ax^2 + bx + c = 0$ when $a = 1$. We consider next the case when $c = 0$,

e.g. solve the equation $3x^2 + 2x = 0$

Since x is common to both terms on the left-hand side we can rewrite this equation

$$x(3x + 2) = 0$$

Then, either $x = 0$ or $3x + 2 = 0$

i.e. $\quad\quad\quad x = 0$ or $3x = -2$

i.e. $\quad\quad\quad x = 0$ or $-\frac{2}{3}$

Exercise 15f

Solve the equations:

1 $x^2 - 2x = 0$ **8** $x^2 + x = 0$

2 $x^2 - 10x = 0$ **9** $3x^2 - 5x = 0$

3 $x^2 + 8x = 0$ **10** $5x^2 - 7x = 0$

4 $2x^2 - x = 0$ **11** $2x^2 + 3x = 0$

5 $4x^2 - 5x = 0$ **12** $8x^2 + 5x = 0$ **15** $7x^2 - 12x = 0$ **18** $x^2 + 4x = 0$

6 $x^2 - 5x = 0$ **13** $x^2 - 7x = 0$ **16** $6x^2 + 7x = 0$ **19** $7x^2 - 2x = 0$

7 $x^2 + 3x = 0$ **14** $3x^2 + 5x = 0$ **17** $12x^2 + 7x = 0$ **20** $14x^2 + 3x = 0$

> **Tip** Start by looking for a common factor, then factorise the left-hand side.

Sometimes a quadratic equation has two answers, or *roots*, that are exactly the same.

Consider $x^2 - 4x + 4 = 0$

then $(x - 2)(x - 2) = 0$

i.e., either $x - 2 = 0$ or $x - 2 = 0$

i.e. $x = 2$ or $x = 2$

i.e. $x = 2$ (twice)

Such an equation involves a *perfect square*. As with any quadratic equation, it has two answers, or roots, but they are equal. We say that such an equation has a repeated root.

Exercise 15g   Solve the equation $x^2 + 14x + 49 = 0$

$$x^2 + 14x + 49 = 0$$

$$(x + 7)(x + 7) = 0$$

Either $x + 7 = 0$ or $x + 7 = 0$

i.e. $x = -7$ (twice)

Solve the equations:

1 $x^2 - 2x + 1 = 0$ **9** $x^2 + 2x + 1 = 0$

2 $x^2 - 10x + 25 = 0$ **10** $x^2 + 20x + 100 = 0$

3 $x^2 - 20x + 100 = 0$ **11** $x^2 + 18x + 81 = 0$

4 $x^2 + 8x + 16 = 0$ **12** $x^2 - 14x + 49 = 0$

5 $x^2 + 6x + 9 = 0$ **13** $x^2 - 22x + 121 = 0$ **17** $x^2 - 12x + 36 = 0$

6 $x^2 - 6x + 9 = 0$ **14** $x^2 + 12x + 36 = 0$ **18** $x^2 - 40x + 400 = 0$

7 $x^2 - 8x + 16 = 0$ **15** $x^2 - x + \frac{1}{4} = 0$ **19** $x^2 - 16x + 64 = 0$

8 $x^2 - 18x + 81 = 0$ **16** $x^2 + 10x + 25 = 0$ **20** $x^2 + \frac{4}{3}x + \frac{4}{9} = 0$

> **Tip** Always begin by trying to factorise.

Frequently $a \neq 1$ in the equation $ax^2 + bx + c = 0$. The next exercise considers quadratic equations for values of a other than 1.

Exercise 15h Solve the equation $5x^2 - 13x - 6 = 0$

$$5x^2 - 13x - 6 = 0$$

$$(5x + 2)(x - 3) = 0$$

(Check that your factors are correct by multiplying out the brackets.)

Either $\quad\quad 5x + 2 = 0 \quad$ or $\quad x + 3 = 0$

i.e $\quad\quad\quad 5x = -2 \quad$ or $\quad x = -3$

i.e. $\quad\quad\quad x = -\frac{2}{5} \quad$ or $\quad -3$

Solve the equations:

1 $2x^2 - 5x + 2 = 0$

2 $2x^2 - 11x + 12 = 0$

3 $2x^2 - 13x + 20 = 0$

4 $3x^2 + 5x + 2 = 0$

5 $2x^2 + 9x - 35 = 0$

6 $3x^2 - 11x + 6 = 0$

7 $3x^2 - 7x + 2 = 0$

8 $2x^2 + 5x - 12 = 0$

9 $3x^2 + 11x + 6 = 0$

10 $5x^2 + 27x + 10 = 0$

11 $6x^2 - x - 2 = 0$

12 $15x^2 + 14x - 8 = 0$

13 $12x^2 - 7x + 1 = 0$

14 $6x^2 - 13x - 5 = 0$

15 $20x^2 + 19x + 3 = 0$

16 $8x^2 - 18x + 9 = 0$

Tip To solve any equation of the form $ax^2 + bx + c = 0$ start by trying to factorise the left-hand side.

17 $12x^2 - 20x - 25 = 0$

18 $4x^2 + 8x + 3 = 0$

19 $12x^2 + 17x + 6 = 0$

20 $10x^2 - 29x - 21 = 0$

Solve the equation $4x^2 - 9 = 0$

This is the difference between two squares.

$$4x^2 - 9 = 0$$

$$(2x - 3)(2x + 3) = 0$$

Either $\quad\quad 2x - 3 = 0 \quad$ or $\quad 2x + 3 = 0$

i.e. $\quad\quad\quad 2x = 3 \quad$ or $\quad 2x = -3$

i.e. $\quad\quad\quad x = \frac{3}{2} \quad$ or $\quad -\frac{3}{2}$

Solve the equations:

21 $16x^2 - 25 = 0$

22 $100x^2 - 81 = 0$

23 $4x^2 - 25 = 0$

24 $9x^2 - 16 = 0$

25 $25x^2 - 144 = 0$

26 $9x^2 - 4 = 0$

27 $81x^2 - 25 = 0$

28 $25x^2 - 4 = 0$

29 $36x^2 - 25 = 0$

30 $4x^2 - 81 = 0$

199

Exercise 15i   Solve the equation $x^2 - x = 12$

$$x^2 - x = 12$$

Subtracting 12 from each side gives

$$x^2 - x - 12 = 0$$

$$(x - 4)(x + 3) = 0$$

Either $x - 4 = 0$ or $x + 3 = 0$

i.e. $x = 4$ or -3

> **Tip** First arrange the equations so that the l.h.s. = 0, then try to factorise.

Solve the equations:

1 $x^2 - x = 30$

2 $x^2 - 6x = 16$

3 $x^2 + 9x = 36$

4 $3x^2 + 4x = 4$

5 $x^2 - x = 6$

6 $x^2 + 6x = 7$

7 $2x^2 + 5x = 3$

8 $5x^2 - 12x = 9$

 **9** $x^2 = 2x + 3$

10 $x^2 = 2x + 24$

11 $x^2 = 12x - 35$

12 $10x^2 = 13x + 3$

13 $x^2 = 3x + 10$

14 $x^2 = 6x - 8$

15 $6x^2 = x + 1$

16 $3x^2 = 13x - 4$

17 $10 = 7x - x^2$

18 $7 = 8x - x^2$

19 $8 = 6x - x^2$

20 $21 = 10x - x^2$

21 $12 = 8x - x^2$

22 $20 = 9x - x^2$

23 $35 = 12x - x^2$

24 $15 = 8x - x^2$

> **Tip** Subtract $2x$ and 3 from each side.

Solve the equation $12x^2 + 10x - 12 = 0$

$$12x^2 + 10x - 12 = 0$$

Taking out the common factor 2 gives

$$(2)(6x^2 + 5x - 6) = 0$$

Since 2 is not zero, $6x^2 + 5x - 6$ must be zero

$\therefore$ $6x^2 + 5x - 6 = 0$

i.e. $(3x - 2)(2x + 3) = 0$

Either $3x - 2 = 0$ or $2x + 3 = 0$

i.e. $3x = 2$ or $2x = -3$

i.e. $x = \frac{2}{3}$ or $-\frac{3}{2}$

Solve the equations:

25 $8x^2 - 4x = 0$

26 $2x^2 - 10x + 12 = 0$

27 $3x^2 - 24x + 36 = 0$

28 $12x^2 + 20x + 8 = 0$

29 $8x^2 + 20x = 12$

30 $3x^2 - 9x = 0$

31 $5x^2 - 15x + 10 = 0$

32 $6x^2 + 18x + 12 = 0$

> **Tip**   Look for common factors.

33 $30x^2 = 39x + 9$

Summary

To solve a quadratic equation by factorising:

a   collect all terms on one side of the equation, i.e. arrange it in the form $ax^2 + bx + c = 0$

b   take out any common factors (these may or may not include x)

c   factorise.

Exercise 15j

Solve the equations:

1 $x^2 - x - 20 = 0$	9. $2x^2 + 3x - 14 = 0$	17 $5x = 3x^2 - 2$
2 $x^2 = 4x - 4$	10 $2x^2 + 12x + 18 = 0$	18 $4 + 11x + 6x^2 = 0$
3 $9x^2 - 1 = 0$	11 $x^2 = 7 - 6x$	19 $7x = 4x^2$
4 $2x^2 + 7x = 0$	12 $4 = 25x^2$	20 $14x - 2 = 24x^2$
5 $x^2 + 13x + 12 = 0$	13 $4x^2 = 25$	21 $6x^2 + 13x - 5 = 0$
6 $1 - 16x^2 = 0$	14 $x^2 + 11x + 18 = 0$	22 $5x + 2 = 3x^2$
7 $x^2 - 6x = 0$	15 $2 - x = 6x^2$	23 $3 + 8x + 4x^2 = 0$
8 $x^2 = 2x + 35$	16 $5x - 2x^2 = 0$	24 $3 - 12x^2 = 0$

Solve the equations:

> **Tip** In questions **25** to **40** multiply out the brackets and arrange each equation in the form $ax^2 + bx + c = 0$. Then try to factorise.

25 $x(x + 1) = 12$	31 $3x(x + 3) = 5x + 4$	37 $(x - 5)(x + 2) = 18$
26 $x(x - 1) = x + 3$	32 $2x(2x - 1) = x^2 + 3x + 2$	38 $(x + 8)(x - 2) = 39$
27 $3x(2x + 1) = 4x + 1$	33 $(x + 2)(x + 3) = 56$	39 $(x + 1)(x + 8) + 12 = 0$
28 $5x(x - 1) = 4x^2 - 4$	34 $(x + 9)(x - 6) = 34$	40 $(x - 1)(x + 10) + 30 = 0$
29 $x(x - 5) = 24$	35 $(x - 2)(x + 6) = 33$	
30 $x(x + 3) = 5(3x - 7)$	36 $(x + 3)(x - 8) + 10 = 0$	

> Solve the equation $x(x - 3)(x - 7) = 0$
>
> $$x(x - 3)(x - 7) = 0$$
>
> Either $x = 0$ or $x - 3 = 0$ or $x - 7 = 0$
>
> i.e. $x = 0$ or   3   or   7

Solve the equations:

41 $x(x-1)(x-2)=0$

42 $x(x-3)(x+4)=0$

43 $x(2x-5)(x-2)=0$

 44 $x^3 - 2x^2 + x = 0$

45 $2x^3 + 9x^2 + 4x = 0$

46 $x(x-6)(x-7)=0$

47 $x(x+2)(x-5)=0$

48 $x(3x+7)(x-5)=0$

49 $4x^3 - 9x = 0$

50 $8x = 6x^2 - x^3$

 **Tip** Take out the common factor, then try to factorise what remains.

Tip Rearrange so that the l.h.s. $= 0$

Investigation

The earliest record of a pair of simultaneous equations that needed solving, where one was a second degree equation and the other linear, is in Babylon c. 2000 BC.

The equations were $x^2 + y^2 = 1000$ and $y = \frac{2x}{3} - 10$. Can you solve them?

Problems

Exercise 15k

I think of a positive number x, square it and then add three times the number I first thought of. If the answer is 54, form an equation in x and solve it to find the number I first thought of.

If the given number is x, its square is x^2 and three times the number is $3x$, adding gives $x^2 + 3x$ which we know is 54.

i.e.
$$x^2 + 3x = 54$$
$$x^2 + 3x - 54 = 0$$
$$(x-6)(x+9) = 0$$

Either $\qquad\qquad x - 6 = 0 \quad \text{or} \quad x + 9 = 0$

i.e. $\qquad\qquad\qquad x = 6 \quad \text{or} \quad -9$

The positive number I first thought of was therefore 6

Tip Read each question carefully; several times if necessary.

1 The square of a number x is 16 more than six times the number. Form an equation in x and solve it.

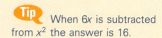
Tip When $6x$ is subtracted from x^2 the answer is 16.

2 When five times a number x is subtracted from the square of the same number, the answer is 14. Form an equation in x and solve it.

3 I think of a number x. If I square it and add it to the number I first thought of the total is 42. Find the number I first thought of.

4 Peter had x marbles. The number of marbles Fred had was six fewer than the square of the number Peter had. Together they had 66 marbles. Form an equation in x and solve it. How many marbles did Fred have?

5 Ahmed is x years old and his father is x^2 years old. If the sum of their ages is 56 years, form an equation in x and solve it to find the age of each.

6 Kathryn is x years old. If her mother's age is two years more than the square of Kathryn's age, and the sum of their ages is 44 years, form an equation in x and solve it to find the ages of Kathryn and her mother.

7 Peter is x years old and his sister is 5 years older. If the product of their ages is 84, form an equation in x and solve it to find Peter's age.

8 Sally is x years old and her sister Ann is 4 years younger. If the product of their ages is 140, form an equation in x and solve it to find Ann's age.

A rectangle is 4 cm longer than it is wide. If it is x cm wide and has an area of 77 cm², form an equation in x and solve it to find the dimensions of the rectangle.

$(x + 4)$ cm

x cm

The area of a rectangle is length × breadth.

$$\text{Area} = (x + 4) \times x \text{ cm}^2$$

But the area is 77 cm²,

$\therefore$ $$(x + 4)x = 77$$

i.e. $$x^2 + 4x = 77$$

$$x^2 + 4x - 77 = 0$$

$$(x - 7)(x + 11) = 0$$

Either $$x - 7 = 0 \quad \text{or} \quad x + 11 = 0$$

i.e. $$x = 7 \quad \text{or} \quad -11$$

The breadth of a rectangle cannot be negative, so we use $x = 7$.

Therefore, the rectangle measures $(7 + 4)$ cm by 7 cm, i.e. 11 cm by 7 cm.

9 A rectangle is x cm long and is 3 cm longer than it is wide. If its area is 28 cm^2, form an equation in x and solve it to find the dimensions of the rectangle.

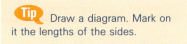

Tip Draw a diagram. Mark on it the lengths of the sides.

10 A rectangle is 5 cm longer than it is wide. If its width is x cm and its area is 66 cm^2 form an equation in x and solve it. Hence find the dimensions of the rectangle.

11 The base of a triangle is x cm long and its perpendicular height is half the length of its base. If the triangle has an area of 25 cm^2, form an equation in x and solve it. What is the height of the triangle?

12 A rectangular lawn measuring 30 m by 20 m is bordered on two adjacent sides by a uniform path of width x m as shown in the diagram.

    a Express in terms of x each of the areas denoted by the letters A, B and C.

    b If the area of the path is 104 m^2 form an equation in x and solve it to find the width of the path.

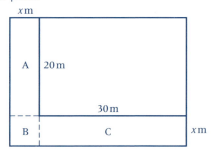

Mixed exercises

Exercise **15I**

1 Find the value of $(x + 4)(x - 3)$ if

    a $x = 1$ **b** $x = 3$ **c** $x = 4$

2 Solve

    a $x(x + 7) = 0$ **b** $4x(2x - 1) = 0$

3 Solve

    a $(x - 3)(x - 8) = 0$ **b** $(x - 2)(5x + 3) = 0$

4 Solve

    a $x^2 - 2x - 35 = 0$ **b** $x^2 - 13x + 40 = 0$

5 Solve

    a $10x^2 - 13x + 4 = 0$ **b** $15x^2 - x - 2 = 0$ **c** $9x^2 - 4 = 0$

6 Solve

    a $7x^2 - 14x = 0$ **b** $4x^2 - 3x = 0$

7 Solve

  a $9x^2 - 3x = 2$ **b** $6x^2 = 17x - 7$

8 Solve

  a $x(x+4) = 45$ **b** $x(x+2) = x + 30$

Exercise 15m

1 Find the value of $(2x-1)(x+2)$ if

  a $x = 0$ **b** $x = \frac{1}{2}$ **c** $x = 2$ **d** $x = -2$

2 Solve

  a $3x(x-2) = 0$ **b** $x(7x+3) = 0$

3 Solve

  a $(x-2)(x+5) = 0$ **b** $(3x-4)(x+2) = 0$ **c** $(2x+3)(2x-3) = 0$

4 Solve

  a $x^2 + x - 6 = 0$ **b** $x^2 + 11x + 30 = 0$

5 Solve

  a $20x^2 + 11x - 3 = 0$ **b** $15x^2 + 41x + 14 = 0$

6 Solve

  a $15x^2 + 10x = 0$ **b** $7x^2 + 3x = 0$

7 Solve

  a $4x^2 - 17x - 15 = 0$ **b** $6x^2 = 11x + 7$

8 Solve

  a $x(x-4) = 32$ **b** $x(2x-1) = 3x + 16$

Exercise 15n

1 Find the value of $(5x+1)(x-3)$ if

  a $x = 2$ **b** $x = 3$ **c** $x = -\frac{1}{5}$

2 Solve

  a $5x(x+7) = 0$ **b** $3x(4x-3) = 0$

3 Solve

  a $(x+4)(x-5) = 0$ **b** $(4x-7)(x+3) = 0$ **c** $(5x-3)(5x+3) = 0$

4 Solve

  a $x^2 - 2x - 15 = 0$ **b** $x^2 + 12x + 32 = 0$

5 Solve

    a $20x^2 + 19x + 3 = 0$ **b** $28x^2 + x - 2 = 0$

6 Solve

    a $12x^2 + 16x = 0$ **b** $3x^2 + 5x = 0$

7 Solve

    a $7x^2 + 5x = 2$ **b** $6x^2 = 7x + 5$

8 Solve

    a $x(x + 6) = 3x + 10$ **b** $x(x + 8) = x + 30$

 Puzzle

A bird table is 3 m from one corner, 4 m from another corner and 5 m from a third corner of a square patio. Find the length of an edge of the patio.

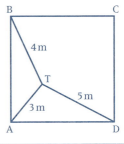

Did you know that Emmy Noether (1882–1935) was the daughter of the outstanding nineteenth century mathematician Max Noether? She was the only woman in a class of over 900 at her father's university. She was a mentor for several female mathematicians including Olga Taussky-Todd, Ruth Stauffer McKee, Marie Weiss and Grace Shover-Quinn.

IN THIS CHAPTER...

you have seen that:

- if $A \times B = 0$ then either $A = 0$ and/or $B = 0$

- the two values of x that satisfy the equation $(x - a)(x - b) = 0$ are $x = a$ and $x = b$

- many quadratic equations can be solved by factorising, e.g. $x^2 - 7x + 12 = 0$ can be rewritten as $(x - 3)(x - 4) = 0$ from which $x = 3$ or $x = 4$ and $4x^2 - 7x = 0$ becomes $x(4x - 7) = 0$, from which $x = 0$ or $x = \frac{7}{4}$

16

ALGEBRAIC FRACTIONS

AT THE END OF THIS CHAPTER...

you should be able to:

1 Reduce a given algebraic fraction to its lowest form.

2 Multiply or divide two algebraic expressions.

3 Find the LCM of a set of algebraic expressions.

4 Add or subtract two algebraic fractions.

5 Solve simple equations containing algebraic fractions.

6 Solve quadratic equations containing algebraic fractions.

When the clock strikes why do
the hands keep working?

BEFORE YOU START

you need to know:
✓ how to cancel fractions
✓ index notation
✓ how to find the LCM of a set of numbers
✓ how to remove brackets
✓ how to factorise a quadratic expression

KEY WORDS

common denominator, denominator, equivalent fraction, factorise, lowest common multiple (LCM), numerator, product, simplify

Simplifying fractions

We simplify a fraction such as $\frac{10}{50}$ by recognising that 10 is a common factor of the numerator and denominator and then cancelling that common factor,

i.e. $\qquad \frac{10}{50} = \frac{{}^{1}\cancel{10}}{5 \times \cancel{10}_{1}} = \frac{1}{5}$

To simplify an algebraic fraction, we do exactly the same: we find and then cancel the common factors of the numerator and denominator.

Note that we do not have to write the number 50 as 5×10 but that when the factors are letters it helps at this stage to put in the multiplication sign.

For example $\qquad$ xy can be written as $x \times y$

and $\qquad$ $2(a+b)$ can be written as $2 \times (a+b)$

Exercise **16a**

Simplify **a** $\dfrac{2xy}{6y}$ $\qquad$ **b** $\dfrac{2a}{a^2 b}$

a $\quad \dfrac{2xy}{6y} = \dfrac{{}^{1}\cancel{2} \times x \times \cancel{y}^{1}}{\cancel{6}_{3} \times \cancel{y}_{1}}$

$\qquad \quad = \dfrac{x}{3}$

b $\quad \dfrac{2a}{a^2 b} = \dfrac{2 \times \cancel{a}^{1}}{{}_{1}\cancel{a} \times a \times b}$

$\qquad \quad = \dfrac{2}{ab}$

Simplify:

1 $\dfrac{2x}{8}$	**4** $\dfrac{a^2}{ab}$	**7** $\dfrac{2ab}{4bc}$	**10** $\dfrac{a^2 b}{abc}$	**13** $\dfrac{b^2}{bd}$	**16** $\dfrac{10x}{15xy}$
2 $\dfrac{ab}{2b}$	**5** $\dfrac{xy}{y^2}$	**8** $\dfrac{6p}{3pq}$	**11** $\dfrac{7a}{14}$	**14** $\dfrac{4}{12x}$	**17** $\dfrac{m^2 n}{kmn}$
3 $\dfrac{p^2}{pq}$	**6** $\dfrac{3}{6a}$	**9** $\dfrac{5p^2 q}{10p}$	**12** $\dfrac{yz}{2x}$	**15** $\dfrac{3pq}{6p}$	**18** $\dfrac{5s^2}{20st}$

Factors

We know that $3 \times 2 = 6$ but neither $3+2$ nor $3-2$ is equal to 6.

We can write a number as the product of its factors but, in general, we cannot write a number as the sum or difference of its factors.

Thus $\qquad \begin{cases} p \text{ and } q \text{ are factors of } pq \\ a \text{ and } (a-b) \text{ are factors of } a\,(a-b) \end{cases}$

but in general $\begin{cases} p \text{ is } not \text{ a factor of } p+q \\ b \text{ is } not \text{ a factor of } a-b \end{cases}$

This means that in the fraction $\dfrac{p+q}{pq}$ we *cannot* cancel q because q is not a factor of the numerator.

Sometimes the common factors in a fraction are not very obvious.

Consider $\dfrac{x-2}{y(x-2)}$

Placing the numerator in brackets and using the multiplication sign gives $\dfrac{(x-2)}{y \times (x-2)}$

Now we can see clearly that $(x-2)$ is a common factor, so

$$\frac{\cancel{(x-2)}^{\,1}}{y \times \underset{1}{\cancel{(x-2)}}} = \frac{1}{y}$$

Exercise 16b

Simplify where possible   **a** $\dfrac{2a(a-b)}{a-b}$ **b** $\dfrac{pq}{p-q}$

a $\dfrac{2a(a-b)}{a-b} = \dfrac{2 \times a \times \cancel{(a-b)}^{\,1}}{\underset{1}{\cancel{(a-b)}}}$ (place the denominator in brackets)

$= 2a$

b $\dfrac{pq}{p-q} = \dfrac{p \times q}{(p-q)}$ which cannot be simplified.

Simplify where possible:

1 $\dfrac{x-y}{x(x-y)}$

2 $\dfrac{st}{s(s-t)}$

3 $\dfrac{2a}{a-b}$

4 $\dfrac{p+q}{2p}$

5 $\dfrac{4x}{8(x-y)}$

6 $\dfrac{3(a+b)}{6ab}$

7 $\dfrac{(p-q)(p+q)}{p+q}$

8 $\dfrac{(4+a)}{(4+a)(4-a)}$

9 $\dfrac{(a-b)}{3(a+b)}$

10 $\dfrac{u-v}{v(u-v)}$

11 $\dfrac{xy}{x(x+y)}$

12 $\dfrac{s-t}{2(s-t)}$

13 $\dfrac{10a}{15(a-b)}$

14 $\dfrac{8(x-y)}{12xy}$

Tip Place brackets round two terms.

15 $\dfrac{s-t}{3s}$

16 $\dfrac{(u+v)(u-v)}{u+v}$

17 $\dfrac{x+y}{2(x-y)}$

18 $\dfrac{s+6}{(s+6)(s-6)}$

Sometimes we have to factorise the numerator and/or the denominator before we look for common factors.

Consider for example $\dfrac{12a - 4b}{3a - b}$

In the numerator there is a common factor of 4, so

$$12a - 4b = 4(3a - b)$$

The fraction becomes $\dfrac{4(3a - b)}{(3a - b)}$

It is now clear that 4 and $(3a - b)$ are factors of $12a - 4b$ so we can cancel

$(3a - b)$, i.e. $\dfrac{4\,\cancel{(3a - b)}^{\,\prime}}{\cancel{(3a - b)}_{\,1}} = 4$

We have been writing expressions such as $4(3a - b)$ in the form $4 \times (3a - b)$, but the multiplication sign is not necessary and we will no longer put it in. If, however, you find that the multiplication sign makes it easier to see individual factors then continue to use it.

Exercise 16c

Simplify $\dfrac{xy - y^2}{3y}$

Start by looking for a common factor in the top.

$$\frac{xy - y^2}{3y} = \frac{\cancel{y}^{\,1}(x - y)}{3\cancel{y}_{\,1}}$$

$$= \frac{x - y}{3}$$

Simplify:

1 $\dfrac{4a}{8a - 2b}$

2 $\dfrac{2pq}{p^2 - pq}$

3 $\dfrac{a - b}{a^2 - ab}$

4 $\dfrac{3a - 6b}{5a - 10b}$

5 $\dfrac{2x - x^2}{3xy}$

6 $\dfrac{a^2}{3a - ab}$

7 $\dfrac{2a - 3b}{6a^2 - 9ab}$

8 $\dfrac{2s^2 - st}{2s - t}$

9 $\dfrac{3a - 6b}{a^2 - 2ab}$

10 $\dfrac{6x}{9x - 3y}$

11 $\dfrac{3ab}{ab + b^2}$

12 $\dfrac{p^2 + pq}{5p}$

13 $\dfrac{2p - 4q}{6p - 12q}$

14 $\dfrac{3a + a^2}{4ab}$

Tip Look for a common factor in the bottom line.

15 $\dfrac{2x - xy}{x^2}$

16 $\dfrac{2x + y}{6xy + 3y^2}$

17 $\dfrac{a^2b - ac}{ab - c}$

18 $\dfrac{p^2 + 2pq}{2p + 4q}$

Simplify $\dfrac{pq - 3q}{p^2 - 6p + 9}$

(The quadratic expression in the bottom line factorises.)

$$\dfrac{pq - 3q}{p^2 - 6p + 9} = \dfrac{q\cancel{(p-3)}^{1}}{(p - 3)\cancel{(p-3)}_{1}}$$

$$= \dfrac{q}{p - 3}$$

Simplify:

19 $\dfrac{a - 4}{a^2 - 6a + 8}$

20 $\dfrac{x - 2}{x^2 - 6x + 8}$

21 $\dfrac{y + 3}{y^2 + 5y + 6}$

22 $\dfrac{2a - 8}{a^2 - a - 12}$

23 $\dfrac{3x + 12}{x^2 + 7x + 12}$

24 $\dfrac{9y - 36}{y^2 - 2y - 8}$

25 $\dfrac{xy - 2y}{x^2 - 4x + 4}$

26 $\dfrac{pq + 5q}{p^2 + 7p + 10}$

27 $\dfrac{st - t}{s^2 - 8s + 7}$

28 $\dfrac{p + 3}{p^2 + 6p + 9}$

29 $\dfrac{x - 5}{x^2 + x - 30}$

30 $\dfrac{2x + 6}{x^2 - x - 12}$

Tip Try to factorise the quadratic expression.

31 $\dfrac{3x - 15}{x^2 - 9x + 20}$

32 $\dfrac{uv + 6v}{u^2 + 12u + 36}$

33 $\dfrac{xy - 7y}{x^2 - 9x + 14}$

Further simplifying

The next exercise contains slightly more complicated factorising.

Remember that $a^2 - b^2 = (a - b)(a + b)$

It sometimes happens that the numerator has a factor $(a - b)$

and the denominator has a factor $(b - a)$

Now $(a - b) = (-1)(b - a)$

therefore a fraction such as $\dfrac{a - b}{b - a}$ can be simplified,

i.e. $\dfrac{a - b}{b - a} = \dfrac{(-1)\cancel{(b-a)}^{1}}{\cancel{(b-a)}_{1}} = -1$

Exercise 16d

Simplify **a** $\dfrac{2x^2 + 5x - 3}{4x^2 - 1}$ **b** $\dfrac{a^2 - 2ab + b^2}{b^2 - ab}$

a (There are quadratic expressions in the top and bottom. Start by factorising them.)

$$\dfrac{2x^2 + 5x - 3}{4x^2 - 1} = \dfrac{\cancel{(2x-1)}^{1}(x + 3)}{\cancel{(2x-1)}_{1}(2x + 1)} = \dfrac{(x + 3)}{(2x + 1)}$$

b $\dfrac{a^2 - 2ab + b^2}{b^2 - ab} = \dfrac{(a-b)(a-b)}{b(b-a)}$

$$= \dfrac{(-1)(b-a)(a-b)}{b(b-a)}$$

$$= \dfrac{(-1)(a-b)}{b}$$

$$= \dfrac{b-a}{b}$$

Simplify:

💬 **1** $\dfrac{x^2 - 9}{2x^2 - 7x + 3}$

9 $\dfrac{a - a^2}{a - 1}$

> **Tip** Start by trying to factorise both the top and the bottom.

2 $\dfrac{4x - 8}{x^2 - 4}$

10 $\dfrac{2y^2 + 5y - 3}{4y^2 - 1}$

3 $\dfrac{4x^2 - 1}{2x^2 - 3x - 2}$

11 $\dfrac{x^2 - 6xy + 9y^2}{x^2 - 3xy}$

> **Tip** Remember that $2 - x = (-1)(x - 2)$.

💬 **4** $\dfrac{2 - x}{x^2 - 4x + 4}$

12 $\dfrac{4x^2 - 7x - 2}{4x^2 - 8x}$

5 $\dfrac{a^2 - b^2}{b^2 - 2ab + a^2}$

13 $\dfrac{2x^2 - 13x + 15}{x^2 - 10x + 25}$

17 $\dfrac{4x^2 + 6x - 4}{x^2 - x - 6}$

6 $\dfrac{2a^2 + ab - b^2}{4a^2 - b^2}$

14 $\dfrac{a - 1}{1 - a^2}$

18 $\dfrac{x^2 - xy - 2y^2}{y^2 + xy}$

7 $\dfrac{3x^2 - xy - 2y^2}{9x^2 - 4y^2}$

15 $\dfrac{ac + bc - ad - bd}{c - d}$

19 $\dfrac{1 - x^2}{3x^2 + 9x + 6}$

8 $\dfrac{x^2 - 5x + 6}{3y - xy}$

16 $\dfrac{2x^2 + 3x - 35}{7 + 5x - 2x^2}$

20 $\dfrac{y + x + xy + y^2}{x^2 + 2xy + y^2}$

Multiplying and dividing fractions

Reminder: The product of two fractions is found by multiplying the numerators and multiplying the denominators,

e.g. $\qquad \dfrac{2}{3} \times \dfrac{4}{5} = \dfrac{2 \times 4}{3 \times 5} = \dfrac{8}{15}$

and $\qquad \dfrac{p}{q} \times \dfrac{(a-b)}{(a+b)} = \dfrac{p(a-b)}{q(a+b)}$

To divide by a fraction, we multiply by its reciprocal,

e.g. $\qquad \dfrac{2}{3} \div \dfrac{3}{4} = \dfrac{2}{3} \times \dfrac{4}{3} = \dfrac{8}{9} \qquad \dfrac{p}{q} \div r = \dfrac{p}{q} \times \dfrac{1}{r} = \dfrac{p}{qr}$

and $\qquad \dfrac{p}{q} \div \dfrac{a}{(a-b)} = \dfrac{p}{q} \times \dfrac{(a-b)}{a} = \dfrac{p(a-b)}{qa}$

Exercise **16e**

Find:

1 $\dfrac{a}{b} \times \dfrac{c}{d}$

5 $\dfrac{a}{b} \div c$

> **Tip** Remember that to divide by a fraction you multiply by its reciprocal.

2 $\dfrac{a}{b} \div \dfrac{c}{d}$

6 $\dfrac{a}{b} \times c$

3 $\dfrac{x - y}{2} \times \dfrac{5}{x}$

7 $\dfrac{(a - b)}{4} \div \dfrac{(a + b)}{3}$

9 $\dfrac{(x - 2)}{3} \div (x + 3)$

4 $\dfrac{x - y}{2} \div \dfrac{5}{x}$

8 $\dfrac{(x - 2)}{3} \times (x + 3)$

10 $\dfrac{p}{q} \div \dfrac{1}{r}$

As is the case in number fractions, it is sometimes possible to simplify before multiplying.

Simplify $\dfrac{ab}{4} \times \dfrac{8}{a^2}$

$$\dfrac{ab}{4} \times \dfrac{8}{a^2} = \dfrac{\cancel{ab}^{\,1}}{\cancel{4}_1} \times \dfrac{\cancel{8}^{\,2}}{\cancel{a^2}_a}$$

$$= \dfrac{2b}{a}$$

Simplify:

11 $\dfrac{2a}{b} \div \dfrac{a^2}{3b^2}$

15 $\dfrac{2p^2}{3} \times \dfrac{q}{4p}$

19 $\dfrac{a^2}{2b} \div 2a$

12 $\dfrac{pq}{6} \times \dfrac{3}{p^2}$

16 $\dfrac{x^2}{4} \div \dfrac{xy}{2}$

20 $\dfrac{a}{b} \times \dfrac{2a}{3b} \div \dfrac{2b}{3a}$

13 $\dfrac{4xy}{3} \times \dfrac{9}{x^2}$

17 $\dfrac{1}{b^2} \div \dfrac{2}{b}$

14 $\dfrac{2ab}{5} \div \dfrac{a}{b}$

18 $\dfrac{7p}{5q} \times \dfrac{10q}{21p^2}$

Simplify $\dfrac{(a^2 - b^2)}{a} \times \dfrac{b}{(a + b)}$

$$\dfrac{(a^2 - b^2)}{a} \times \dfrac{b}{(a + b)} = \dfrac{(a - b)\cancel{(a + b)}^{\,1}b}{a\cancel{(a + b)}_1}$$

$$= \dfrac{b(a - b)}{a}$$

(Notice that we leave the answer in factor form.)

Simplify $(x^2 - 3x - 4) \div (x - 4)$

$\left(\text{Remember that } 7 \div 4 \text{ can be written } \frac{7}{4}; \text{ similarly } (x^2 - 3x - 4) \div (x - 4) \text{ can be written } \frac{x^2 - 3x - 4}{x - 4}\right)$

$$(x^2 - 3x - 4) \div (x - 4) = \frac{x^2 - 3x - 4}{x - 4}$$

$$= \frac{\cancel{(x - 4)}(x + 1)}{\cancel{(x - 4)}}$$

$$= x + 1$$

Simplify:

21 $\dfrac{(b - 2)}{4} \times \dfrac{1}{(b^2 - 4b + 4)}$

22 $\dfrac{(x^2 - 4)}{3} \times \dfrac{6}{(x + 2)}$

23 $\dfrac{(a^2 - 9)}{2} \div \dfrac{(a - 3)}{4}$

24 $\dfrac{(3b - 6)}{5} \div \dfrac{(b - 2)}{10}$

25 $(x^2 - 6x + 9) \div (x - 3)$

26 $\dfrac{1}{(x - 2)} \times (x^2 - 5x + 6)$

27 $(x - 4) \div (x^2 - 6x + 8)$

28 $(2x - 4) \times \dfrac{1}{(x^2 + 2x - 8)}$

29 $\dfrac{(x^2 - 4)}{5} \times \dfrac{3}{(x^2 + 8x + 12)}$

> **Tip** Remember to try to factorise quadratic expressions and that division by a fraction is the same as multiplying by its reciprocal.

30 $\dfrac{(4x^2 - 9)}{3} \div \dfrac{(6x + 9)}{2}$

31 $(3x - 6) \div (3x^2 - 4x - 4)$

32 $\dfrac{(2x^2 - 5x + 3)}{2} \times \dfrac{(x - 1)}{(x^2 - 2x + 1)}$

33 $(4x^2 - 1) \div (12x^2 + 8x + 1)$

34 $\dfrac{(ab + a^2)}{b} \times \dfrac{b}{a + b}$

35 $\dfrac{a - b^2}{c} \div \dfrac{b^2 - ab}{c^2}$

36 $\dfrac{(x^2 - 5x + 4)}{(x + 2)} \times \dfrac{(x^2 - 4)}{(x - 1)}$

Investigation

Multiply 32547891 by 6. Compare your answer with the original number. What do you notice?

Investigate similar relationships. For example, is there an 8-digit number using every digit from 1 to 9 except 7 which, when multiplied by 7, gives a 9-digit answer which uses every digit from 1 to 9 once?

Lowest common multiple

Before we can simplify $\frac{2}{3} + \frac{1}{5}$ we must change both $\frac{2}{3}$ and $\frac{1}{5}$ into equivalent fractions with the same denominator. This common denominator must contain both 3 and 5 as factors; there are many numbers we could choose but 15 is the lowest such number, i.e. 15 is the *lowest common multiple* (LCM) of 3 and 5.

To simplify $\dfrac{3}{x} + \dfrac{2}{y}$ we follow the same pattern. We need a common denominator with both x and y as factors. Again there are many we could use, but the simplest is xy; this is the LCM of x and y.

Exercise 16f

Find the LCM of ab and c

The LCM is abc

Find the LCM of:

1 p, q	**3** $2, 3, 5$	**5** x, y, wz	**7** v, uw
2 r, st	**4** a, b, c	**6** a, d	**8** $3, 7, 8$

Find the LCM of   **a** $4, 10$ **b** ab, a^2 **c** $2x, 6x$

a $4 = 2 \times 2$ and $10 = 2 \times 5$

(The LCM is the *lowest* number that 4 and 10 divide into exactly, so the factors it must include are

2×2 from 4 and 5 from 10

The factor of 2 from 10 is not needed as 2 is already included.)

$\therefore$ the LCM of 4 and 10 is $2 \times 2 \times 5 = 20$

b $ab = a \times b$ and $a^2 = a \times a$

$\therefore$ the LCM is $a \times b \times a = a^2 b$

c $2x = 2 \times x$ and $6x = 2 \times 3 \times x$

$\therefore$ the LCM is $2 \times 3 \times x = 6x$

Find the LCM of:

9 x, xy	**12** $x^2, 2xy$	**15** $3p, p^2$	**18** $2x, 3x$	**21** $6, 4, 10$	**24** $4x, 6x$
10 $x^2, 2x$	**13** ab, bc	**16** $5a, ab$	**19** $4x, 8x$	**22** a, ab, a^2	**25** $3y, 5y$
11 $pq, 3p$	**14** s, st	**17** $3pq, q^2$	**20** $6a, 9a$	**23** $10x, 15x$	**26** $2x, 3x, 4x$

Addition and subtraction of fractions

To add or subtract fractions we first have to change them into equivalent fractions with a common denominator.

Thus to find $\frac{2}{3} + \frac{1}{5}$, we choose a common denominator of 15 which is the LCM of 3 and 5.

Now $\qquad \frac{2}{3} = \frac{2 \times 5}{3 \times 5} = \frac{10}{15}$ and $\frac{1}{5} = \frac{1 \times 3}{5 \times 3} = \frac{3}{15}$

Therefore $\qquad \frac{2}{3} + \frac{1}{5} = \frac{10 + 3}{15} = \frac{13}{15}$

To simplify $\frac{3}{x} + \frac{2}{y}$ we follow the same pattern:

xy is the LCM of x and y.

$$\frac{3}{x} = \frac{3 \times y}{x \times y} = \frac{3y}{xy} \quad \text{and} \quad \frac{2}{y} = \frac{2 \times x}{y \times x} = \frac{2x}{xy}$$

$\therefore \qquad \frac{3}{x} + \frac{2}{y} = \frac{3y + 2x}{xy}$

Exercise 16g

Simplify $\dfrac{1}{2a} + \dfrac{1}{b}$

($2ab$ is the LCM of $2a$ and b)

$$\frac{1}{2a} + \frac{1}{b} = \frac{(1) \times (b) + (1) \times (2a)}{2ab} = \frac{b + 2a}{2ab}$$

Simplify $\dfrac{3}{4x} - \dfrac{1}{6x}$

($12x$ is the LCM of $4x$ and $6x$)

$$\frac{3}{4x} - \frac{1}{6x} = \frac{(3) \times (3) - (1) \times (2)}{12x} = \frac{7}{12x}$$

Simplify:

1 $\dfrac{1}{x} + \dfrac{1}{y}$

2 $\dfrac{3}{p} + \dfrac{2}{q}$

3 $\dfrac{2}{s} - \dfrac{1}{t}$

4 $\dfrac{3}{a} + \dfrac{1}{2b}$

5 $\dfrac{1}{3x} - \dfrac{2}{5y}$

6 $\dfrac{1}{a} + \dfrac{5}{2b}$

7 $\dfrac{2}{x} - \dfrac{3}{y}$

8 $\dfrac{4}{3p} + \dfrac{2}{q}$

9 $\dfrac{3}{x} - \dfrac{2}{y}$

10 $\dfrac{5}{7a} + \dfrac{3}{4b}$

11 $\dfrac{1}{2x} + \dfrac{1}{3x}$

12 $\dfrac{2}{5x} - \dfrac{3}{7x}$

13 $\dfrac{2}{y} - \dfrac{3}{4y}$

14 $\dfrac{3}{8p} - \dfrac{1}{4p}$

15 $\dfrac{1}{a} + \dfrac{5}{8a}$

16 $\dfrac{1}{3x} - \dfrac{1}{7x}$

17 $\dfrac{4}{7x} - \dfrac{2}{5x}$

18 $\dfrac{1}{y} - \dfrac{2}{3y}$

> **Tip** To simplify the sum or difference of two fractions you must put them over a common denominator.

Simplify $\dfrac{4a}{3b} - \dfrac{b}{6a}$

$(3b = 3 \times b \quad \text{and} \quad 6a = 2 \times 3 \times a, \quad \therefore \quad \text{LCM} = 6ab)$

$$\dfrac{4a}{3b} - \dfrac{b}{6a} = \dfrac{(4a) \times (2a) - (b) \times (b)}{6ab}$$

$$= \dfrac{8a^2 - b^2}{6ab}$$

Simplify:

19 $\dfrac{1}{2a} + \dfrac{3}{4b}$

25 $\dfrac{s}{t^2} + \dfrac{s^2}{2t}$

20 $\dfrac{a}{2b} - \dfrac{a^2}{b^2}$

26 $\dfrac{5}{2a} + \dfrac{2}{3ab}$

21 $\dfrac{3}{x} - \dfrac{4}{xy}$

27 $\dfrac{1}{x^2} + \dfrac{2}{3x}$

31 $\dfrac{5}{7x} - \dfrac{3}{14xy}$

34 $\dfrac{7}{9p} - \dfrac{5}{6q}$

22 $\dfrac{2}{p^2} - \dfrac{3}{2p}$

28 $\dfrac{2y}{3x} - \dfrac{3x}{2y}$

32 $\dfrac{9}{a^2} - \dfrac{3}{2ab}$

35 $\dfrac{a^2}{b^2} + \dfrac{4a}{5b}$

23 $\dfrac{3a}{4b} + \dfrac{b}{6a}$

29 $\dfrac{5}{8x} + \dfrac{2}{4y}$

33 $\dfrac{3x}{2y} - \dfrac{3y}{2x}$

36 $\dfrac{7}{5pq} + \dfrac{8}{15q}$

24 $\dfrac{5}{2p} - \dfrac{3}{4q}$

30 $\dfrac{p}{3q} + \dfrac{p^2}{q^2}$

Tip Always begin by finding the LCM of the denominator.

Exercise 16h

Simplify $\dfrac{x-2}{3} - \dfrac{x-4}{2}$

$$\dfrac{(x-2)}{3} - \dfrac{(x-4)}{2} = \dfrac{2(x-2) - 3(x-4)}{6}$$

$$= \dfrac{2x - 4 - 3x + 12}{6}$$

$$= \dfrac{-x + 8}{6}$$

$$= \dfrac{8 - x}{6}$$

(Notice that we placed brackets round the numerators *before* putting the fractions over a common denominator. This ensured that each numerator was kept together and that the signs were not confused.)

Simplify:

1 $\dfrac{x+2}{5} + \dfrac{x-1}{4}$

3 $\dfrac{2x-1}{3} + \dfrac{x+2}{5}$

5 $\dfrac{x+3}{7} - \dfrac{x+2}{5}$

7 $\dfrac{2x-1}{7} - \dfrac{x-2}{5}$

2 $\dfrac{x+3}{4} - \dfrac{x+1}{3}$

4 $\dfrac{2x+3}{4} - \dfrac{x-2}{6}$

6 $\dfrac{x+4}{5} + \dfrac{x-1}{2}$

8 $\dfrac{3x+1}{14} - \dfrac{2x+3}{21}$

217

Remember that $\frac{1}{5}(x-2) = \frac{(x-2)}{5}$

💬 **9** $\frac{1}{7}(2x-3) - \frac{1}{3}(4x-2)$

Tip This is the same as $\frac{2x-3}{7} - \frac{4x-2}{3}$

10 $\frac{1}{4}(5x-1) - \frac{1}{3}(2x-3)$

11 $\frac{5-2x}{3} + \frac{4-3x}{2}$

12 $\frac{1}{4}(3-x) + \frac{1}{6}(1-2x)$

15 $\frac{3-x}{2} + \frac{1-2x}{6}$

18 $\frac{1}{9}(4-x) - \frac{1}{6}(2+3x)$

13 $\frac{1}{8}(5x+4) - \frac{1}{3}(4x-1)$

16 $\frac{2+5x}{8} - \frac{3-4x}{6}$

14 $\frac{2x+3}{5} - \frac{3x-2}{4}$

17 $\frac{1}{5}(4-3x) + \frac{1}{10}(3-x)$

Simplify $\frac{2(x+1)}{3} - \frac{3(x-2)}{5}$

$$\frac{2(x+1)}{3} - \frac{3(x-2)}{5} = \frac{5 \times 2(x+1) - 3 \times 3(x-2)}{15}$$

$$= \frac{10(x+1) - 9(x-2)}{15}$$

(Now multiply out the brackets.)

$$= \frac{10x + 10 - 9x + 18}{15}$$

$$= \frac{x+28}{15}$$

Simplify:

19 $\frac{4(x+2)}{3} + \frac{2(x-1)}{5}$

22 $\frac{5(2x-1)}{2} - \frac{4(x+3)}{5}$

25 $\frac{2(3x-1)}{5} + \frac{4(2x-3)}{15}$

20 $\frac{3(x-1)}{4} + \frac{2(x+1)}{3}$

23 $\frac{3(x-1)}{2} + \frac{3(x+4)}{7}$

26 $\frac{3(x-2)}{5} - \frac{7(x-4)}{6}$

21 $\frac{2(x-2)}{3} - \frac{3(x-1)}{7}$

24 $\frac{7(x-3)}{3} - \frac{2(x+5)}{9}$

Simplify $\frac{2}{x} - \frac{1}{x+2}$

$$\frac{2}{x} - \frac{1}{(x+2)} = \frac{(2)(x+2) - (1)(x)}{x(x+2)}$$

$$= \frac{2x+4-x}{x(x+2)} \quad \text{(Multiplying out the brackets}$$
$$\text{in the numerator.)}$$

$$= \frac{x+4}{x(x+2)}$$

(Notice that we placed the two-term denominator in brackets. Notice also that we left the common denominator in factorised form.)

Simplify:

 27 $\dfrac{2}{a}+\dfrac{1}{a+3}$

30 $\dfrac{2}{2x+1}-\dfrac{3}{4x}$

Tip The LCM is $a(a+3)$

28 $\dfrac{4}{x+2}+\dfrac{2}{x}$

31 $\dfrac{3}{a}+\dfrac{2}{a+4}$

29 $\dfrac{3}{x-4}+\dfrac{1}{2x}$

32 $\dfrac{3}{x-1}+\dfrac{4}{x}$

33 $\dfrac{3}{2x+1}+\dfrac{1}{3x}$

34 $\dfrac{5}{2x+3}-\dfrac{2}{5x}$

Investigation

You now know how to express the sum of two fractions such as $\dfrac{1}{x+1}+\dfrac{1}{x-1}$ as a single fraction,

i.e. $\dfrac{1}{x+1}+\dfrac{1}{x-1}=\dfrac{(x-1)+(x+1)}{(x-1)(x+1)}=\dfrac{2x}{(x-1)(x+1)}$

Therefore starting with a single fraction such as $\dfrac{2x}{(x-1)(x+1)}$ it must be possible to reverse the process and express it as the sum (or difference) of two fractions.

Investigate how $\dfrac{2}{(x+1)(x-1)}$ can be expressed as the sum or difference of two fractions.

If you think you have found a method that will work with any such fraction, try it by expressing $\dfrac{4}{(x+2)(2x-1)}$ as the sum or difference of two fractions.

Mixed questions

Exercise 16i

Simplify:

1 $\dfrac{2}{a}-\dfrac{b}{c}$

5 $\dfrac{3}{4x}-\dfrac{2}{3x}$

9 $\dfrac{1}{x}-\dfrac{3}{x+1}$

13 $\dfrac{2}{5x}\times\dfrac{3}{4x}$

2 $\dfrac{pq}{r}\times\dfrac{r^3}{p^2}$

6 $\dfrac{x+3}{x^2+5x+6}$

10 $\dfrac{a^2}{bc}\div\dfrac{a}{b^2}$

14 $\dfrac{x+4}{5}+\dfrac{2x-1}{10}$

3 $\dfrac{x+2}{4}+\dfrac{x-5}{3}$

7 $\dfrac{p^2-pq}{q^2-p^2}$

11 $\dfrac{2}{5x}\div\dfrac{3}{4x}$

15 $\dfrac{x+4}{5}\times\dfrac{2x-1}{10}$

4 $\dfrac{a^2+ab}{a^2-b^2}$

8 $\dfrac{4}{x^2}-\dfrac{2}{3x}$

12 $\dfrac{2}{5x}+\dfrac{3}{4x}$

16 $\dfrac{5}{4x}+\dfrac{5}{6x}$

17 $\dfrac{5}{4x} \times \dfrac{5}{6x}$

19 $\dfrac{1}{3x} + \dfrac{6}{x-1}$

21 $\dfrac{3}{2a} - \dfrac{2}{a-1}$

23 $\dfrac{3}{4y-3} \div \dfrac{y}{4y-3}$

18 $\dfrac{5}{4x} \div \dfrac{5}{6x}$

20 $\dfrac{1}{3x} \times \dfrac{6}{x-1}$

22 $\dfrac{3}{2a} \times \dfrac{2}{a-1}$

24 $\dfrac{3}{4y-3} - \dfrac{4y}{4y-3}$

Solving equations with fractions

Remember that when solving an equation we *must* keep the equality true. This means that if we alter the size of one side of the equation then we must alter the other side in the same way.

Consider the equation $\dfrac{1}{x} + \dfrac{1}{2x} = \dfrac{5}{6}$

If we choose to multiply each side by the LCM of the denominators, we can remove all fractions from the equation.

The LCM of x, $2x$ and 6 is $6x$.

Multiplying each side by $6x$ gives

$$6x\left(\dfrac{1}{x} + \dfrac{1}{2x}\right) = 6x \times \dfrac{5}{6}$$

$\therefore \quad \dfrac{6x}{1} \times \dfrac{1}{x} + \dfrac{6x}{1} \times \dfrac{1}{2x} = \dfrac{6x}{1} \times \dfrac{5}{6}$

$\therefore \quad 6 + 3 = 5x$

$$9 = 5x$$

$$\tfrac{9}{5} = x \quad \text{i.e.} \quad x = 1\tfrac{4}{5}$$

Exercise 16j

Solve the following equations:

1 $\dfrac{1}{2} + \dfrac{4}{x} = 1$

5 $\dfrac{1}{2x} + \dfrac{2}{x} = \dfrac{1}{4}$

9 $\dfrac{1}{x} - \dfrac{1}{2} = \dfrac{3}{2x}$

10 $\dfrac{3}{2x} + \dfrac{2}{5} = \dfrac{5}{x}$

2 $\dfrac{2}{3} - \dfrac{1}{x} = \dfrac{13}{15}$

6 $\dfrac{3}{x} + \dfrac{3}{10} = \dfrac{9}{10}$

Tip The LCM is 15x.

3 $\dfrac{3}{4} - \dfrac{2}{x} = \dfrac{5}{12}$

7 $\dfrac{3}{8} - \dfrac{2}{x} = \dfrac{1}{6}$

4 $\dfrac{1}{x} - \dfrac{1}{3x} = \dfrac{1}{2}$

8 $\dfrac{3}{2x} + \dfrac{1}{4x} = \dfrac{1}{3}$

Tip The LCM is 6x.

Solve the equation $\dfrac{x-2}{4} - \dfrac{x-3}{6} = 2$

$$\frac{(x-2)}{4} - \frac{(x-3)}{6} = 2$$

Multiply each side by 12

$$12\left[\frac{(x-2)}{4} - \frac{(x-3)}{6}\right] = 12 \times 2$$

$$\therefore \qquad \frac{\overset{3}{\cancel{12}}}{1} \times \frac{(x-2)}{\cancel{4}_1} - \overset{2}{\cancel{12}} \times \frac{(x-3)}{\cancel{6}_1} = 24$$

$$\therefore \qquad 3(x-2) - 2(x-3) = 24$$

$$3x - 6 - 2x + 6 = 24$$

$$x = 24$$

Solve the following equations:

11 $\dfrac{x+2}{4} + \dfrac{x-3}{2} = \dfrac{1}{2}$

16 $\dfrac{x+3}{5} + \dfrac{x-2}{4} = \dfrac{3}{10}$

12 $\dfrac{x}{4} - \dfrac{x+3}{3} = \dfrac{1}{2}$

17 $\dfrac{2}{3} - \dfrac{x+1}{9} = \dfrac{5}{6}$

13 $\dfrac{x}{5} + \dfrac{x+1}{4} = \dfrac{8}{5}$

18 $\dfrac{x+3}{4} - \dfrac{x}{2} = 5$

14 $\dfrac{2x}{5} - \dfrac{x-3}{8} = \dfrac{1}{10}$

19 $\dfrac{3x}{20} + \dfrac{x-2}{8} = \dfrac{3}{10}$

21 $\dfrac{2x-1}{7} + \dfrac{3x-3}{4} = \dfrac{1}{7}$

15 $\dfrac{x-4}{3} - \dfrac{x+1}{4} = \dfrac{1}{6}$

20 $\dfrac{x+3}{7} - \dfrac{x-4}{3} = 1$

22 $\dfrac{2x}{9} - \dfrac{3x+2}{4} = \dfrac{7}{12}$

Tip Put brackets around the numerators before you multiply each side by the LCM.

Sometimes a fractional equation is a quadratic equation in disguise.

Solve the equation $\dfrac{x-4}{2} + \dfrac{4}{x} = 1$

$$\frac{(x-4)}{2} + \frac{4}{x} = 1$$

Multiply each side by $2x$ $\qquad 2x\left[\frac{(x-4)}{2} + \frac{4}{x}\right] = 2x \times 1$

$$\frac{\overset{1}{\cancel{2x}}}{1} \times \frac{(x-4)}{\cancel{2}_1} + \frac{\overset{2}{\cancel{2x}}}{1} \times \frac{4}{\cancel{x}_1} = 2x$$

$$x(x-4) + 8 = 2x$$

$$x^2 - 4x + 8 = 2x$$

(This contains an x^2 term so it is a quadratic equation. Therefore we collect all terms on one side.)

Take $2x$ from each side $\qquad\qquad x^2 - 6x + 8 = 0$

$\therefore \qquad\qquad\qquad\qquad\qquad (x - 2)(x - 4) = 0$

$\therefore \qquad\qquad\qquad\qquad$ either $x - 2 = 0$ or $x - 4 = 0$

$\therefore \qquad\qquad\qquad\qquad\qquad\qquad x = 2$ or $\qquad x = 4$

Solve the equations in questions 23 to 42:

> **Tip** Start by multiplying both sides by the LCM of the denominators.

23 $\dfrac{x+5}{2} + \dfrac{1}{x} = 1$

24 $\dfrac{2}{x} + \dfrac{x+1}{3} = 2$

25 $\dfrac{x+2}{2} + \dfrac{2}{x} + 1 = 0$

26 $\dfrac{x+12}{3} + \dfrac{3}{x} = 2$

27 $\dfrac{x+9}{2} - \dfrac{2}{x} = 3$

28 $x + 8 + \dfrac{9}{x} = 2$

29 $\dfrac{2-x}{4} + \dfrac{1}{2x} = \dfrac{3}{4x}$

30 $\dfrac{1}{x} - \dfrac{2}{x-2} = 3$

31 $\dfrac{1}{3x} - \dfrac{x-2}{4} = \dfrac{1}{6}$

32 $\dfrac{2}{x+3} + x = 0$

Not all of the following questions give quadratic equations:

33 $\dfrac{1}{2x} + \dfrac{1}{4x} = \dfrac{1}{6}$

34 $\dfrac{x+2}{3} - \dfrac{3x}{2} = \dfrac{1}{5}$

35 $\dfrac{x-1}{2} + \dfrac{1}{x} = 1$

36 $\dfrac{2x-1}{3} - \dfrac{3x}{2} = 2$

37 $\dfrac{x+4}{3} + 2 = \dfrac{x}{4}$

38 $\dfrac{x+1}{x} + \dfrac{1}{2} = 4$

39 $\dfrac{x}{5} + \dfrac{1}{x+1} = 1$

40 $\dfrac{1}{2}(x-1) + \dfrac{1}{3}(2x-3) = 2$

41 $\dfrac{4x+1}{2} - \dfrac{1}{2x} = \dfrac{1}{2}$

42 $\dfrac{x-3}{x} = 0$

Mixed exercises

Exercise 16k

1 Simplify:

  a $\dfrac{ab^2}{2ab}$
  b $\dfrac{a(a+b)}{a+b}$
  c $\dfrac{a^2-b^2}{a+b}$

2 Simplify:

  a $\dfrac{1}{x} + \dfrac{1}{3x}$
  b $\dfrac{1}{x} \times \dfrac{1}{3x}$
  c $\dfrac{1}{x} \div \dfrac{1}{3x}$

3 Solve the equations:

  a $\dfrac{x+1}{3} - \dfrac{x-1}{2} = 3$
  b $\dfrac{x+1}{3} - \dfrac{1}{x} = 1$

4 **a** Simplify $\dfrac{1}{2}(x-1) + \dfrac{1}{3}(x-2)$

  b Solve the equation $\dfrac{1}{2}(x-1) + \dfrac{1}{3}(x-2) = \dfrac{1}{4}$

Exercise 16l

1 Simplify:

   a $\dfrac{6xy}{3y^2}$
      b $\dfrac{2x(x-y)}{4x^2}$
      c $\dfrac{x^2+4x+3}{x+1}$

2 Simplify:

   a $\dfrac{1}{2p}-\dfrac{1}{3p}$
      b $(x^2-2x)\div x$
      c $\dfrac{2y^2}{3}\times\dfrac{9}{4xy}$

3 Solve the equations:

   a $\dfrac{x}{4}+\dfrac{3x-2}{6}=\dfrac{1}{3}$
      b $\dfrac{x-5}{4}+\dfrac{1}{2x}=\dfrac{4}{x}$

4 a Simplify $\dfrac{x-2}{4}+\dfrac{3}{x}$

   b Solve the equation $\dfrac{x}{2}-\dfrac{2x-4}{3}=\dfrac{1}{4}$

Exercise 16m

1 Simplify:

   a $\dfrac{uv^2}{wu^2v}$
      b $\dfrac{a-b}{(b-a)(b-2a)}$
      c $\dfrac{x^2}{3x-x^2}$

2 Simplify:

   a $\dfrac{3s}{t}\div\dfrac{1}{6st}$
      b $\dfrac{x^2-4}{3}\times\dfrac{6}{x+2}$
      c $\dfrac{3}{4x-1}-\dfrac{2}{x}$

3 Solve the equations:

   a $\dfrac{x-2}{4}-\dfrac{x-3}{5}=\dfrac{3}{10}$
      b $\dfrac{x-2}{4}+\dfrac{1}{2x}=\dfrac{1}{4}$

4 a Simplify $\tfrac{1}{2}(x-2)-\tfrac{1}{3}(x-3)$

   b Solve the equation $\tfrac{1}{2}(x-2)-\tfrac{1}{3}(x-3)=5$

5 Solve the equations:

   a $\dfrac{x}{5}-\dfrac{x}{8}=3$
      b $\dfrac{x}{3}-\dfrac{x}{7}=4$
      c $\dfrac{x}{6}-\dfrac{x}{11}=5$

   What pattern do you notice in the above?

6 Solve for x:

$\dfrac{x}{a}-\dfrac{x}{b}=b-a$

MATHS IS OUT THERE

On what day of the week were you born? Can you answer this question? There is a formula which you may use to help you.

If d = day of the month, y = year, and m = month,

$$w = d + 2m + [3(m + 1)/5] + y + [y/4] - [y/100] + [y/400] + 2$$

January is taken as the 13th month of the previous year, and February as the 14th month. All other months are given their regular number.

The numbers in the 'square' brackets denote 'the greatest whole number less than the number'

e.g. $[7.6] = 7$; $[18.39] = 18$.

When you have a value for w, divide it by 7. Your remainder is the day of the week. Sunday is the first day and Saturday is day 0.

On what day of the week did New Year's Day fall in 1982?

$d = 1$, $m = 13$, $y = 1981$ (using the previous year for January) we have,

$$w = 1 + 2(13) + [3(14)/5] + 1981 + [1981/4] - [1981/100] + [1981/400] + 2$$
$$= 1 + 26 + [8.4] + 1981 + [495.25] - [19.81] + [4.9525] + 2$$
$$= 1 + 26 + 8 + 1981 + 495 - 19 + 4 + 2 = 2498$$

$2498/7 = 356$ remainder 6. The day of the week is therefore Friday.

Try the following exercises:

1. Find the day of the week on which you were born.
2. On what day of the week will your birthday fall in the year 2010?

IN THIS CHAPTER...

you have seen that:

- algebraic expressions can often be simplified by cancelling factors that are common to the numerator and denominator

- some expressions need factorising before a common factor becomes apparent

- two fractions can be multiplied together by multiplying their numerators and multiplying their denominators

- to divide by a fraction, turn it upside down and multiply

- to add or subtract algebraic fractions, express each fraction as an equivalent fraction with the LCM of all the denominators as the denominator

- to solve equations with fractions, multiply every term by the LCM of the denominators. The resulting equation should be of a type you are familiar with

17

QUADRATIC EQUATIONS 2

AT THE END OF THIS CHAPTER...

you should be able to:

1   Add a number to an expression of the form $ax^2 + bx$ to form a perfect square.

2   Solve a quadratic equation by completing the square

3   Recall and use the formula for solving a quadratic equation.

4   Find the roots of a quadratic equation from a graph.

MATHS IS OUT THERE

Around 400 BC, the ancient Babylonians and Chinese used completing the square to solve quadratic equations with positive roots, but did not have a general formula. Euclid produced a more abstract geometrical method around 300 BC.

BEFORE YOU START

you need to know:
- ✓ that expressions of the form $x^2 + 2ax + a^2$ can be expressed as $(x + a)^2$
- ✓ how to substitute numbers into a formula
- ✓ how to work with algebraic fractions
- ✓ how to plot a graph

KEY WORDS   perfect square, roots, surd

Completing the square

The numbers 9 and 25 can be written as 3^2 and 5^2, whereas 7 and 11 cannot be written as the square of another exact number. Because 9 and 25 can be written in this way they are called *perfect squares*.

Other examples are $\frac{9}{4}$ which is $(\frac{3}{2})^2$ and $\frac{121}{169}$ which is $(\frac{11}{13})^2$.

In a similar way, because we can write $x^2 + 2x + 1$ as $(x + 1)^2$ and $4x^2 + 12x + 9$ as $(2x + 3)^2$, we say that $x^2 + 2x + 1$ and $4x^2 + 12x + 9$ are perfect squares.

Exercise 17a

Express $x^2 + 12x + 36$ in the form $(x + a)^2$.

$$x^2 + 12x + 36 = (x + 6)(x + 6)$$
$$= (x + 6)^2$$

Express each of the following expressions in the form $(x + a)^2$.

1 $x^2 + 6x + 9$

2 $a^2 + 4a + 4$

3 $p^2 - 10p + 25$

4 $s^2 - 12s + 36$

5 $x^2 - 5x + \frac{25}{4}$

6 $b^2 + 3b + \frac{9}{4}$

7 $x^2 + 9x + \frac{81}{4}$

8 $a^2 - a + \frac{1}{4}$

9 $x^2 - \frac{1}{2}x + \frac{1}{16}$

10 $x^2 + 8x + 16$

11 $x^2 + x + \frac{1}{4}$

12 $x^2 + \frac{2}{3}x + \frac{1}{9}$

13 $p^2 + 18p + 81$

14 $a^2 - \frac{4}{5}a + \frac{4}{25}$

15 $t^2 - \frac{3}{2}t + \frac{9}{16}$

16 $x^2 + 2bx + b^2$

17 $x^2 - 2cx + c^2$

18 $x^2 + \frac{b}{a}x + \frac{b^2}{4a^2}$

Express $4x^2 + 12x + 9$ in the form $(ax + b)^2$

$$4x^2 + 12x + 9 = (2x + 3)(2x + 3)$$
$$= (2x + 3)^2$$

Express each of the following expressions in the form $(ax + b)^2$

19 $9x^2 + 18x + 9$

20 $4x^2 - 12x + 9$

21 $100x^2 - 60x + 9$

22 $9x^2 - 24x + 16$

23 $4x^2 - 4x + 1$

24 $25x^2 + 20x + 4$

25 $9x^2 - 6x + 1$

26 $4x^2 + 2x + \frac{1}{4}$

27 $\frac{9}{4}x^2 + 2x + \frac{4}{9}$

Forming a perfect square

To make $x^2 + 6x$ into a perfect square we must add 9 to it.

Then $\qquad x^2 + 6x + 9 = (x + 3)^2$

and to make $x^2 - 3x$ into a perfect square we must add $\frac{9}{4}$ to it.

Then $\qquad x^2 - 3x + \frac{9}{4} = (x - \frac{3}{2})^2$

More generally, to make $x^2 + px$ into a perfect square we must half the coefficient of x

$\left(\text{i.e. } \dfrac{p}{2}\right)$, square it $\left(\text{i.e. } \dfrac{p^2}{4}\right)$ and add this to $x^2 + px$.

Then $x^2 + px + \dfrac{p^2}{4}$ is a perfect square and can be written $(x + \frac{p}{2})^2$

Exercise 17b   What must be added to $x^2 + 6x$ to make it a perfect square?

The coefficient of x is 6

Half of this is 3

The square of 3 is 9

9 must be added to $x^2 + 6x$ to make it into a perfect square.

What must be added to each of the following expressions to make it a perfect square?

1   $x^2 + 4x$ 4 $p^2 - 14p$ 7 $c^2 + 7c$ 10 $x^2 + x$

2   $a^2 + 8a$ 5 $x^2 - 3x$ 8 $b^2 - \frac{1}{2}b$ 11 $x^2 + 2hx$

3   $x^2 - 12x$ 6 $x^2 + 20x$ 9 $a^2 - \frac{3}{2}a$ 12 $x^2 + \dfrac{b}{a}x$

Now consider $9x^2 + 12x$. If we want to make this into a perfect square we first take out the factor 9,

i.e. $9x^2 + 12x = 9(x^2 + \frac{12}{9}x)$

$$= 9(x^2 + \tfrac{4}{3}x)$$

To complete the square within the bracket find the coefficient of x (i.e. $\frac{4}{3}$), find half of it ($\frac{2}{3}$), square this value ($\frac{4}{9}$), and add it to the expression within the bracket.

Then $9(x^2 + \tfrac{4}{3}x + \tfrac{4}{9}) = 9x^2 + 12x + 4$

$$= (3x + 2)^2$$

which is in the form $(ax + b)^2$

Exercise 17c

What must be added to each of the following expressions to make it a perfect square?

1   $9x^2 + 12x$ 4 $100x^2 - 60x$ 7 $49a^2 - 28a$

2   $4x^2 + 12x$ 5 $25x^2 - 20x$ 8 $\dfrac{a^2}{4} - 2a$

3   $36a^2 + 60a$ 6 $4x^2 + 20x$ 9 $\frac{4}{9}a^2 - \frac{2}{3}a$

Quadratic equations

Now consider the equation $(x + 1)^2 = 4$

If we take the square root of each side we get

$$x + 1 = \pm\sqrt{4}$$
$$= \pm 2$$

If $x + 1 = 2$
$$x = 1$$

and if $x + 1 = -2$
$$x = -3$$

These two values of x satisfy the equation $(x + 1)^2 = 4$.

Exercise 17d

Solve the equations:

1 $(x + 1)^2 = 9$	**6** $(x - 1)^2 = 25$	**11** $(x + 3)^2 = 25$
2 $(x - 2)^2 = 16$	**7** $(x + 2)^2 = 49$	**12** $(x - 9)^2 = 36$
3 $(x - 3)^2 = 25$	**8** $(x - 5)^2 = 16$	**13** $(x + 1)^2 = \frac{1}{4}$
4 $(x + 6)^2 = 100$	**9** $(x - 7)^2 = 4$	**14** $(x - 2)^2 = \frac{9}{4}$
5 $(x + 7)^2 = 1$	**10** $(x + 4)^2 = 16$	**15** $(x - \frac{1}{2})^2 = \frac{25}{4}$

Solve the equation $(2x + 3)^2 = 4$

Taking the square root of each side we get

$$2x + 3 = \pm 2$$

i.e
$$2x + 3 = 2 \quad \text{or} \quad 2x + 3 = -2$$
$$2x = -1 \quad \text{or} \quad 2x = -5$$
$$x = -\tfrac{1}{2} \quad \text{or} \quad x = -\tfrac{5}{2}$$

Solve the equations:

16 $(2x - 1)^2 = 16$	**20** $(7x + 2)^2 = 100$	**24** $(4x - 3)^2 = 1$
17 $(3x + 2)^2 = 25$	**21** $(2x + 1)^2 = 36$	**25** $(9x - 5)^2 = 4$
18 $(5x - 1)^2 = 36$	**22** $(3x - 4)^2 = 49$	**26** $(5x + 3)^2 = 16$
19 $(3x - 4)^2 = 1$	**23** $(5x + 2)^2 = 25$	**27** $(7x - 5)^2 = 81$

Solution of quadratic equations by completing the square

In a previous chapter, we were able to solve some quadratic equations by factorising the left-hand side and using the fact that if $A \times B = 0$ then either $A = 0$ or $B = 0$ (or A and B are both zero).

There are other equations which cannot be used in this way because the left-hand side will not factorise. We can solve these equations by expressing them in the form $(x + a)^2 = c$ and taking the square root of both sides.

This is called the method of *completing the square*.

Consider the equation $x^2 - 6x + 2 = 0$

Subtract 2 from each side $x^2 - 6x = -2$

Complete the square on the LHS by adding 9 and add the same quantity to the RHS.

$$x^2 - 6x + 9 = -2 + 9$$

$$(x - 3)^2 = 7$$

Take the square root of each side $\quad x - 3 = \pm 2.646 \ldots$

$$x = 3 \pm 2.646 \ldots$$

$$= 5.646 \ldots \text{ or } 0.354 \ldots$$

i.e. $x = 5.65$ or 0.35 correct to 2 d.p.

These values of x are called the *roots* of the equations.

Exercise 17e   Solve the equation $x^2 + 5x - 3 = 0$ by completing the square.

$$x^2 + 5x - 3 = 0$$

$$x^2 + 5x = 3$$

$$x^2 + 5x + \tfrac{25}{4} = 3 + \tfrac{25}{4}$$

$$(x + \tfrac{5}{2})^2 = \frac{12 + 25}{4}$$

$$(x + \tfrac{5}{2})^2 = \frac{37}{4}$$

$$x + \tfrac{5}{2} = \pm \frac{\sqrt{37}}{2}$$

$$x = -\frac{5}{2} \pm \frac{6.083 \ldots}{2}$$

$$= \frac{1.083 \ldots}{2} \text{ or } \frac{-11.083 \ldots}{2}$$

$$= 0.541 \ldots \text{ or } -5.541 \ldots$$

i.e. $x = 0.54$ or -5.54 correct to 2 d.p.

Solve the following equations by completing the square.

1 $x^2 + 4x = 5$	**5** $x^2 - 4x + 1 = 0$	**9** $x^2 - 7x + 5 = 0$
2 $x^2 - 6x = 7$	**6** $x^2 + 8x = 3$	**10** $x^2 - x - 4 = 0$
3 $x^2 + 10x = 11$	**7** $x^2 - 4x = 9$	**11** $x^2 + 9x - 3 = 0$
4 $x^2 + 8x + 3 = 0$	**8** $x^2 + 9x + 4 = 0$	**12** $x^2 + 8x + 4 = 0$

Previously, when solving linear equations, we have frequently divided both sides by a non-zero number. When solving a quadratic equation by completing the square *always* divide both sides by the non-zero coefficient of x^2.

Solve the equation $2x^2 + 6x - 5 = 0$ by completing the square.

$$2x^2 + 6x - 5 = 0$$

$$x^2 + 3x - \frac{5}{2} = 0$$

$$x^2 + 3x = \frac{5}{2}$$

$$x^2 + 3x + \frac{9}{4} = \frac{5}{2} + \frac{9}{4}$$

$$(x + \tfrac{3}{2})^2 = \frac{10 + 9}{4}$$

$$(x + \tfrac{3}{2})^2 = \frac{19}{4}$$

$$x + \frac{3}{2} = \pm \frac{\sqrt{19}}{2}$$

$$x = -\frac{3}{2} \pm \frac{4.359\ldots}{2}$$

$$x = \frac{1.359\ldots}{2} \text{ or } \frac{-7.359\ldots}{2}$$

i.e. $x = 0.68$ or -3.68 correct to 2 d.p.

Solve the following equations by completing the square.

13 $2x^2 + 6x = 9$	**17** $3x^2 + 12x - 8 = 0$	**21** $4x^2 - 7x - 3 = 0$
14 $6x^2 - 12x = 5$	**18** $3x^2 - 5x = 1$	**22** $6x^2 - 5x - 1 = 0$
15 $4x^2 + 8x = 3$	**19** $5x^2 - 5x = 4$	**23** $7x^2 + 7x - 4 = 0$
16 $2x^2 - 3x - 4 = 0$	**20** $5x^2 + 8x + 2 = 0$	**24** $3x^2 - 9x = 2$

Puzzle

To prove that $2 = 3$

$$4 - 10 = 9 - 15$$

So $4 - 10 + \frac{25}{4} = 9 - 15 + \frac{25}{4}$

Expressing each side as a perfect square gives

$$(2 - \tfrac{5}{2})^2 = (3 - \tfrac{5}{2})^2$$

Taking the square root of each side

$$2 - \tfrac{5}{2} = 3 - \tfrac{5}{2}$$

Add $\tfrac{5}{2}$ to both sides

Therefore $2 = 3$. Where is the flaw in this argument?

Solution of quadratic equations by formula

If we apply the method of completing the square to the general quadratic equation $ax^2 + bx + c = 0$, where a, b and c are positive or negative numbers, we can establish a formula for solving the equation.

Consider the general equation $ax^2 + bx + c = 0$

Divide both sides by a
$$x^2 + \frac{b}{a}x + \frac{c}{a} = 0,$$

Subtract $\dfrac{c}{a}$ from each side
$$x^2 + \frac{b}{a}x = -\frac{c}{a}$$

Complete the square on the LHS and add the same quantity to the RHS
$$x^2 + \frac{b}{a}x + \frac{b^2}{4a^2} = -\frac{c}{a} + \frac{b^2}{4a^2}$$

Therefore
$$\left(x + \frac{b}{2a}\right)^2 = \frac{-4ac + b^2}{4a^2} = \frac{b^2 - 4ac}{4a^2}$$

Take square roots of each side
$$x + \frac{b}{2a} = \pm \frac{\sqrt{b^2 - 4ac}}{2a}$$

Subtract $\dfrac{b}{2a}$ from each side
$$x = -\frac{b}{2a} \pm \frac{\sqrt{b^2 - 4ac}}{2a}$$

i.e.
$$x = \frac{-b \pm \sqrt{b^2 - 4ac}}{2a}$$

This is called the *formula* for solving quadratic equations. It gives values of x, or roots of the equation, for any given values of a, b and c (provided that $b^2 - 4ac$ is not negative and, clearly, $a \neq 0$).

Remember that a is the coefficient of x^2

b is the coefficient of x

c is the constant number term

Since the two values of x are

$$-\frac{b}{2a} + \frac{\sqrt{b^2 - 4ac}}{2a} \quad \text{and} \quad -\frac{b}{2a} - \frac{\sqrt{b^2 - 4ac}}{2a}$$

the sum of the two roots is always $\left(-\dfrac{b}{2a}\right) + \left(-\dfrac{b}{2a}\right) = -\dfrac{b}{a}$

This provides a useful check that your answers are correct.

Exercise 17f Use the formula to solve the equation $x^2 - 9x - 2 = 0$ giving your answers correct to 2 decimal places.

$$x^2 - 9x - 2 = 0$$

$$a = 1, \ b = -9, \ c = -2$$

$$x = \frac{-b \pm \sqrt{b^2 - 4ac}}{2a}$$

$$= \frac{-(-9) \pm \sqrt{(-9)^2 - 4(1)(-2)}}{2 \times 1}$$

$$= \frac{9 \pm \sqrt{81 + 8}}{2} = \frac{9 \pm \sqrt{89}}{2}$$

$$= \frac{9 \pm 9.433\ldots}{2}$$

$$= \frac{18.433\ldots}{2} \quad \text{or} \quad \frac{-0.433\ldots}{2}$$

$$= 9.216\ldots \quad \text{or} \quad -0.216\ldots$$

$$\therefore \quad x = 9.22 \quad \text{or} \quad -0.22 \quad \text{(correct to 2 d.p.)}$$

Check: Sum of roots is $9.22 + (-0.22) = 9$

and $\dfrac{-b}{a} = \dfrac{-(-9)}{1} = 9$ confirming the results.

 Tip Note that if we leave the answers as

$$x = \frac{9 + \sqrt{89}}{2} \quad \text{and} \quad x = \frac{9 - \sqrt{89}}{2}$$

they are *exact* answers in square root form. Answers that are given correct to a given number of decimal places are approximate answers.

Use the formula to solve the following equations, giving answers correct to 2 decimal places.

1 $x^2 + 6x + 3 = 0$ **8** $x^2 + 10x - 15 = 0$ **15** $x^2 - 6x + 6 = 0$

2 $x^2 + 7x + 4 = 0$ **9** $x^2 + 6x - 6 = 0$ **16** $x^2 - 5x - 5 = 0$

3 $x^2 + 5x + 5 = 0$ **10** $x^2 + 9x - 1 = 0$ **17** $x^2 - 5x + 2 = 0$

4 $x^2 + 7x - 2 = 0$ **11** $x^2 + 3x - 5 = 0$ **18** $x^2 - 3x + 1 = 0$

5 $x^2 + 4x - 3 = 0$ **12** $x^2 + 4x - 7 = 0$ **19** $x^2 - 9x - 2 = 0$

6 $x^2 + 9x + 12 = 0$ **13** $x^2 - 4x + 2 = 0$ **20** $x^2 - 4x - 9 = 0$

7 $x^2 + 8x + 13 = 0$ **14** $x^2 - 7x + 3 = 0$ **21** $x^2 + 7x - 2 = 0$

For questions **22** to **25** give your answers in square root form.

22 $x^2 - 4x - 3 = 0$ **24** $x^2 + 8x + 5 = 0$

23 $x^2 - 7x - 3 = 0$ **25** $x^2 - 2x - 4 = 0$

Solve the equation $3x^2 + 7x - 2 = 0$ giving your answer correct to 2 decimal places.

$$3x^2 + 7x - 2 = 0$$
$$a = 3, \quad b = 7, \quad c = -2$$

$$x = \frac{-b \pm \sqrt{b^2 - 4ac}}{2a}$$

$$= \frac{-7 \pm \sqrt{7^2 - 4(3)(-2)}}{2 \times 3}$$

$$= \frac{-7 \pm \sqrt{49 + 24}}{6}$$

$$= \frac{-7 \pm \sqrt{73}}{6}$$

$$= \frac{-7 \pm 8.544\ldots}{6}$$

$$= \frac{1.544\ldots}{6} \quad \text{or} \quad \frac{-15.544\ldots}{6}$$

$$= 0.257\ldots \quad \text{or} \quad -2.590\ldots$$

$$= 0.26 \quad \text{or} \quad -2.59 \quad \text{(correct to 2 d.p.)}$$

Check: Sum of roots is $0.26 + (-2.59) = -2.33$

and $\dfrac{-b}{a} = \dfrac{-7}{3} = -2.33$ (correct to 2 d.p.) confirming the results.

Solve the equations, giving answers correct to 2 decimal places.

26 $2x^2 + 7x + 2 = 0$

27 $2x^2 + 7x + 4 = 0$

28 $5x^2 + 9x + 2 = 0$

29 $2x^2 - 7x + 4 = 0$

30 $4.2x^2 - 7.5x + 1 = 0$

31 $3x^2 + 7x + 3 = 0$

32 $4x^2 + 7x + 1 = 0$

33 $5x^2 - 9x + 2 = 0$

34 $3x^2 + 5x - 3 = 0$

35 $3.7x^2 + 8.5x - 2 = 0$

36 $5a^2 - 2a - 1 = 0$

37 $40b^2 + 90b + 21 = 0$

38 $2S^2 - 4S - 1 = 0$

39 $\frac{1}{2}x^2 - \frac{1}{3}x - 1 = 0$

40 $0.7R^2 + 1.2R + 0.3 = 0$

41 $10n^2 + 50n - 3 = 0$

42 $500p^2 + 75p - 23 = 0$

43 $\frac{2}{3}x^2 + \frac{3}{4}x + \frac{1}{9} = 0$

44 $0.2A^2 + A + 0.5 = 0$

45 $\frac{1}{2}N^2 + 3N - \frac{1}{8} = 0$

46 Use the formula to solve the equation $4x^2 - 12x + 9 = 0$.
Comment on the number of different solutions and on the value of $b^2 - 4ac$.

47 Repeat question **46** for the equation

a $25x^2 + 10x + 1 = 0$ **b** $9x^2 - 42x + 29 = 0$.

How can you test whether or not a quadratic equation has two identical solutions?

48 **a** Use the formula to solve the equation $2x^2 + 9x - 5 = 0$.

What do you notice about the value of $b^2 - 4ac$?

b Solve the equation $2x^2 + 9x - 5 = 0$ by factorising. What can you conclude about an equation if the value of $b^2 - 4ac$ is a perfect square?

49 **a** Use the formula to try to solve the equation $x^2 + 2x + 3 = 0$.

b What is the relationship between the result in part **a** and the value of $b^2 - 4ac$?

c Use the results of this question and questions **46** to **48** to deduce a relationship between the number of solutions to a quadratic equation and the value of $b^2 - 4ac$.

Quadratic equations do not always arise arranged in the order $ax^2 + bx + c = 0$. They should, however, be arranged in this order before the formula is used.

Exercise **17g**

Solve the equation $4x^2 = 7x + 1$ giving your answers correct to 2 decimal places.

$$4x^2 = 7x + 1$$

Tip First arrange the equation in the form $ax^2 + bx + c = 0$

$$4x^2 - 7x - 1 = 0$$

$$a = 4, \ b = -7, \ c = -1$$

$$x = \frac{-b \pm \sqrt{b^2 - 4ac}}{2a}$$

$$= \frac{-(-7) \pm \sqrt{(-7)^2 - 4(4)(-1)}}{2 \times 4}$$

$$= \frac{7 \pm \sqrt{49 + 16}}{8} = \frac{7 \pm \sqrt{65}}{8}$$

$$= \frac{7 \pm 8.062\ldots}{8} = \frac{15.062\ldots}{8} \quad \text{or} \quad \frac{-1.062\ldots}{8}$$

$$= 1.882\ldots \quad \text{or} \quad -0.132\ldots$$

$$= 1.88 \quad \text{or} \quad -0.13 \quad \text{(correct to 2 d.p.)}$$

Check: Sum of roots is $1.88 + (-0.13) = 1.75$

and $\dfrac{-b}{a} = \dfrac{-(-7)}{4} = \dfrac{7}{4} = 1.75$

confirming the results.

Solve the following equations giving your answers correct to 2 decimal places.

1	$2x^2 = 8x + 11$	7	$2x^2 = 3x + 1$
2	$4x^2 = 8x + 3$	8	$4x^2 = 5 - 3x$
3	$3x^2 = 3 - 5x$	9	$3x^2 + 2 = 9x$
4	$5x^2 = x + 3$	10	$6x^2 - 9x = 4$
5	$4x^2 + 2 = 7x$	11	$2x^2 = 5x + 5$
6	$3x^2 = 12x + 2$	12	$3x^2 + 4x = 1$

13	$4x^2 = 4x + 1$
14	$3x^2 + 7x = 2$
15	$5x^2 = 5x - 1$
16	$8x^2 = x + 1$
17	$3.2x^2 = 5.4x + 1.2$
18	$7.6x^2 = 3.4x + 0.6$

In the next exercise, some of the equations can be solved by factorisation. Try this method first. If factors cannot be found, use the formula.

Exercise 17h

For questions **1** to **9** solve the equations, giving answers correct to 2 decimal places when necessary.

1	$2x^2 + 3x - 2 = 0$	4	$2x^2 + 3x - 3 = 0$	7	$2x^2 - 3x - 3 = 0$
2	$3x^2 + 6x + 2 = 0$	5	$3x^2 - 8x + 2 = 0$	8	$8x^2 + 10x - 3 = 10$
3	$6x^2 + 7x + 2 = 0$	6	$3x^2 - 8x - 3 = 0$	9	$6x^2 + 7x - 2 = 0$

For questions **10** to **18** solve the equations, giving answers exactly, in surd form when necessary.

10	$7x^2 + 8x - 2 = 0$	14	$20x^2 = 3 - 11x$
11	$5x^2 - 3x - 1 = 0$	15	$3x^2 - 14x + 15 = 0$
12	$3x^2 = 7x - 2$	16	$5x^2 + 8x + 2 = 0$
13	$11x^2 + 12x + 3 = 0$	17	$2x^2 = 7x + 3$
		18	$2x^2 + 9x = 5$

> **Tip** A surd is an expression such as $\sqrt{2}$, $\sqrt{19}$.

Harder expansions and equations

Solving some problems leads to equations of the form

$$(x + 5)\left(\frac{4}{x} - 1\right) = 7$$

To solve such equations we must first expand the left-hand side.

Exercise 17i

Expand $(x + 2)\left(\dfrac{10}{x} - 3\right)$

$$(x + 2)\left(\frac{10}{x} - 3\right) = x \times \frac{10}{x} + x \times (-3) + 2 \times \frac{10}{x} + 2 \times (-3)$$

$$= 10 - 3x + \frac{20}{x} - 6 = 4 - 3x + \frac{20}{x}$$

Expand

1 $(x+5)\left(\dfrac{6}{x}+1\right)$

4 $(x-7)\left(3-\dfrac{5}{x}\right)$

7 $\left(\dfrac{4}{x}-3\right)(5x+1)$

2 $(x+3)\left(\dfrac{10}{x}-3\right)$

5 $(x-4)\left(\dfrac{12}{x}-5\right)$

8 $\left(5-\dfrac{6}{x}\right)(3+2x)$

3 $(x-4)\left(\dfrac{3}{x}+5\right)$

6 $\left(\dfrac{5}{x}-2\right)(2x+1)$

9 $\left(3-\dfrac{2}{x}\right)(x-3)$

Solve the equation $(x-7)\left(\dfrac{20}{x}+3\right)=100$

$(x-7)\left(\dfrac{20}{x}+3\right)=100$ Expand the brackets.

$\cancel{x}\times\dfrac{20}{\cancel{x}_1} + x\times 3 - 7\times\dfrac{20}{x} - 7\times 3 = 100$

$20+3x-\dfrac{140}{x}-21=100$ Collect like terms on the LHS.

$3x-\dfrac{140}{x}-1=100$ Subtract 100 from both sides.

$3x-\dfrac{140}{x}-101=0$ Multiply every term by x.

$3x^2-140-101x=0$ Arrange in the form $ax^2+bx+c=0$

$3x^2-101x-140=0$ Factorise $\Rightarrow$ $(3x+4)(x-35)=0$

Either $3x+4=0$ or $x-35=0$ $\Rightarrow$ $3x=-4$ or $x=35$

i.e. $x=-\dfrac{4}{3}$ or 35

Solve the equations.

10 $(x-5)\left(\dfrac{8}{x}-9\right)=0$

11 $(x+5)\left(\dfrac{14}{x}-5\right)=0$

> **Tip** Expand the brackets then multiply each term by the appropriate expression to get rid of any fractions.

12 $(x+5)\left(\dfrac{6}{x}+1\right)=24$

15 $(2x+7)\left(\dfrac{20}{x}+9\right)=0$

13 $(x+4)\left(\dfrac{120}{x}-3\right)=144$

16 $(x-4)\left(\dfrac{12}{x}-5\right)=96$

14 $(x+7)\left(\dfrac{12}{x}-5\right)=0$

17 $(x-3)\left(\dfrac{12}{x}-7\right)=70$

a Rewrite the equation $\dfrac{6}{x}+\dfrac{3}{x+1}=4$ in the form $ax^2+bx+c=0$

b Hence solve the equation $\dfrac{6}{x}+\dfrac{3}{x+1}=4$

a $\qquad\qquad \dfrac{6}{x}+\dfrac{3}{x+1}=4$ Multiply each term by x and by $x+1$

$\dfrac{6}{\cancel{x}}\times\cancel{x}\times(x+1)+\dfrac{3}{\cancel{x+1}}\times x\times\cancel{(x+1)}=4\times x\times(x+1)$

$\Rightarrow \qquad\qquad 6x+6+3x=4x^2+4x$

$\qquad\qquad\qquad 9x+6=4x^2+4x$

$\qquad\qquad\qquad 0=4x^2-5x-6$

$\qquad\qquad 4x^2-5x-6=0$

b Factorising $\quad 4x^2-5x-6=0$ gives $(4x+3)(x-2)=0$

Either $4x+3=0$ or $x-2=0$

$4x=-3$ or $x=2$ $\Rightarrow$ $x=-\frac{3}{4}$ or 2

In questions **18** to **23**

a Express each equation in the form $ax^2+bx+c=0$

b Solve the given equation.

18 $x-\dfrac{12}{x}=1$

21 $x+\dfrac{2}{x}=7$

> **Tip** Start by getting rid of fractions.

19 $x+\dfrac{2}{x}=11$

22 $\dfrac{2}{x}+\dfrac{1}{x+1}=4$

20 $x-\dfrac{5}{x}=3$

23 $\dfrac{5}{x}-2x=5$

Solve the equation $\dfrac{1}{x+1} + \dfrac{2}{x-3} = 4$ giving your answers

correct to 2 decimal places.

$$\frac{1}{x+1} + \frac{2}{x-3} = 4$$

Multiply both sides by $(x+1)(x-3)$

$$\frac{(x+1)(x-3)}{x+1} + \frac{2(x+1)(x-3)}{x-3} = 4(x+1)(x-3)$$

$$(x-3) + 2(x+1) = 4(x+1)(x-3)$$

$$x - 3 + 2x + 2 = 4(x^2 - 2x - 3)$$

$$3x - 1 = 4x^2 - 8x - 12$$

i.e. $\quad 4x^2 - 11x - 11 = 0$

$$a = 4, \quad b = -11, \quad c = -11$$

$$x = \frac{-b \pm \sqrt{b^2 - 4ac}}{2a}$$

$$= \frac{11 \pm \sqrt{121 + 176}}{8} = \frac{11 \pm \sqrt{297}}{8}$$

$$= \frac{11 \pm 17.233\ldots}{8} = \frac{28.233\ldots}{8} \quad \text{or} \quad \frac{-6.233\ldots}{8}$$

$$= 3.529\ldots \quad \text{or} \quad -0.779\ldots$$

i.e. $\quad x = 3.53 \quad \text{or} \quad -0.78 \quad$ (correct to 2 d.p.)

Solve the following equations; give answers correct to 2 decimal places where necessary.

24 $\dfrac{3}{x-1} + \dfrac{2}{x+2} = \dfrac{7}{2}$

28 $\dfrac{4}{x-4} + \dfrac{4}{x-2} = 3$

32 $\dfrac{3}{x+2} - \dfrac{1}{x+4} = 2$

25 $\dfrac{4}{x} + \dfrac{5}{x+2} = 1$

29 $\dfrac{1}{x-3} - \dfrac{1}{x-1} = \dfrac{1}{4}$

33 $\dfrac{2}{x+5} + \dfrac{3}{x-2} = 4$

26 $\dfrac{2}{x-3} + \dfrac{4}{x+6} = 1$

30 $\dfrac{4}{x+1} + \dfrac{2}{3x-2} = 1$

34 $\dfrac{3}{x-1} - \dfrac{2}{x+3} = 1$

27 $\dfrac{3}{x-1} + \dfrac{2x}{x-2} = 0$

31 $\dfrac{2x}{x+20} - \dfrac{1}{x+1} = 0$

35 $\dfrac{1}{x-3} + \dfrac{5}{x+3} = 1$

Problems

A rectangular lawn measuring 20 m by 15 m is surrounded by a path x m wide. If the area of the path is 74 m² form an equation in x and solve it to find the width of the path.

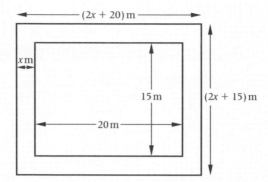

The lawn and path together form a rectangle measuring $(2x + 20)$ m by $(2x + 15)$ m

∴ area of lawn and path together $= (2x + 20)(2x + 15)$ m²

while the area of the lawn is 20×15 m² $= 300$ m²

∴ area of path is $(2x + 20)(2x + 15) - 300$ m²

But the area of the path is given as 74 m².

Equating the two values for the area of the path gives the equation

$$(2x + 20)(2x + 15) - 300 = 74$$

i.e. $\quad 4x^2 + 30x + 40x + 300 - 300 = 74$

$$4x^2 + 70x - 74 = 0 \qquad \text{Divide each term by 2.}$$

$$2x^2 + 35x - 37 = 0$$

$$(2x + 37)(x - 1) = 0$$

i.e.   either $2x + 37 = 0$ or $x - 1 = 0$

$$2x = -37 \quad \text{or} \quad x = 1$$

$$x = -\tfrac{37}{2} \quad \text{or} \quad 1$$

Since the width of the path must be positive the only acceptable solution is $x = 1$

∴ the width of the path is 1 m

> **Tip** Check by reading the question and seeing if a path of width 1 m fits the information given.

The following questions may lead to quadratic equations that do not factorise but always check first whether a quadratic equation will factorise before using the formula. If the answer is not rational give it correct to 3 significant figures.

1. The sum of two numbers is 10 and the sum of their squares is 80. Find them.

2. The sum of two numbers is 9 and the difference between their squares is 60. Find them.

3. Find a number such that the sum of the number and its reciprocal is 20. In this case give the answers correct to 2 decimal places.

4. One side of a rectangle is 3 cm longer than another. Find the sides if the area of the rectangle is 20 cm².

5. The parallel sides of a trapezium are $(x - 2)$ cm and $(x + 4)$ cm long. If the distance between the parallel sides is x cm and the area of the trapezium is 42 cm² find its dimensions.

6. A rectangular block is 2 cm wider than it is high and twice as long as it is wide. If its total surface area is 190 cm² find its dimensions.

7. Sally is x years old. Her mother's age is $(x^2 - 4)$ years and her father is 6 years older than her mother. If the combined age of all three is 76 years, form an equation in x and solve it. How old is her father?

8.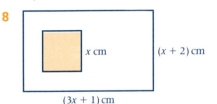

   A square of side x cm is removed from a rectangular piece of cardboard measuring $(3x + 1)$ cm by $(x + 2)$ cm. If the area of card remaining is 62 cm² form an equation in x and solve it to find the dimensions of the original card.

9. Mrs Brown bought x grapefruit for $4.20. If the price of each grapefruit is increased by 5 c she can buy 2 fewer grapefruit for the same money.

   a. Express the price, in cents, of one grapefruit in terms of x.

   b. Use your answer to part a to write down the price, in terms of x, of a grapefruit at the increased price.

   c. Hence show that x satisfies the equation $\left(\dfrac{420}{x} + 5\right)(x - 2) = 420$

   d. Solve the equation to find the number of grapefruit bought and the price of each one.

10. Alan bought x copies of a reference book for $240. If he had waited for the summer sale, when the books were reduced by $2 each, he could have bought 4 more books for the same amount.

    a. Express the original price of a book in terms of x.

    b. Express the sale price of a book in terms of x.

    c. Express the number of books that could be bought at the sale price in terms of x.

    d. Use your answers to parts b and c to form the equation $(x + 4)\left(\dfrac{240}{x} - 2\right) = 240$

    e. Solve the equation given in part d. How many books did Alan buy and how much did each one cost?

A coach is due to reach its destination 30 kilometres away at a certain time. Its start is delayed by 18 minutes, but by increasing the average speed by 5 km/h the driver arrives on time. How long did the journey actually take? What was the intended average speed?

Let the intended average speed be x km/h.

The information can then be set out in table form taking care to work in compatible units.

	Speed in km/h	Distance in km	Time in hours
Intended journey	x	30	$\dfrac{30}{x}$
Actual journey	$x + 5$	30	$\dfrac{30}{x+5}$

Since the actual time is 18 minutes, i.e. $\frac{3}{10}$ hour, shorter than the intended time, then

$$\frac{30}{x} - \frac{30}{x+5} = \frac{3}{10}$$

Multiply both sides by $10x(x + 5)$

$$\frac{30 \times 10\!\!\not{x}(x+5)}{\not{x}} - \frac{30 \times 10x\cancel{(x+5)}}{\cancel{x+5}} = \frac{3 \times 10\!\!\not{x}(x+5)}{\cancel{10}}$$

$$300(x + 5) - 300x = 3x(x + 5)$$
$$100(x + 5) - 100x = x^2 + 5x$$
$$100x + 500 - 100x = x^2 + 5x$$
$$0 = x^2 + 5x - 500$$

i.e. $\qquad x^2 + 5x - 500 = 0$

$$(x + 25)(x - 20) = 0$$

$\therefore \qquad\qquad\qquad x = -25 \quad \text{or} \quad 20$

But -25 is unacceptable as the average speed has to be positive.

$\therefore \qquad\qquad\qquad x = 20$

i.e. the intended speed is 20 km/h and the time actually taken is

$\dfrac{30}{20 + 5}$ hours $= \dfrac{30}{25}$ hours i.e. 1 hour 12 minutes.

---

11 When its average speed increases by 10 m.p.h. the time taken for a car to make a journey of 105 miles is reduced by 15 minutes. Find the original average speed of the car.

12 Find the price of potatoes per kilogram if, when the price rises by 5 c per kg, I can buy 1 kg less for $2.10.

13 Tickets are available for a concert at two prices, the dearer ticket being $3 more than the cheaper one. Find the price of each ticket if a youth group can buy ten more of the cheaper tickets than the dearer tickets for $180.

14 From a piece of wire 42 cm long, a length 10x cm is cut off and bent into a rectangle whose length is one and a half times its width. The remainder is bent to form a square. If the combined area of the rectangle and square is 63 cm² find their dimensions.

15 The members of a club hire a coach for the day at a cost of $210. Seven members withdraw which means that each member who makes the trip must pay an extra $1. How many members originally agreed to go?

16 Find the original price of oranges if, when the price of each orange rises by 4 c, Jean can buy 5 fewer oranges for $6.

17 Two footpaths connect Antley and Berry. The distance between the two villages along one footpath is 15 miles while the distance between them along the other path is 18 miles. Len takes the shorter route and, while walking at x m.p.h., takes $\frac{1}{2}$ hour longer than Mandy. Mandy takes the longer route and walks at 1 m.p.h. faster than Len. Find the speed at which each walks and the time each of them takes.

18 George walks at x km/h, which is 2 km per hour faster than Liam, and in consequence takes 40 minutes less to walk 8 km. Find the speed at which each walks.

Solving quadratic equations graphically

We saw in Chapter 10 that, if we draw the graph of $y = ax^2 + bx + c$, we can read the values of x where the curve crosses the x-axis. The curve crosses the x-axis where $y = 0$, i.e. where $ax^2 + bx + c = 0$, so those values of x are the solutions of the quadratic equation $ax^2 + bx + c = 0$

Therefore we can solve a quadratic equation, such as $5x^2 - 2x - 8 = 0$, by first drawing the graph of $y - 5x^2 - 2x - 8$.

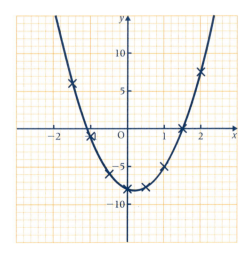

As accurately as we can read from the graph, the curve cuts the x-axis where $x = -1.1$ and $x = 1.5$.

Therefore the roots of the equation $5x^2 - 2x - 8 = 0$ are -1.1 and 1.5.

(The roots of an equation are the values of x that satisfy the equation.)

Note that the accuracy of graphical solutions depends on how accurately the graph is drawn and on the scale used. The larger the scale, the more accurate the solutions.

Exercise 17k

Draw graphs to find the roots of the following equations.

Use scales of 1 cm $\equiv$ 1 unit on both axes for values of x between -4 and 4.

Give answers to 1 decimal place.

1 $x^2 - x - 1 = 0$ **2** $2x^2 + 3x - 3 = 0$ **3** $4 - 2x - x^2 = 0$ **4** $x(2x - 3) = 4$

Investigation

a Use the formula to try to find solutions to the equation $x^2 - 4x + 8 = 0$.

Explain why the method breaks down.

What can you deduce about the equation $x^2 - 4x + 8 = 0$?

b Is there another method, not using the formula, from which you can reach the same deduction about the equation $x^2 - 4x + 8 = 0$?

c Repeat parts **a** and **b** for the equation $2x^2 + 3x + 4 = 0$.

d Make a deduction about any quadratic equation of the form $ax^2 + bx + c = 0$ when a is positive and c is negative.

e Find a simple test that will show whether a quadratic equation has two different roots or the same root twice (these are called repeated roots) or no real roots.

The first known use of an algebraic formula, giving negative and positive solutions to a quadratic equation was by Brahmagupta, who lived in India in the 7th century.

IN THIS CHAPTER...

you have seen that:

- you can add a number to $x^2 + bx$ to make a perfect square; the number is $\frac{b^2}{4}$, giving $x^2 + bx + \frac{b^2}{4} = \left(x + \frac{b}{2}\right)^2$

- the solutions of the equation $ax^2 + bx + c = 0$ are given by the formula

$$x = \frac{-b \pm \sqrt{b^2 - 4ac}}{2a}$$

- you can solve the equation $ax^2 + bx + c = 0$ by drawing the graph of $y = ax^2 + bx + c$ and finding where the curve cuts the x-axis

18 POLYGONS

AT THE END OF THIS CHAPTER...

you should be able to:

1   Classify polygons in terms of their number of sides.

2   Identify regular polygons.

3   State the sum of the exterior or interior angles of a given polygon with n sides
    as $360°$ or $(n - 2)180°$
    respectively.

4   Calculate the size of an exterior or interior angle of a regular polygon.

5   Use the formula for the sum of the exterior angles of a polygon and that for interior to solve problems.

6   Make patterns using regular polygons that tessellate.

Lance: What do polygons sing?

Jo: Go on, tell me.

Lance: Polly Wolly Doodle all the day.

BEFORE YOU START

you need to know:
✓   about triangles and their angle properties
✓   about quadrilaterals and their angle properties
✓   properties of the special quadrilaterals

KEY WORDS

dodecahedron, equilateral, exterior angle, hexagon, interior angle, isosceles, octagon, parallelogram, pentagon, polygon, polyhedra, quadrilateral, rectangle, regular, rhombus, square, tessellate, triangle

Polygons

A polygon is a plane (flat) figure formed by three or more points joined by line segments. The points are called vertices (singular-'vertex'). The line segments are called sides.

This is a nine-sided polygon.

Some polygons have names which you already know:

a three-sided polygon is a triangle

a four-sided polygon is a quadrilateral

a five-sided polygon is a pentagon

a six-sided polygon is a hexagon

an eight-sided polygon is an octagon

Regular polygons

A polygon is called regular when all its sides are the same length *and* all its angles are the same size. The polygons below are all regular:

Exercise *18a*

State which of the following figures are regular polygons. Give a brief reason for your answer:

1 Rhombus

2 Square

3 Rectangle

4 Parallelogram

5 Isosceles triangle

6 Right-angled triangle

7 Equilateral triangle

8 Circle

Make a rough sketch of each of the following polygons. (Unless you are told that a polygon is regular, you must assume that it is *not* regular.)

9 A regular quadrilateral

10 A hexagon

11 A triangle

12 A regular triangle

13 A regular hexagon

14 A pentagon

15 A quadrilateral

16 A ten-sided polygon

When the vertices of a polygon all point outwards, the polygon is convex.

Sometimes one or more of the vertices point inwards, in which case the polygon is concave.

convex polygon

concave polygon

In this chapter we consider only convex polygons.

Interior angles

The angles enclosed by the sides of a polygon are the *interior angles*. For example,

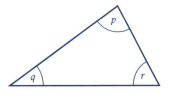

p, *q* and *r* are the interior angles of the triangle,

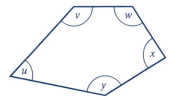

u, *v*, *w*, *x* and *y* are the interior angles of the pentagon.

The exterior angles

If we produce (extend) one side of a polygon, an angle is formed outside the polygon. It is called an *exterior angle*.

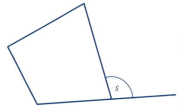

For example, *s* is an exterior angle of the quadrilateral.

If we produce all the sides in order we have all the exterior angles.

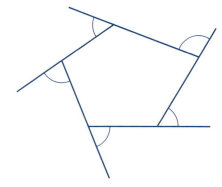

Exercise 18b

1 What is the sum of the interior angles of any triangle?

2 What is the sum of the interior angles of any quadrilateral?

3 In triangle ABC, find

    a the size of each marked angle

    b the sum of the exterior angles.

4 ABCD is a parallelogram. Find

    a the size of each marked angle

    b the sum of the exterior angles.

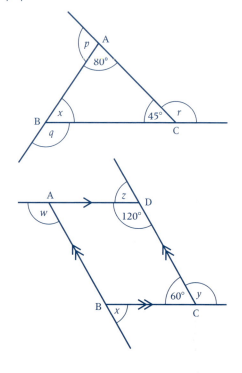

5 In a triangle ABC, write down the value of

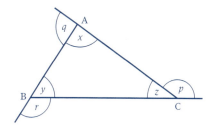

   a $x + q$

   b the sum of all six marked angles

   c the sum of the interior angles

   d the sum of the exterior angles.

6 Draw a pentagon. Produce the sides in order to form the five exterior angles. Measure each exterior angle and then find their sum.

7 Construct a regular hexagon of side 5 cm. (Start with a circle of radius 5 cm and then with your compasses still open to a radius of 5 cm, mark off points on the circumference in turn.) Produce each side of the hexagon in turn to form the six exterior angles.

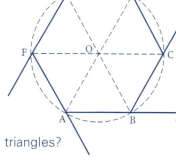

If O is the centre of the circle, joining O to each vertex forms six triangles:

   a What kind of triangle is each of these triangles?

   b What is the size of each interior angle in these triangles?

   c Write down the value of $A\hat{B}C$.

   d Write down the value of $C\hat{B}G$.

   e Write down the value of the sum of the six exterior angles of the hexagon.

The sum of the exterior angles of a polygon

In the last exercise, we found that the sum of the exterior angles is 360° in each case. This is true of any polygon, whatever its shape or size.

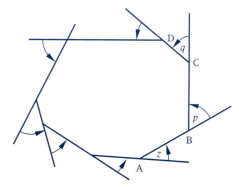

Consider walking round this polygon. Start at A and walk along AB. When you get to B you have to turn through angle p to walk along BC. When you get to C you have to turn through angle *q* to walk along CD, . . . and so on until you return to A. If you then turn through angle *z* you are facing in the direction AB again. You have now turned through each exterior angle and have made just one complete turn, i.e.

the sum of the exterior angles of a polygon is 360°.

Exercise **18c**   Find the size of the angle marked *p*.

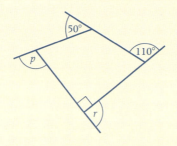

$$p + r + 110° + 50° = 360° \quad \text{(sum of exterior angles of a polygon)}$$

but $\qquad\qquad\qquad r = 90° \quad$ (angles on a straight line)

∴ $\qquad\qquad\qquad p = 360° - 90° - 110° - 50°$

$$p = 110°$$

In each case find the size of the angle marked *p*:

1

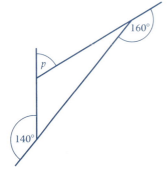

3

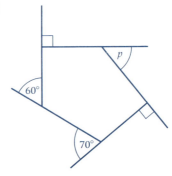

2

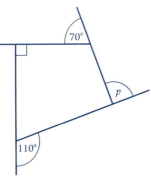

4

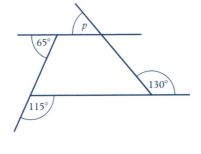

5

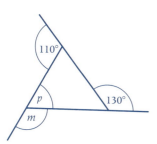

8

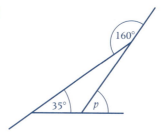

6

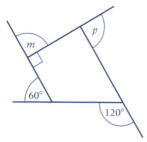

9

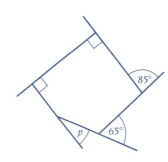

7

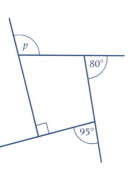

10

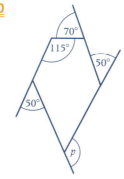

In questions **11** and **12** find the value of x.

11

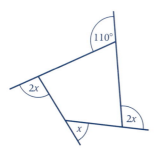

12

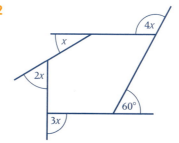

13 The exterior angles of a hexagon are x, $2x$, $3x$, $4x$, $3x$ and $2x$. Find the value of x.

14 Find the number of sides of a polygon if each exterior angle is

   a 72°   **b** 45°.

The exterior angle of a regular polygon

If a polygon is regular, all its exterior angles are the same size. We know that the sum of the exterior angles is 360°, so the size of one exterior angle is easily found; we just divide 360° by the number of sides of the polygon, i.e.

in a *regular* polygon with n sides, the size of an exterior angle is $\dfrac{360°}{n}$

Exercise 18d   Find the size of each exterior angle of a 24-sided regular polygon.

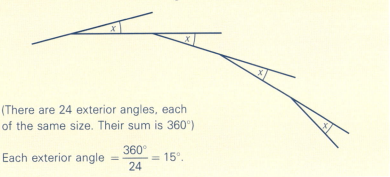

(There are 24 exterior angles, each of the same size. Their sum is 360°)

Each exterior angle $= \dfrac{360°}{24} = 15°$.

Find the size of each exterior angle of a regular polygon with:

1	10 sides	**4**	6 sides	**7**	9 sides
2	8 sides	**5**	15 sides	**8**	16 sides
3	12 sides	**6**	18 sides	**9**	20 sides

The sum of the interior angles of a polygon

Consider an octagon:

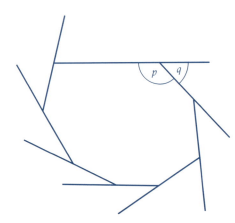

At each vertex there is an interior angle and an exterior angle and the sum of these two angles is 180° (angles on a straight line), i.e. $p + q = 180°$ at each one of the eight vertices.

Therefore, the sum of the interior angles and exterior angles together is

$$8 \times 180° = 1440°$$

The sum of the eight exterior angles is $360°$.

Therefore, the sum of the interior angles is

$$1440° - 360° = 1080°$$

Exercise 18e   Find the sum of the interior angles of a 14-sided polygon.

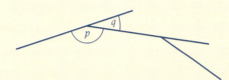

At each vertex $p + q = 180°$

If there are 14 sides there are 14 pairs of exterior and interior angles.

∴ sum of interior angles and exterior angles is

$$14 \times 180° = 2520°$$

∴ sum of interior angles $= 2520° - 360°$

$$= 2160°$$

Find the sum of the interior angles of a polygon with:

1   6 sides   **4**   4 sides   **<u>7</u>**   18 sides

2   5 sides   **5**   7 sides   **<u>8</u>**   9 sides

3   10 sides   **6**   12 sides   **<u>9</u>**   15 sides

Formula for the sum of the interior angles

If a polygon has n sides, then the sum of the interior and exterior angles together is $n \times 180° = 180n°$ so the sum of the interior angles only is $180n° - 360°$ which, as $360° = 180° \times 2$, can be written as $180°(n - 2)$,

i.e.   in a polygon with n sides, the sum of the interior angles is

$$(180n - 360)° \quad \text{or} \quad (n - 2)180°$$

Exercise *18f*

1 Find the sum of the interior angles of a polygon with

    a 20 sides   **b** 16 sides   **c** 11 sides.

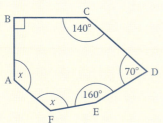

In the hexagon ABCDEF, the angles marked x are equal. Find the value of x.

The sum of the interior angles is $180° \times 6 - 360° = 1080° - 360° = 720°$

$$\therefore 90° + 140° + 70° + 160° + 2x = 720°$$
$$460° + 2x = 720°$$
$$2x = 260°$$
$$x = 130°$$

In each of the following questions find the size of the angle(s) marked x:

2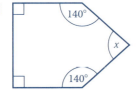

5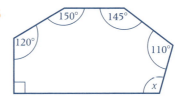

3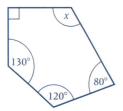

6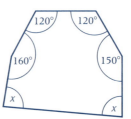

4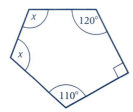

7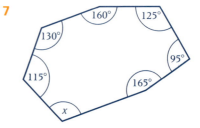

Find the size of each interior angle of a regular nine-sided polygon.

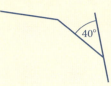

(As the polygon is regular, all the exterior angles are equal and all the interior angles are equal.)

Method 1 Sum of exterior angles = 360°

∴ each exterior angle = 360° ÷ 9 = 40°

∴ each interior angle = 180° − 40° = 140°

Method 2 Sum of interior angles = 180° × 9 − 360° = 1260°

∴ each interior angle = 1260° ÷ 9 = 140°

Find the size of each interior angle of:

8 A regular pentagon.

9 A regular hexagon.

10 A regular octagon.

11 A regular ten-sided polygon.

12 A regular 12-sided polygon.

13 A regular 20-sided polygon.

14 How many sides has a regular polygon if each exterior angle is

  a 20°   **b** 15°?

15 How many sides has a regular polygon if each interior angle is

  a 150°   **b** 162°

> **Tip** Find the exterior angle first.

16 Is it possible for each exterior angle of a regular polygon to be

  a 30° **b** 40° **c** 50° **d** 60° **e** 70° **f** 90°

In those cases where it is possible, give the number of sides.

17 Is it possible for each interior angle of a regular polygon to be

  a 90° **b** 120° **c** 180° **d** 175° **e** 170° **f** 135°

In those cases where it is possible, give the number of sides.

18 Construct a regular pentagon with sides 5 cm long.

19 Construct a regular octagon of side 5 cm.

> **Tip** Find the size of each interior angle, then use your protractor.

Puzzle

Arrange ten counters in such a way as to form five rows with four counters only in each row.

Mixed problems

ABCDE is a pentagon, in which the interior angles at A and D are each $3x°$ and the interior angles at B, C and E are each $4x°$. AB and DC are produced until they meet at F.

Find $B\hat{F}C$.

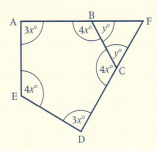

Sum of the interior angles of a pentagon $= 180° \times 5 - 360°$
$$= 540°$$

$\therefore$ $\quad\quad\quad\quad\quad\quad 3x + 4x + 3x + 4x + 4x = 540$
$$18x = 540$$
$$x = 30$$

$\therefore$ $\quad\quad\quad\quad A\hat{B}C = 120°$ $\quad$ and $\quad$ $B\hat{C}D = 120°$

$\therefore$ $\quad\quad\quad\quad y = 60°$

$\therefore$ $\quad\quad\quad\quad B\hat{F}C = 180° - 2 \times 60°$ $\quad$ (angle sum of $\triangle$BFC)

$$= 60°$$

In questions **1** to **10** find the value of x:

1

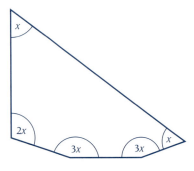

3

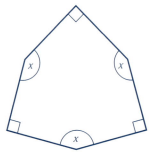

2

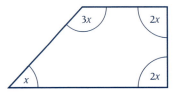

4

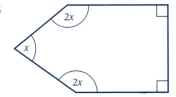

5

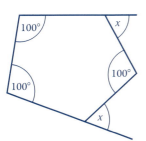

8

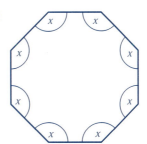

6

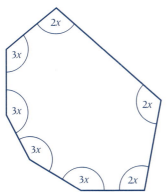

9

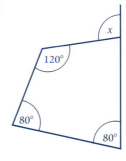

7

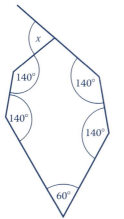

10

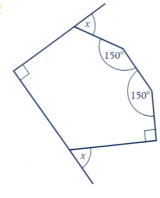

11 ABCDE is a regular pentagon.

OA = OB = OC = OD = OE.

Find the size of each angle at O.

12 ABCDEFGH is a regular octagon. O is a point in the middle of the octagon such that O is the same distance from each vertex. Find $A\hat{O}B$.

Tip Draw a diagram.

13 ABCDEF is a regular hexagon. AB and DC are produced until they meet at G. Find $B\hat{G}C$.

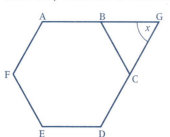

14 ABCDE is a regular pentagon. AB and DC are produced until they meet at F. Find $B\hat{F}C$.

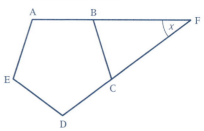

ABCDEF is a regular hexagon.
Find $A\hat{D}B$.

As ABCDEF is regular, the exterior angles are all equal.

Each exterior angle $= 360° \div 6 = 60°$

$\therefore$ each interior angle $= 180° - 60° = 120°$

$\triangle$ BCD is isosceles (BC = DC).

$\therefore$ $C\hat{B}D = B\hat{D}C = 30°$ (angle sum of $\triangle$ BCD)

AD is a line of symmetry for the hexagon.

$\therefore$ $E\hat{D}A = C\hat{D}A = 60°$

$\therefore$ $A\hat{D}B = 60° - 30°$

                                              $- 30°$

In questions **15** to **20**, each polygon is regular. Give answers correct to one decimal place where necessary:

15

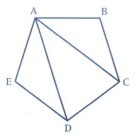

Find **a** $A\hat{C}B$ **b** $D\hat{A}C$

16

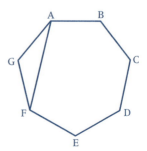

Find **a** $A\hat{G}F$ **b** $G\hat{A}F$

17

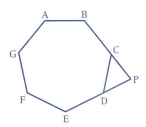

Find CP̂D

19

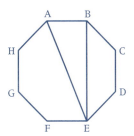

Find AÊB

18

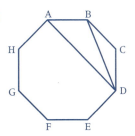

Find **a** CB̂D   **b** BD̂A

20

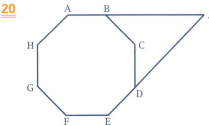

Find BĴD

 Puzzle

This flag is to be coloured red, white and blue.

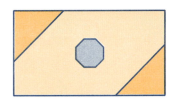

Adjoining regions must have different colours.

How many different flags are possible?

Pattern making with regular polyhedra

Regular hexagons fit together without leaving gaps,
to form a flat surface. We say that they *tessellate*.

The hexagons tessellate because each interior angle of
a regular hexagon is 120°, so three vertices fit together
to make 360°.

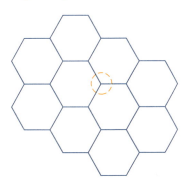

Exercise 18h

1   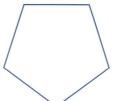 This is a pattern using regular octagons.

They do not tessellate:

> **Tip** Make sure that you have a sharp pencil.

   a Explain why they do not tessellate.

   b What shape is left between the four octagons?

   c Continue the pattern. (Trace one of the shapes above, cut it out and use it as a template.)

2   Trace this regular pentagon and use it to cut out a template:

   a Will regular pentagons tessellate?

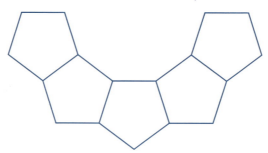

   b

   Use your template to copy and continue this pattern until you have a complete circle of pentagons. What shape is left in the middle?

   c Make up a pattern using pentagons.

3

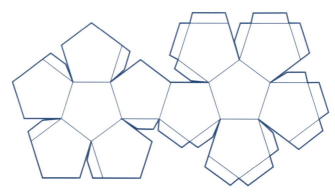

   Use your template from question 2 to copy this net on to thick paper. Cut it out and fold along the lines. Stick the edges together using the flaps. You have made a regular dodecahedron.

4   Apart from the hexagon, there are two other regular polygons that tessellate. Which are they, and why?

5 Regular hexagons, squares and equilateral triangles can be combined to make interesting patterns. Some examples are given below:

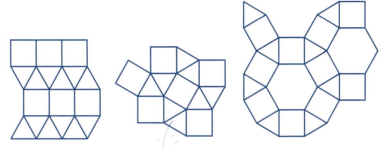

Copy these patterns and extend them. (If you make templates to help you, make each shape of side 2 cm.)

6 Make some patterns of your own using the shapes in question 5.

MATHS IS OUT THERE

Did you know that there are only five regular convex polyhedra that can be made from regular shapes?

They are:

- a tetrahedron which uses 4 equilateral triangles,
- an octahedron which uses 8 equilateral triangles,
- an icosahedron which uses 20 equilateral triangles,
- a cube which uses 4 squares,
- a dodecahedron which uses 5 regular pentagons.

These five solids are called the Platonic solids.

IN THIS CHAPTER...

you have seen that:

- the sum of the exterior angles of any polygon is 360°

- if a polygon is regular (i.e. equal sides) the exterior angles are equal and the size of each one is 360° ÷ the number of sides

- for a polygon with n sides the sum of the interior angles is $(180n - 360)°$ or, in a slightly more useful form, $(n - 2)180°$

- some regular polygons tessellate, i.e. they fit together without leaving gaps

19 CIRCLES

AT THE END OF THIS CHAPTER...

you should be able to:

1 Identify the centre, radius, diameter, chord, arc, segment and sector of a circle.

2 Identify angles subtended by the same arc of a circle and state the relationship between them.

3 Identify a cyclic quadrilateral and state its properties.

4 State the relationship between an angle at the centre and an angle at the circumference subtended by the same arc.

5 State the size of an angle in a semicircle.

6 Use ruler and a pair of compasses to construct the circumcircle of a given triangle.

Teacher: What do you call a squashed rectangle?

Student: A cute angle.

BEFORE YOU START

you need to know:
- ✓ the angle properties of a triangle
- ✓ the angle sum of a quadrilateral
- ✓ the properties of angles formed by a pair of parallel lines and a transversal
- ✓ the properties of isosceles and equilateral triangles
- ✓ the properties of the special quadrilaterals
- ✓ the meaning of supplementary angles
- ✓ how to construct triangles using a ruler and a pair of compasses

KEY WORDS

arc, bisect, chord, circumcircle, circumference, cyclic quadrilateral, diameter, exterior angle, interior angle, major arc, minor arc, perpendicular bisector, radius, sector, segment, semicircle, subtend, supplementary angles, trapezium, vertex

The basic facts

First we will revise some of the facts we already know about the circle.

Every point on a circle is the same distance from its centre. This distance is called the *radius* of the circle.

The *circumference* of the circle is the length of the curve.

A straight line joining any two points on the circle is called a *chord*.

Any chord passing through the centre of a circle is called a *diameter*.

We will now learn some new facts and definitions.

Any part of the circumference is called an arc. If the arc is less than half the circumference it is called a *minor arc*; if it is greater than half the circumference it is called a *major arc*.

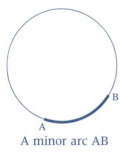

A minor arc AB

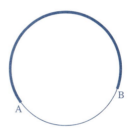

A major arc AB

The shaded area is enclosed by two radii and an arc. It looks like a slice of cake and is called a *sector*.

A chord divides a circle into two regions called segments. The larger region is called a *major segment* and the smaller region is called a *minor segment*.

A diameter divides a circle into two equal halves called semicircles.

Exercise **19a**

1 Name six chords in this diagram. Is any one of these chords a diameter? If so, name it.

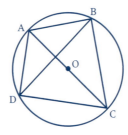

Copy the following diagrams. They do not have to be identical:

2 DC divides the circle into two segments. Shade the minor segment.

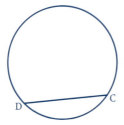

4 AB divides the circle into two segments. Shade the major segment.

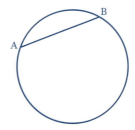

3 BC divides the circle into two segments. Shade the major segment.

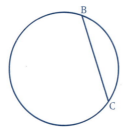

5 AD divides the circle into two segments. Shade the minor segment.

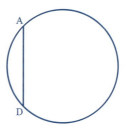

The angle subtended by an arc

Consider a triangle ABC whose vertices A, B and C lie on a circle. The angle BAC is said to stand on the minor arc BC. We say that BC *subtends* an angle BAC at A which is on the circumference.

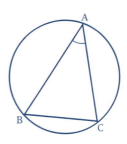

Similarly the angle ABC stands on the arc AC, or AC subtends an angle ABC at B, and the angle ACB stands on the arc AB, or AB subtends an angle ACB at C.

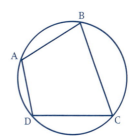

If the four vertices of a quadrilateral ABCD all lie on a circle, we say that the quadrilateral is *cyclic*, i.e. ABCD is a *cyclic quadrilateral*.

Exercise **19b**

Questions **1** to **6** refer to the following diagram:

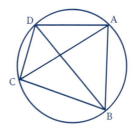

1 What arc does $\widehat{DBC}$ stand on?

2 What arc does $\widehat{BDC}$ stand on?

3 Name the two angles at the circumference standing on arc AB.

4 Name the two angles at the circumference standing on the arc BC.

5 What arc subtends $\widehat{DBA}$?

6 What arc subtends $\widehat{ACB}$?

Questions **7** to **12** refer to the following diagram:

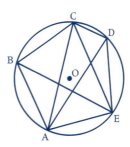

7 What arc does $\widehat{BCE}$ stand on?

8 What arc does $\widehat{CAD}$ stand on?

9 What arc subtends $\widehat{CAE}$?

10 What arc subtends $\widehat{DAB}$?

11 What angles stand on
  a the arc AB   **b** the arc BC?

12 What angles stand on
  a the arc AE   **b** the arc CE?

Discovering relationships between angles

Exercise **19c**

Copy the following diagrams making them at least twice as large. For each diagram measure the angles denoted by the letters:

1

3

2

4

First Fact

The results for questions **1** to **4** show that

angles standing on the same arc of a circle and in the same segment are equal.

Exercise **19d**   Find the angle denoted by the letter *x*.

$x = 44°$ (angles in the same segment of a circle)

Find the angles denoted by the letters:

1

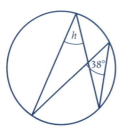

2

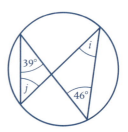

3

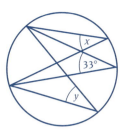

4

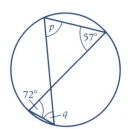

Find the angle denoted by the letter k.

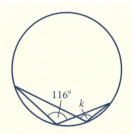

$k = 116°$ (angles in the same segment)

5

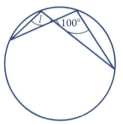

7

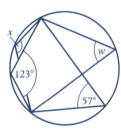

6

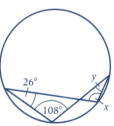

8

Exercise **19e**

Copy the following diagrams making them at least twice as large. They need not be identical. For each diagram measure the angles denoted by the letters. Hence find a relationship between x and y:

1

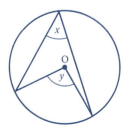

2

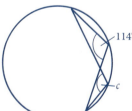

3

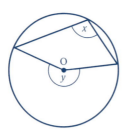

4

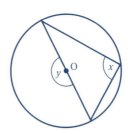

Second Fact

These results show that

> the angle which the arc of a circle subtends at the centre is equal to twice the angle it subtends at any point on the remaining circumference

Exercise **19f**   Find the angle denoted by the letter *d*.

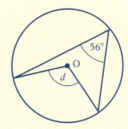

Angle *d* and the angle of 56° are both standing on the same arc: *d* is at the centre of the circle.

so $d = 2 \times 56° = 112°$

($\angle$ at centre $= 2 \times \angle$ at circumference)

Find the angles denoted by the letters. O denotes the centre of the circle:

1

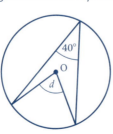

3

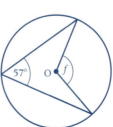

2

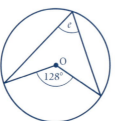

4

Find the angle denoted by the letter *f*.

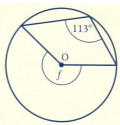

Angle *f* and the angle of 113° are both standing on the same arc; *f* is at the centre of the circle.

so $f = 2 \times 113° = 226°$ (∠at centre = 2 × ∠at circumference)

5

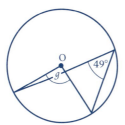

7

6

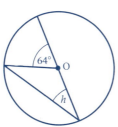

8

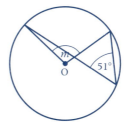

Investigation

Cut squares of side 10 cm from two sheets of plain A4 paper. On the interior of each square draw a circle of radius 5 cm and then cut the square into four cards as shown in the diagram.

a The cards are laid down side by side so that they are touching. The smallest number of arcs needed to make a closed shape is four.

In how many different ways can four arcs be placed together to give a closed shape? One example is drawn for you. Which shape has

i the largest area   **ii** the smallest area?

    b Decide whether or not it is possible to make a closed shape if the number of arcs used is **i** 5   **ii** 6   **iii** 7   **iv** 8, etc.

    c How many different shapes can be made if 6 arcs are used? How many of these shapes have

       i line symmetry

       ii rotational symmetry

       iii both line and rotational symmetry?

    b For shapes that use 6 arcs investigate which arrangement of the arcs encloses

       i the maximum area   **ii** the minimum area.

    e Write down any generalisation you can make about what happens if the number of arcs used is odd or even, or which way the arcs must be arranged to give the greatest or least enclosed area.

Exercise 19g

Copy the following diagrams making them at least twice as large. For each diagram measure the angles denoted by p and q. What do you notice about their sum?

1 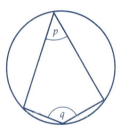   **2**    **3**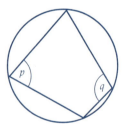

Third Fact

These results show that

    the opposite angles of a cyclic quadrilateral are supplementary.

Exercise 19h

Find the angle denoted by the letter p.

All four vertices are on the circumference so it is a cyclic quadrilateral

    ∴ $p + 132° = 180°$ (opp. ∠s cyclic quad. supplementary)

    so $p = 180° - 132° = 48°$

Find the angles denoted by the letters.

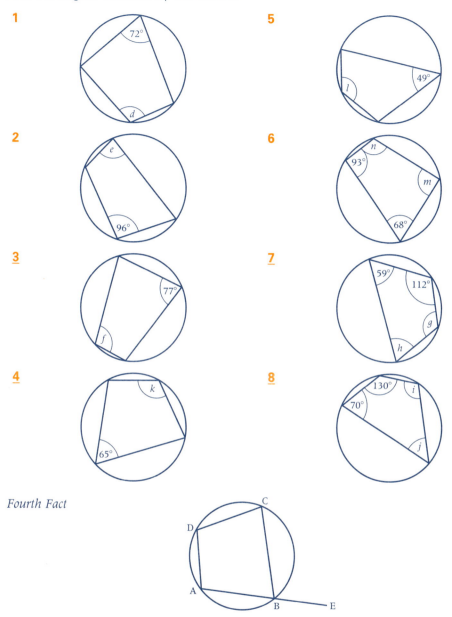

1 72° d

5 49° l

2 e 96°

6 n 93° m 68°

3 77° f

7 59° 112° g h

4 k 65°

8 130° i 70° j

Fourth Fact

If the side AB of the cyclic quadrilateral ABCD is produced to E, the angle CB̂E is called an exterior angle of quadrilateral ABCD.

Then $\quad\quad\quad$ AB̂C + CB̂E = 180° $\quad\quad$ (∠s on a str. line)

and $\quad\quad\quad$ AB̂C + AD̂C = 180° $\quad\quad$ (opp. ∠s cyclic quadrilateral)

Hence $\quad\quad\quad$ CB̂E = AD̂C

i.e. $\quad\quad$ any exterior angle of a cyclic quadrilateral is equal to the opposite interior angle.

Exercise 19i

Copy each of the following diagrams making them at least twice as large. For each diagram measure the angles denoted by the letters. What result do they confirm?

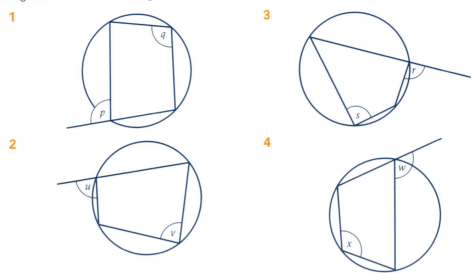

1

3

2

4

In questions **5** to **8** find the angles denoted by the letters:

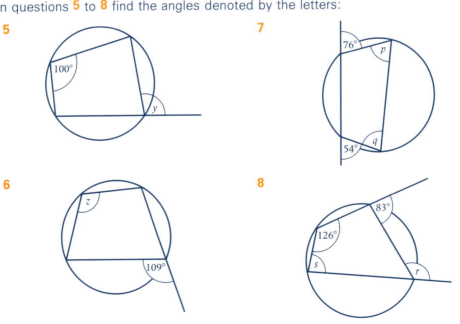

5

7

6

8

Exercise 19j

This exercise brings together the results used in the last few exercises. In some questions more than one of those results is required. It begins by revising some of the results for 'angles in a triangle' and 'parallel lines'.

In the following diagrams, O denotes the centre of the circle.

Find the angles denoted by the letters:

> **Tip** Copy the diagram and mark the facts given. Then mark the sizes of any other angles that you can find until you can 'see' the sizes of the angles you need.

1

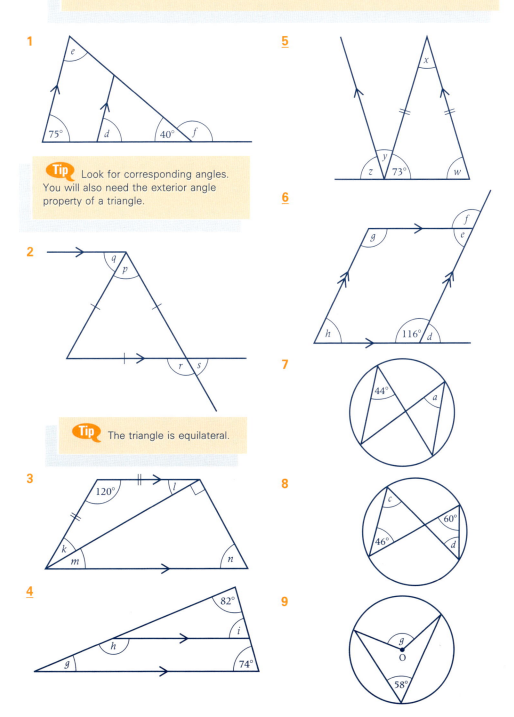

> **Tip** Look for corresponding angles. You will also need the exterior angle property of a triangle.

2

> **Tip** The triangle is equilateral.

3

4

5

6

7

8

9

10

39°

O

b

11

34° 52°

e f

12

h

O

144°

13

40° 37°

O

m l n

14

58°

110°

s

r

15

y

92°

30° z

O

x

16

25°

c

e

d

17

64°

O

i

h

18

m

O

l

n 54°

19

O 34°

v

w u

x

20

l

O

62°

n k

m 40°

Exercise 19k

Copy the following diagrams making yours at least twice as big. Measure the angles denoted by the letters. What result do these values show?

1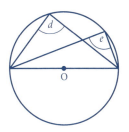

2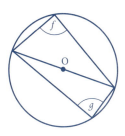

3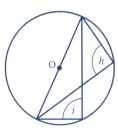

Fifth Fact

The angle in a semicircle is a right angle.

Exercise 19l

Find the angles denoted by the letters. The centre of the circle is marked O.

1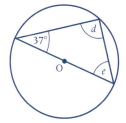

3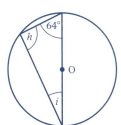

5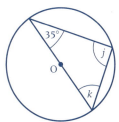

2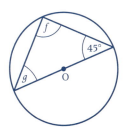

4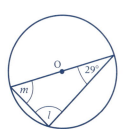

6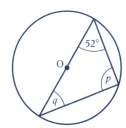

Find the angles denoted by the letters.

$p = 90°$ ($\angle$ in a semicircle)

$q = 33°$ ($\angle$ s of a triangle)

$r = 90°$ ($\angle$ in a semicircle)

$s + t = 90°$ ($\angle$ s of a triangle)

But $s = t$ (isosceles triangle)

$\therefore$ $s = t = 45°$

> **Tip** In these questions you will need all the angle facts. Copy the diagram and fill in the sizes of all angles you can find until you can 'see' the sizes of the angles you need.

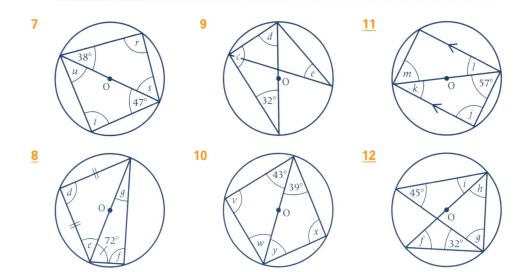

7 8 9 10 11 12

Construction of the perpendicular bisector

Reminder

Any point on the perpendicular bisector of the line joining two points A and B is equidistant from those two points.

LMN is the perpendicular bisector of the line joining AB. For any point P on this line PA = PB.

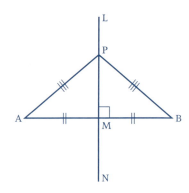

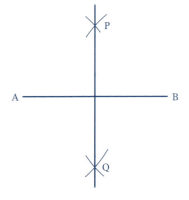

Use the following method to construct the perpendicular bisector of the line joining two points A and B.

With centre A and radius more than half AB, draw arcs above and below AB. *Keeping the radius the same*, move the point of your compasses to B and draw arcs above and below AB so that they intersect the first two arcs at P and Q. Join P to Q. PQ is the perpendicular bisector of AB. Any point on PQ is equidistant from A and B.

To construct the circumcircle of a given triangle

The circumcircle of a given triangle is the circle that passes through its vertices.

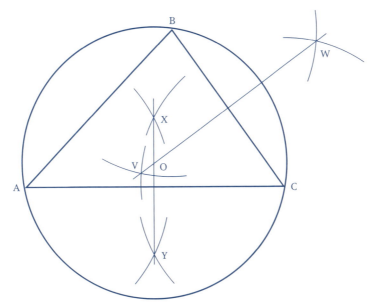

Draw a triangle ABC similar in size and shape to the triangle shown in the diagram. Use the method given above to construct the perpendicular bisector of AC and mark it XY. Next turn the page round so that BC is at the bottom and use the same method to construct the perpendicular bisector of BC. Mark this line VW. Any point on XY is equidistant from A and C and any point on VW is equidistant from B and C.

Mark O, the point of intersection of XY and VW. The point O is equidistant from all three vertices A, B and C.

With centre O and radius OA, draw a circle. If you have done your constructions carefully, you will find that this circle passes through A, B and C. This circle is called the circumcircle of triangle ABC and its centre O is called the circumcentre.

Now that you know exactly what to do, draw another triangle of a similar size and repeat this construction. When you have finished, the triangle and the circumcircle should stand out clearly from all construction lines.

Exercise *19m*

1 Draw a triangle of any size and construct its circumcircle.

Tip You need a sharp pencil.

2 Construct a triangle PQR in which PQ = 11 cm, PR = 10 cm and QR = 9 cm. Construct the circumcircle of this triangle. Measure its radius.

3 Construct a triangle XYZ in which XY = 12.5 cm, YZ = 7.5 cm and $X\hat{Y}Z = 60°$. Construct the circumcircle to this triangle. Measure

   a the length of XZ   **b** the radius of the circumcircle.

4 Construct a triangle ABC in which AB = 12.5 cm, BC = 7.5 cm and AC = 10 cm. Find the position of the circumcentre O and hence draw the circumcircle. What do you notice about the position of O? What value would you now expect $A\hat{C}B$ to have? Give reasons. Check your result by measuring $A\hat{C}B$ with a protractor.

5 Construct a triangle DEF in which EF = 8.8 cm, $D\hat{E}F = 30°$ and $D\hat{F}E = 45°$. Construct the perpendicular bisectors of all three sides and hence draw the circumcircle. What do you notice

   a about the three perpendicular bisectors

   b about the circumcentre of an obtuse angled triangle?

6 Construct a trapezium ABCD in which AB = 10.2 cm, BC = AD = 5.2 cm, $A\hat{B}C = 60°$, $D\hat{A}B = 60°$, $A\hat{D}C = 120°$ and $B\hat{C}D = 120°$. Construct the circumcircle to triangle ABC. Does this circle pass through any other particular point? Can you give a reason for what has happened? Will this happen for every trapezium?

Mixed exercise

Find the angles denoted by the letters:

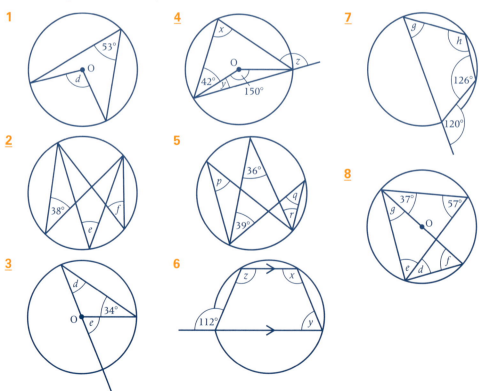

Did you know that if you draw any three intersecting circles, the chords common to any two of the circles intersect at one point?

Try it and see for yourself.

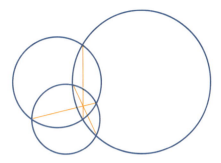

IN THIS CHAPTER...

you have seen that:

- angles standing on the same arc and in the same segment of a circle are equal

- the angle subtended by an arc at the centre of a circle is twice the angle subtended by the same arc at the circumference

- a cyclic quadrilateral has all four vertices on the circumference of a circle

- the opposite angles of a cyclic quadrilateral are supplementary

- an exterior angle of a cyclic quadrilateral is equal to the interior opposite angle

- the angle in a semicircle is a right angle

- the circumcircle of a triangle passes through the three vertices of that triangle

- you can find the centre of the circumcircle of a triangle by constructing the perpendicular bisectors of two sides of the triangle; the centre is where these intersect.

In questions **1** to **10**, choose the letter for the correct answer.

1 The mean of six scores is 12 and the mean of four other scores is 17. What is the mean of the ten scores.

    A 13.5   **B** 14   **C** 14.5   **D** 29

2 $(3x + 4)(2x − 5) =$

    A $6x^2 − 20$ **B** $6x^2 − 15x − 20$ **C** $6x^2 − 7x − 20$ **D** $6x^2 + 8x − 20$

3 A solution of $(2x + 5)(3x − 4) = 0$ is $x =$

    A $−5$ **B** $−\frac{5}{2}$ **C** $\frac{3}{4}$ **D** 4

4 A minimum value of $7 + (x + 3)^2$ is

    A $−3$ **B** 0 **C** 3 **D** 7

5 $x^2 − 2x + 5 =$

    A $(x − 1)^2$ **B** $(x − 1)^2 + 4$ **C** $(x − 1)^2 + 6$ **D** $x^2 − 1$

6 The parallel sides of a trapezium are 10 cm and 15 cm in length. The distance between the parallel sides is 5 cm. The area of the trapezium is

    A $30.2\,\text{cm}^2$ **B** $30\,\text{cm}^2$ **C** $62.5\,\text{cm}^2$ **D** $125\,\text{cm}^2$

7 Each interior angle of a regular 5-sided polygon measures

    A $144°$ **B** $108°$ **C** $72°$ **D** $36°$

8 In a cyclic quadrilateral PQRS, the angle at P measures 70 degrees. What is the measure of the angle at R in degrees?

    A 20 **B** $70°$ **C** $110°$ **D** 140

9 $5x^2 − 3x + 2 =$

    A $(5x + 1)(x − 3)$ **B** $(5x − 2)(x + 1)$ **C** $(5x − 3)(x − 1)$ **D** none of these

10 A bus travels 10 km at 15 km/h and 20 km at 45 km/h. The average speed of the bus is

    A $30\,\text{km/h}$ **B** $\frac{1}{2}\,\text{km/h}$ **C** $18\,\text{km/h}$ **D** $27\,\text{km/h}$

11 For the first $2\frac{1}{2}$ hours of a 182 km journey the average speed was 60 km per hour. The average speed for the remainder of the journey was 64 km per hour. Calculate the average speed for the entire journey.

12 a The length of a rectangle is 10 cm more than its width. The area of the rectangle is 45 cm². Find the length of the rectangle.

b

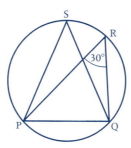

In the above diagram PR is the diameter of the circle. Given $Q\hat{R}P = 30°$ and $SP = SQ$, find, giving reasons for your answers, **i** $P\hat{S}Q$ **ii** $Q\hat{S}R$ **iii** $S\hat{Q}P$.

13 a Factorise completely **i** $2x^3 - 72xy^2$ **ii** $a^2 - 2ab - 2bc + ac$ **iii** $6x^2 - 5x - 4$

b Solve by the method of completing the squares, $2x^2 + 7x + 1 = 0$. Give your answers to two decimal places.

14 Solve the equation $x^2 + 4x + 2 = 0$.

a by drawing a graph using 1 cm = 1 unit on each axis for $-4 \leqslant x \leqslant 1$

b by completing the square

15 a Simplify $\dfrac{3}{x - 1} + \dfrac{5}{x - 2}$

b Solve the equation $\dfrac{2}{x} - \dfrac{2}{x - 3} = 3$

AT THE END OF THIS CHAPTER...

you should be able to:

1 Reduce a given ratio to its simplest form.

2 Compare two ratios in terms of size.

3 Express a given ratio in the form $n : 1$, where n is written correct to a stated number of significant figures.

4 Find the ratio of two quantities measured in different units.

5 Calculate the missing quantity, given two equivalent ratios.

6 Use ratios to solve problems in sharing.

7 Express quantities which are in direct or inverse proportion in terms of ratio.

8 Solve problems involving direct or inverse proportion.

9 Use the unitary method to solve problems.

The Golden Ratio

Early Greek mathematicians were fascinated by a rectangle, the ratio of whose sides is such that if a square is cut off one end, you are left with a rectangle of the same shape as the original, i.e. the ratios of their corresponding sides are equal. This process can be repeated as often as you like.

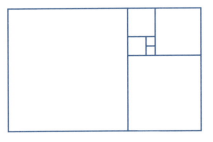

Because such a rectangle is thought to be pleasing to the eye, the ratio of the sides is also called the divine proportion.

<table>
<tr><td>BEFORE YOU START</td><td>you need to know:</td></tr>
</table>

| **BEFORE YOU START** | you need to know:
✓ how to work with fractions and decimals
✓ the units of length and the relationships between them
✓ the units of mass and the relationships between them
✓ how to find the areas of squares and triangles
✓ how to solve equations involving fractions |

| **KEY WORDS** | denominator, direct proportion, inverse proportion, lowest common multiple, proportional, ratio, reciprocal |

The first few exercises are to remind you about ratios. Remember that ratios compare the sizes of related quantities.

Simplifying ratios

A ratio can be divided or multiplied throughout by the same number.

Exercise **20a**   Give the ratios   **a** $225 : 105$ **b** $\frac{4}{5} : \frac{7}{8} : \frac{1}{2}$ in their simplest forms

a $225 : 105 = 45 : 21$ (dividing by 5)

$\qquad\qquad\;\; = 15 : 7$ (dividing by 3)

b   40 is the lowest common multiple of the denominators so multiply by 40 to get rid of the fractions

$\frac{4}{5} : \frac{7}{8} : \frac{1}{2} = \overset{8}{\cancel{40}} \times \frac{4}{\cancel{5}} : \overset{5}{\cancel{40}} \times \frac{7}{\cancel{8}} : \overset{20}{\cancel{40}} \times \frac{1}{\cancel{2}}$

$\qquad\qquad = 32 : 35 : 20$

Give the following ratios in their simplest form:

1 $12 : 18$	**4** $320 : 480$	**7** $3.2 : 7.2$
2 $3 : 6 : 9$	**5** $288 : 128 : 144$	**8** $\frac{1}{2} : \frac{5}{6} : \frac{2}{3}$
3 $3.5 : 2.5$	**6** $\frac{1}{2} : \frac{3}{4} : \frac{1}{4}$	**9** $36 : 54 : 18$

Which ratio is larger, $3 : 2$ or $14 : 9$?

(To compare the sizes of the ratios, we write them as fractions and then convert them to equivalent fractions with a common denominator.)

$$\frac{3}{2} = \frac{27}{18} \quad \text{and} \quad \frac{14}{9} = \frac{28}{18}$$

The second ratio is the larger.

Which ratio is the larger?

10 $6:11$ or $2:5$ **12** $20:3$ or $31:4$

11 $15:4$ or $11:3$ **13** $2:7$ or $5:16$

Express the following ratios in the form $n:1$, giving n correct to three significant figures where necessary:

14 $3:2$ **17** $30:11$ **20** $4:3$

15 $12:5$ **18** $3:5$ **21** $3:4$

16 $6:7$ **19** $21:8$ **22** $10:7$

> **Tip** To express $a:b$ in the form $n:1$, divide throughout by b, e.g. $4:5 = \frac{4}{5}:1 = 0.8:1$

Mixed units and problems

If we are asked to compare two quantities expressed in different units, we need to change one or both so that the two quantities are in the same unit. It is easier to change to smaller units (where multiplication is required) rather than to larger units (where division is required).

Exercise **20b**

Simplify the following ratios:

1 $45\,\text{cm} : 0.1\,\text{m}$ **4** $32\,\text{g} : 2\,\text{kg}$

2 $42\,\text{c} : \$1.05$ **5** $450\,\text{mg} : 1\,\text{g}$

3 $340\,\text{m} : 1.2\,\text{km}$ **6** $2.2\,\text{t} : 132\,\text{kg}$

> **Tip** Express both quantities in the same unit.

> Find the ratio of 14 c per gram to \$120 per kilogram.
>
> (In order to compare we will use both prices in \$ per kg.)
>
> $$14\,\text{c per g} = 14\,000\,\text{c per kg}$$
> $$= \$140\,\text{per kg}$$
> $$14\,\text{c per g} : \$120\,\text{per kg} = \$140\,\text{per kg} : \$120\,\text{per kg}$$
> $$= 140 : 120$$
> $$= 7 : 6$$

Find the ratios of the following prices:

  7 4 c per kilogram to \$38 per tonne.

> **Tip** First express 4 c per kg as a price per tonne.

  8 6 c each to 70 c per dozen.

9 \$16.20 per metre to 15 c per centimetre.

> **Tip** First find the cost per dozen.

10 72 c for twenty to 4 c each.

Give the ratio of the cost of 6 m of material at $2.40 per metre to the cost of 8 m at $2.20 per metre.

We need to find the cost of each length of material first.

$$\text{First cost} = \$6 \times 2.40 = \$14.40$$
$$\text{Second cost} = \$8 \times 2.20 = \$17.60$$
$$\text{Ratio of costs} = \$14.40 : \$17.60$$
$$= 144 : 176$$
$$= 18 : 22 = 9 : 11$$

11 In a school of 1029 pupils, 504 are girls. What is the ratio of the number of boys to the number of girls?

12 I spend $3.60 on groceries and $2.40 on vegetables. What is the ratio of the cost of

  a groceries to vegetables

  b vegetables to groceries

  c groceries to the total?

13 One rectangle has a length of 6 cm and a width of $4\frac{1}{2}$ cm. A second rectangle has a length of 9 cm and a width of $2\frac{1}{2}$ cm. Find the ratios of

  a their lengths

  b their widths

  c their perimeters

  d their areas.

14 Find the ratio of the cost of 12 m^2 of carpet at $7.20 per m^2 to the cost of 50 carpet tiles at $2.40 per tile.

15 Find the ratios of the following areas

  a B : A   **b** C : B

  c E : A + B   **d** E : D

  e E : C + D   **f** C : whole square

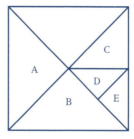

16 The areas of the small triangles are equal. Find the ratios of the following areas:

  a A : whole figure

  b A : A + B + C + D

  c B + E + F + G : whole figure

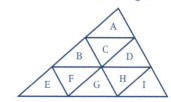

Finding missing quantities

If we are given the ratio $x : 6 = 3 : 5$ we may write this as $\frac{x}{6} = \frac{3}{5}$ and solve the equation for x.

It does not matter in which order we compare the two quantities as long as we are consistent. If the ratio is $4 : x = 5 : 3$ we may rewrite it as $x : 4 = 3 : 5$.

Exercise 20c Find x if **a** $x : 6 = 13 : 15$ **b** $4 : x = 3 : 2$

a $\qquad \dfrac{x}{6} = \dfrac{13}{15}$

$\qquad \overset{1}{\cancel{6}} \times \dfrac{x}{\cancel{6}_1} = \overset{2}{\cancel{6}} \times \dfrac{13}{\cancel{15}_5}$ (multiply both sides by 6)

$\qquad x = \dfrac{26}{5} = 5.2$

b $\qquad x : 4 = 2 : 3$ (rearrange the ratio so that x comes first)

$\qquad \dfrac{x}{4} = \dfrac{2}{3} \Rightarrow 4 \times \dfrac{x}{4} = 4 \times \dfrac{2}{3}$ (multiply both sides by 4)

$\qquad x = \dfrac{8}{3} = 2\dfrac{2}{3}$

Find x in questions **1** to **6**:

1 $x : 5 = 2 : 9$ **3** $x : 6 = 5 : 4$ **5** $3 : 8 = 9 : x$

2 $x : 3 = 1 : 7$ **4** $5 : x = 7 : 2$ **6** $15 : 2 = x : 3$

Complete the following ratios:

7 $4 : = 3 : 7$ **9** $3 : = 5 : 2$ **11** $6 : 5 = 4 :$

8 $: 5 = 6 : 11$ **10** $9 : 5 = : 4$ **12** $: 12 = 5 : 7$

In a town, the ratio of the number of males to the number of females is $80 : 81$. There are 9680 males. How many females are there? What is the total population?

Suppose there are x females,

then $x : 9680 = 81 : 80$ (Notice the change of order so that x comes first.)

$$\frac{x}{9680} = \frac{81}{80}$$

$$\overset{1}{\cancel{9680}} \times \frac{x}{\cancel{9680}_1} = \overset{121}{\cancel{9680}} \times \frac{81}{\cancel{80}_1}$$

$$x = 9801$$

There are 9801 females and the total population is 19 481.

13 The numbers of Mr and Mrs James' grandsons and granddaughters are in the ratio $4 : 3$. There are nine granddaughters. How many grandsons are there? What is the ratio of the number of granddaughters to the number of grandchildren?

14 The ratio of the number of cats to the number of dogs owned by the children in one year group in a school is $5 : 3$. There are 95 cats. How many cats and dogs are there altogether?

15 The ratio of the lengths of two rectangles is $6 : 5$. The length of the second is 8.4 cm. What is the length of the first?

16 The ratio of the numbers of orange flowers to white flowers in a garden is $6 : 11$. There are 144 orange flowers. How many white flowers are there?

Division in a given ratio

Share 72 c amongst three people so that their shares are in the ratio 3 : 4 : 5.

There are 12 portions (i.e. $3 + 4 + 5$).

First share is 3 out of 12 portions, i.e. $\frac{3}{12}$ of 72 c $= \frac{3}{12} \times 72 \, c = 18 \, c$

Second share is 4 out of 12 portions, i.e. $\frac{4}{12}$ of 72 c $= \frac{4}{12} \times 72 \, c = 24 \, c$

Third share $= \frac{5}{12} \times 72 \, c = 30 \, c$

Check: $18 \, c + 24 \, c + 30 \, c = 72 \, c$

1 Divide $45 into two parts in the ratio 4 : 5

2 Divide 96 m into two parts in the ratio 9 : 7

3 Divide 5 kg into three parts in the ratio 1 : 2 : 5

4 Divide seven hours into three parts in the ratio 1 : 5 : 8

5 There are 32 children in a class. The ratio of the number of boys to the number of girls is 9 : 7. How many boys and how many girls are there?

6 The angles of a triangle are in the ratio 6 : 5 : 7. Find the sizes of the three angles.

7 Share the contents of a box containing 30 chocolates amongst Anne, Mary and Sue in the ratio 3 : 4 : 3. How many chocolates will each get?

8 A marksman fires at a target and the ratio of hits to misses is 11 : 4. He fires 90 times. How many hits does he score and how many times does he miss?

Mixed questions

1 Simplify the ratio 324 : 252

2 Divide 72 m into two parts in the ratio 5 : 7

3 Complete the ratio 4 : 7 = 3 :

4 Find x if $4 : x = 9 : 5$

5 The ratio of two lengths is 4 : 5. If the first length is 22 cm what is the second?

6 If $p : q = 2 : 3$, what is $5p : 2q$?

7 Find the ratio of 2 c per gram to $2.12 per kilogram.

8 Simplify $4\frac{2}{3} : 3\frac{1}{2}$

9 In a block of flats, 24 have two bedrooms and 32 have one bedroom. Give the ratio of the number of two-bedroomed flats to the number of one-bedroomed flats.

10 Simplify the ratio 3.2 : 4.8

Simple direct proportion

If we know the cost of *one* article, we can easily find the cost of ten similar articles, or if we know that someone is paid for *one* hour's work, we can find what the pay is for five hours.

Exercise 20f   If $1\,cm^3$ of lead weighs $11.3\,g$, what is the weight of
  a $6\,cm^3$ **b** $0.8\,cm^3$?

$$1\,cm^3 \text{ weighs } 11.3\,g$$

a $6\,cm^3$ weigh $11.3 \times 6\,g = 67.8\,g$

b $0.8\,cm^3$ weigh $11.3 \times 0.8\,g = 9.04\,g$

---

1 The cost of $1\,kg$ of sugar is $60\,c$. What is the cost of

  a $3\,kg$ **b** $12\,kg$?

2 In one hour an electric fire uses $1\frac{1}{2}$ units. Find how much it uses in

  a four hours   **b** $\frac{1}{2}$ hour.

3 One litre of petrol takes a car $18\,km$. At the same rate, how far does it travel on

  a four litres   **b** 6.6 litres?

4 A knitting pattern states that, at the correct tension, five rows measure $1\,cm$. How many rows must be knitted to measure

  a $7\,cm$ **b** $8.4\,cm$?

5 The cost of $1\,kg$ of tomatoes is $\$2.25$. Find the cost of

  a $\frac{1}{2}\,kg$ **b** $2.4\,kg$.

We can reverse the process and, for instance, find the cost of *one* article if we know the cost of three similar articles.

Exercise 20g $18\,cm^3$ of copper weigh $162\,g$. What is the weight of $1\,cm^3$?

$$18\,cm^3 \text{ weigh } 162\,g$$

$$1\,cm^3 \text{ weighs } \frac{162}{18}\,g = 9\,g$$

---

1 Six pens cost $\$7.20$. What is the cost of one pen?

2 A car uses eight litres of petrol to travel $124\,km$. At the same rate, how far can it travel on one litre?

3 A man walks steadily for three hours and covers $13\,km$. How far does he walk in one hour?

4 Dress material costs $\$2.60$ for $4\,m$. What is the cost of $1\,m$?

5 A carpet costs $\$92.40$. Its area is $12\,m^2$. What is the cost of $1\,m^2$?

We can use the same process even if the quantities are not a whole number of units.

The mass of $0.6\,cm^3$ of a metal is $3\,g$. What is the mass of $1\,cm^3$?

The mass of $0.6\,cm^3$ is $3\,g$

The mass of $1\,cm^3$ is $\dfrac{3}{0.6}\,g = 5\,g$

6 The cost of $2.8\,m$ of material is $11.76. What is the cost of $1\,m$?

7 $8.6\,m^2$ of carpet cost $71.38. What is the cost of $1\,m^2$?

8 The cost of running a refrigerator for 3.2 hours is $4.8\,c$. What is the cost of running the refrigerator for one hour?

9 A bricklayer takes 0.8 hour to build a wall $1.2\,m$ high. How high a wall (of the same length) could he build in one hour?

10 A piece of webbing is $12.4\,cm$ long and its area is $68.2\,cm^2$. What is the area of a piece of this webbing that is $1\,cm$ long?

Direct proportion

If two varying quantities are always in the same ratio, they are said to be *directly proportional* to one another (or sometimes simply *proportional*).

For example, when buying pens which each cost the same amount, the total cost is proportional to the number of pens. The ratio of the cost of 11 pens to the cost of 14 pens is $11 : 14$, and if we know the cost of 11 pens, we can find the cost of 14 pens.

One method for solving problems involving direct proportion uses ratio, another uses the ideas in the last two exercises. This method is called the *unitary* method because it makes use of the cost of *one* article or the time taken by *one* man to complete a piece of work.

Exercise 20h   If the mass of $16\,cm^3$ of a metal is $24\,g$, what is the mass of $20\,cm^3$?

First Method (using ratios)

Let the mass of $20\,cm^3$ be x grams

Then $x : 24 = 20 : 16$ (The ratio of the masses = the ratio of the volumes)

$$\frac{x}{24} = \frac{20}{16}$$

$$\overset{1}{24} \times \frac{x}{\underset{1}{24}} = \overset{3}{24} \times \frac{\overset{10}{20}}{\underset{\underset{1}{2}}{16}}$$

$$x = 30$$

The mass of $20\,cm^3$ is $30\,g$.

Second Method (unitary method)

(Write the first sentence so that it ends with the quantity you want, i.e. the mass.)

$16 \, \text{cm}^3$ has a mass of $24 \, \text{g}$.

$1 \, \text{cm}^3$ has a mass of $\dfrac{24}{16} \, \text{g}$. (There is no need to work out the value of $\dfrac{24}{16}$ yet.)

$20 \, \text{cm}^3$ has a mass of $\overset{10}{\cancel{20}} \times \dfrac{\overset{3}{\cancel{24}}}{\cancel{16}_{1}} \, \text{g} = 30 \, \text{g}$

1 At a steady speed a car uses four litres of petrol to travel $75 \, \text{km}$. At the same speed how much petrol is needed to travel $60 \, \text{km}$?

2 A hiker walked steadily for four hours, covering $16 \, \text{km}$. How long did he take to cover $12 \, \text{km}$?

3 An electric fire uses $7\frac{1}{2}$ units in three hours. How many units does it use in five hours?

4 How long does the same electric fire take to use 9 units?

5 A taxi journey of $30 \, \text{km}$ costs $\$36$. At the same rate per kilometre

   a what would be the cost of travelling $25 \, \text{km}$

   b how far could you travel for $\$42$?

6 It costs $\$108$ to turf a lawn of area $63 \, \text{m}^2$. How much would it cost to turf a lawn of area $56 \, \text{m}^2$?

7 A machine in a soft drinks factory fills 840 bottles in six hours. How many could it fill in five hours?

8 A $6 \, \text{kg}$ bag of beans costs $198 \, \text{c}$. At the same rate, what would an $8 \, \text{kg}$ bag cost?

9 A knitting pattern states that the correct tension is such that 55 rows measure $10 \, \text{cm}$. How many rows should be knitted to give $12 \, \text{cm}$?

10 A scale model of a ship is such that the mast is $9 \, \text{cm}$ high and the mast of the original ship is $12 \, \text{m}$ high. The length of the original ship is $27 \, \text{m}$. How long is the model ship?

Either method will work, whether the numbers are complicated or simple. Even if the question is about something unfamiliar, it is sufficient to know that the quantities are proportional.

In a spring balance, the extension in the spring is proportional to the load. If the extension is $2.5 \, \text{cm}$ when the load is 8 newtons, what is the extension when the load is 3.6 newtons?

Ratio Method

Let the extension be $x \, \text{cm}$.

$$x : 2.5 = 3.6 : 8$$

$$\frac{x}{2.5} = \frac{3.6}{8}$$

$$2.5 \times \frac{x}{2.5} = 2.5 \times \frac{3.6}{8}$$

$$x = 1.125$$

The extension is $1.125 \, \text{cm}$.

Unitary Method

If a load of 8 newtons gives an extension of 2.5 cm, then a load of 1 newton gives an extension of $\frac{2.5}{8}$ cm.

∴ a load of 3.6 newtons gives an extension of

$$3.6 \times \frac{2.5}{8} \text{ cm} = 1.125 \text{ cm}$$

11 It costs $196 to hire scaffolding for 42 days. How much would it cost to hire the same scaffolding for 36 days at the same rate per day?

12 The rates of currency exchange published in the newspapers on a certain day showed that 14 pounds could be exchanged for 49 dollars. How many dollars could be obtained for 112 pounds?

13 At a steady speed, a car uses 15 litres of petrol to travel 164 km. At the same speed, what distance could be travelled if six litres were used?

14 If a 2 kg bag of sugar contains 9×10^6 crystals, how many crystals are there in

  a 5 kg   **b** 1.8 kg   **c** 0.03 kg?

15 The current flowing through a lamp is proportional to the voltage across the lamp. If the voltage across the lamp is ten volts the current is 0.6 amp. What voltage is required to make a current of 0.9 amp flow?

16 The amount of energy carried by an electric current is proportional to the number of coulombs. If five coulombs carry 19 joules of energy, how many joules are carried by 6.5 coulombs?

17 A recipe for date squares uses the following quantities:

Ingredients	Costs
125 g of brown sugar	500 g cost $1.52
75 g of oats	750 g cost $2.04
75 g of flour	$1\frac{1}{2}$ kg cost $1.76
100 g of margarine	250 g cost 72 c
100 g of dates	250 g cost $1.68
Pinch of bicarbonate of soda	–
Squeeze of lemon juice	4 c

Find the cost of making these date squares as accurately as possible, then give your answer correct to the nearest cent.

18 A do-it-yourself enthusiast makes a base for a table.

Materials	Costs
4 legs each 30 cm long	2 m cost $9.60
4 stretchers each 70 cm long	3 m cost $6.30
4 stretchers each 35 cm long	2 m cost $4.50
3 pieces each 80 cm long	3 m cost $18.90
$\frac{3}{4}$ litre of varnish	1 litre costs $14.40
12 screws	20 screws cost $2.40

What is the total cost of the materials that are actually used?

Puzzle

Roger makes two tables. They are the same height. The top of the first table is 0.5 m by 1 m. The top of the second table is twice the area of the first.

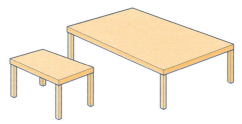

The wood for the table tops costs $4.50 a square metre and the legs cost $2.50 each.

What is the ratio of the costs of the tables?

Inverse proportion

Some quantities are not directly proportional to one another, although there is a connection between them. As one increases in size, the other may decrease, so that the reciprocal, or inverse, of the second is proportional to the first.

Suppose, for example, that a certain amount of food is available for several days. If each person eats the same amount each day, the more people there are, the shorter is the time that the food will last. The number of days the food will last is *inversely proportional* to the number of people eating it.

Exercise 20i

In this exercise, assume that the rates are constant.

Four bricklayers can build a certain wall in ten days. How long would it take five bricklayers to build it?

Ratio Method
Suppose it takes five bricklayers x days to build it.

(Five bricklayers will take a shorter time so we use the inverse ratio.)

$$\frac{x}{10} = \frac{4}{5}$$

$$\overset{1}{\cancel{10}} \times \frac{x}{\cancel{10}_{1}} = \overset{2}{\cancel{10}} \times \frac{4}{\cancel{5}_{1}}$$

$$x = 8$$

It would take them 8 days.

Unitary Method

Four bricklayers take 10 days.

One bricklayer would take 40 days.

Five bricklayers would take $\dfrac{40}{5}$ days = 8 days

1 Eleven taps fill a tank in three hours. How long would it take to fill the tank if only six taps are working?

2 Nine children share out equally the chocolates in a large tin and get eight each. If there were only six children, how many would each get?

3 The length of an essay is 174 lines with an average of 14 words per line. If it is rewritten with an average of 12 words per line, how many lines will be needed?

4 A field of grass feeds 24 cows for six days. How long would the same field feed 18 cows?

5 The dimensions of a block of stamps are 30 cm wide by 20 cm high. The same number of stamps could also have been arranged in a block 24 cm wide. How high would this second block be?

6 A batch of bottles was packed in 25 boxes taking 12 bottles each. If the same batch had been packed in boxes taking 15 each, how many boxes would be filled?

7 When knitting a scarf 48 stitches wide, one ball of wool will give a length of 18 cm. If there had been 54 stitches instead, how long a piece would the same ball give?

8 In a school, 33 classrooms are required if each class has 32 pupils. How many classrooms would be required if the class size was reduced to 22?

9 A factory requires 42 machines to produce a given number of articles in 63 days. How many machines would be required to produce the same number of articles in 54 days?

Mixed questions

Exercise 20j

Some of the following questions cannot be answered because the quantities are neither in direct nor in inverse proportion. In these cases write 'There is no answer'. For those questions that can be solved, give answers correct to three significant figures where necessary:

1 The list of exchange rates states that US $1 = 126 yen and US $1 = TT $6.

 a How many TT dollars can 315 yen be exchanged for?

 b How many yen are 100 TT dollars worth?

2 A man earned $41.20 for an eight-hour day. How much would he earn at the same time for a 38-hour week?

3 A typist typed 3690 words in $4\frac{1}{2}$ hours. How long would it take to type 2870 words at the same rate?

4 At the age of twelve, a boy is 1.6 m tall. How tall will he be at the age of eighteen?

5 A ream of paper (500 sheets) is 6.2 cm thick. How thick is a pile of 360 sheets of the same paper?

6 If I buy balloons at 14 c each, I can buy 63 of them. If the price of a balloon increases to 18 c, how many can I buy for the same amount of money?

7 A boy's mark for a test is 18 out of a total of 30 marks. If the test had been marked out of 40 what would the boy's mark have been?

8 Twenty-four identical mathematics text books occupy 60 cm of shelf space. How many books will fit into 85 cm?

9 A lamp post 4 m high has a shadow 3.2 m long cast by the sun. A man 1.8 m high is standing by the lamp post. At the same moment, what is the length of his shadow?

10 A contractor decides that he can build a barn in nine weeks using four men. If he employs two more men, how long will the job take? Assume that all the men work at the same rate.

11 A girl twelve years old gained 27 marks in a competition. How many marks did her six-year-old sister gain?

12 For a given voltage, the current flowing is inversely proportional to the resistance. When the current flowing is 2.5 amps the resistance is 0.9 ohm. What is the current when the resistance is 1.5 ohms?

Mixed exercises

Exercise 20k

1 Simplify the ratio 7.35 : 2.45

2 Complete the ratio   : 9 = 2 : 5

3 Divide 56 m into three parts in the ratio 1 : 2 : 4

4 A car uses seven litres of petrol for a 100 km journey. At the same rate, how far could it go on eight litres?

5 Eight typists together could complete a task in five hours. If all the typists work at the same rate, how long would six typists take?

6 Simplify the ratio $7\frac{1}{2} : 2\frac{1}{2} : 1\frac{1}{4}$

7 The ratio of the numbers of eleven-year-olds to twelve-year-olds in a class is 8 : 3. There are 24 eleven-year-olds. How many twelve-year-olds are there?

8 Give the ratio 6 : 5 in the form n : 1

Investigation

Sweets at a 'Pick and Mix' counter are sold by weight at $0.56 per 100 grams.

x	20	50	100	200	500	1000
y						

a If x grams cost y cents, copy and complete this table giving values of y corresponding to values of x.

b Use a scale of 1 cm for 50 units on the x-axis and a scale of 2 cm for 50 units on the y-axis to plot these points on a graph.

c What do you notice about these points? Can you use graph to find the cost of 162 grams?

d The cost and weight of these sweets are directly proportional. Investigate the graphical relationship between other quantities that are directly proportional. What do you notice? Is this always true?

e Extend your work to investigate the graphical relationship between two quantities that are inversely proportional.

Did you know that if you are given the ratio of two numbers and you know either their sum, or their difference or their product, you can find the two numbers? Try it.

IN THIS CHAPTER...

you have seen that:

- to compare the size of two quantities as a ratio, both quantities must be in the same units

- ratios can be simplified by multiplying or dividing all parts of the ratio by the same number

- when two varying quantities are directly proportional, they are always in the same ratio

- when two varying quantities are inversely proportional, the reciprocal of one of them is proportional to the other

21 PYTHAGORAS' THEOREM

AT THE END OF THIS CHAPTER...

you should be able to:

1   State the relationship between the hypotenuse and the other sides of a right-angled triangle.

2   Calculate the length of one side of a right-angled triangle, given the other two sides.

3   Identify Pythagorean triples.

4   Use Perigal's dissection.

5   Determine whether or not a triangle is right-angled given the lengths of the sides.

Did you know that the square root sign ($\sqrt{}$) comes from the first letter of the word *radix* which was the Latin word for root?

BEFORE YOU START

you need to know:
- ✓ how to work with decimals
- ✓ the properties of the special quadrilaterals
- ✓ how to recognise right-angled triangles

KEY WORDS   chord, converse, hypotenuse, significant figures, symmetry

Squares and square roots

The following exercise revises the finding of squares and square roots. Remember that you need a rough estimate first.

Exercise 21a   Use a calculator to find the squares of

a 2.3	**b** 23	**c** 2300	**d** 0.023

a $2.3^2 \approx 2 \times 2 = 4$

   $2.3^2 = 5.29$

b $23^2 \approx 20 \times 20 = 400$

   $23^2 = 529$

c $2300^2 \approx 2000 \times 2000 = 4\,000\,000$

   $2300^2 = 5\,290\,000$

d $0.023^2 \approx 0.02 \times 0.02 = 0.0004$

   $0.023^2 = 0.000\,529$

Always find a rough value of the square of a number because it is easy to press the wrong button on a calculator. Check that your calculator answer is sensible.

Find the squares of the following numbers, giving your answers correct to four significant figures where necessary:

1 6.2	**5** 0.71	**9** 3.12	**13** 5210
2 13.7	**6** 0.059	**10** 0.0312	**14** 52.1
3 242	**7** 0.0017	**11** 9.2	**15** 0.521
4 2780	**8** 312	**12** 92	**16** 0.0521

Find the square roots of the following numbers, giving your answers correct to four significant figures:

17 9.87	**23** 96		
18 19.9	**24** 321		
19 124	**25** 2.62		
20 96 800	**26** 0.062		
21 0.0482	**27** 0.000 78	**29** 0.461	**31** 461
22 0.004 82	**28** 0.5	**30** 4.61	**32** 0.000 461

Tip To estimate a square root, pair off the numbers each way from the decimal point.

Pythagoras' theorem

Pythagoras was a native of Samos who travelled frequently to Egypt for the purpose of education. The Egyptians are believed to have known this theorem many years before Pythagoras was born. It is very likely that the Egyptian priests explained this theorem to Pythagoras. Indeed it is claimed that Pythagoras offered a sacrifice to the Muses when the Egyptian priests explained to him the properties of the right-angled triangle.

We can show that the properties involve a relationship between the lengths of the three sides.

Exercise 21b

First we will collect some evidence. Bear in mind that, however accurate your drawing, it is not perfect.

Construct the triangles in questions **1** to **6** and in each case measure the third side, the hypotenuse.

1

6 cm

8 cm

4

5 cm

12 cm

2

6 cm

10 cm

5

7 cm

9 cm

3

5 cm

8 cm

6

7 cm

12 cm

7 In each of the questions 1 to 6, find the squares of the lengths of the three sides. Write the squares in ascending order (i.e. the smallest first). Can you see a relation between the first two squares and the third square?

Pythagoras' theorem

If your drawings are reasonably accurate you will find that by adding the squares of the two shorter sides you get the square of the hypotenuse.

$$AB^2 = 16$$

$$BC^2 = 9$$

$$AC^2 = 25$$

$$25 = 16 + 9$$

so $$AC^2 = AB^2 + BC^2$$

C

5 cm

3 cm

A

4 cm

B

This result is called Pythagoras' theorem, which states that in a right-angled triangle the square of the hypotenuse is equal to the sum of the squares of the other two sides.

Practical work

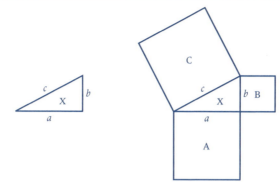

Draw any right-angled triangle and draw the square on each of the three sides. Mark the four areas A, B, C and X as shown in the diagram. Cut out one of each shape and make another three triangles indentical to X. Arrange the shapes in two different ways as shown below. Sketch the two arrangements and mark in as many lengths as possible with a, b or c.

a What can you say about the areas of these two diagrams? Justify your answer.

b If the four triangles marked X are removed from each diagram, what can you say about the areas that remain? What relation does this give for

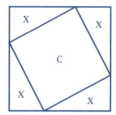 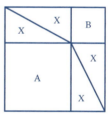

i areas A, B and C   **ii** lengths a, b and c?

Finding the hypotenuse

Give your answers correct to 3 s.f.

In △PQR, $\hat{R} = 90°$, PR = 7 cm and QR = 6 cm.

Find PQ.

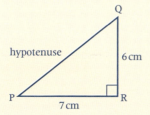

$PQ^2 = PR^2 + QR^2$ (Pythagoras' theorem)

$= 7^2 + 6^2$

$= 49 + 36 = 85$

$PQ = \sqrt{85}$ $(9.---)$

$PQ = 9.22$ cm   correct to 3 s.f.

In the following right-angled triangles find the required lengths.

> **Tip** Start each question by labelling the hypotenuse. Then write down Pythagoras' Theorem in terms of the sides of this triangle.

1 Find AC.

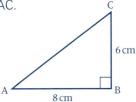

2 Find PR.

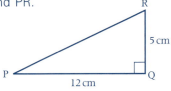

3 Find MN.

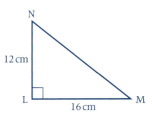

4 Find AC.

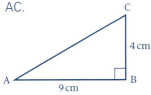

5 Find LN.

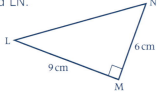

6 Find QR.

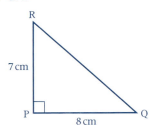

7 Find AC.

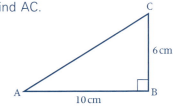

8 Find EF.

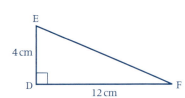

9 Find QR.

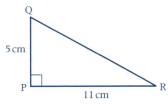

10 Find YZ.

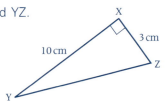

11 In $\triangle ABC$, $\hat{C} = 90°$, $AC = 2$ cm and $BC = 3$ cm. Find AB.

> **Tip** Start by drawing the triangle.

12 In $\triangle DEF$, $\hat{E} = 90°$, $DE = 7$ cm and $EF = 9$ cm. Find DF.

13 In △ABC, $\hat{A} = 90°$, $AB = 4\,m$ and $AC = 5\,m$. Find BC.

14 In △PQR, $\hat{Q} = 90°$, $PQ = 11\,m$ and $QR = 3\,m$. Find PR.

15 In △XYZ, $\hat{X} = 90°$, $YX = 12\,cm$ and $XZ = 2\,cm$. Find YZ.

In △XYZ, $\hat{Z} = 90°$, $XZ = 5.3\,cm$ and $YZ = 3.6\,cm$.
Find XY.

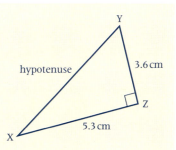

$$XY^2 = XZ^2 + ZY^2 \quad \text{(Pythagoras' theorem)}$$

$$= 5.3^2 + 3.6^2 \qquad\qquad 5.3^2 \approx 5 \times 5 = 25$$

$$= 28.09 + 12.96 \qquad\qquad 3.6^2 \approx 4 \times 4 = 16$$

$$= 41.05$$

$$XY = \sqrt{41.05} = 6.407\ldots \qquad (\sqrt{41.05} = 6.----)$$

Length of $XY = 6.41\,cm$ correct to 3 s.f.

16 Find AC.

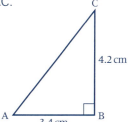

17 Find AC.

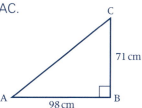

18 Find XY.

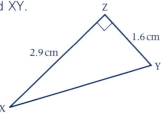

19 Find QR.

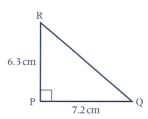

20 Find PR.

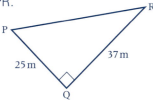

21 Find DF.

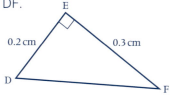

22 In △ABC, $\hat{B} = 90°$, $AB = 7.9\,cm$, $BC = 3.5\,cm$. Find AC.

23 In △PQR, $\hat{Q} = 90°$, $PQ = 11.4\,m$, $QR = 13.2\,m$. Find PR.

24 In △XYZ, $\hat{Z} = 90°$, XZ = 1.23 cm, ZY = 2.3 cm. Find XY.

25 In △ABC, $\hat{C} = 90°$, AC = 32 cm, BC = 14.2 cm. Find AB.

26 In △PQR, $\hat{P} = 90°$, PQ = 9.6 m, PR = 8.8 m. Find QR.

27 In △DEF, $\hat{F} = 90°$, DF = 10.1 cm, EF = 6.4 cm. Find DE.

The 3, 4, 5 triangle

You will have noticed that, in most cases when two sides of a right-angled triangle are given and the third side is calculated using Pythagoras' theorem, the answer is not a rational number. There are a few special cases where all three sides are rational numbers.

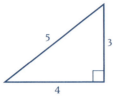

The simplest one is the 3, 4, 5 triangle. Any triangle similar to this has sides in the ratio 3 : 4 : 5 so whenever you spot this case you can find the missing side very easily.

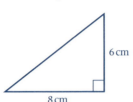

For instance, in the triangle opposite, $6 = 2 \times 3$ and $8 = 2 \times 4$. The triangle is similar to the 3, 4, 5 triangle, so the hypotenuse is 2×5 cm, that is, 10 cm.

The other triangle with exact sides which might be useful is the 5, 12, 13 triangle.

Sets of numbers like { 3, 4, 5 } and { 5, 12, 13 } are called *Pythagorean triples*.

Exercise 21d In △ABC, $\hat{B} = 90°$, AB = 20 cm and BC = 15 cm. Find AC.

Notice that BC = 3 × 5 cm and AB = 4 × 5 cm so the sides about the right angle are in the ratio 3 : 4.

ABC is therefore a '3, 4, 5 triangle'.

so   AC = 5 × 5 cm (3, 4, 5 △)

   = 25 cm

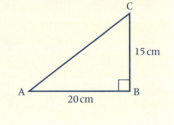

In each of the following questions, decide whether the triangle is similar to the 3, 4, 5 triangle or to the 5, 12, 13 triangle or to neither. Find the hypotenuse, using the method you think is easiest.

1

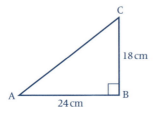

2

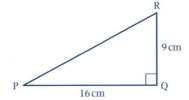

3

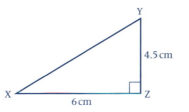

6

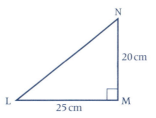

4

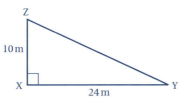

7

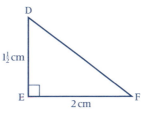

5

8

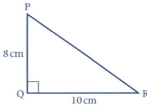

Investigation

Find more Pythagorean triples.

Tip The two larger numbers are always consecutive whole numbers.

Finding one of the shorter sides

If we are given the hypotenuse and one other side we can find the third side.

Exercise **21e**

In $\triangle$ABC, $\hat{B} = 90°$, AB = 7 cm and AC = 10 cm. Find BC.

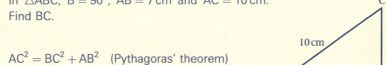

$AC^2 = BC^2 + AB^2$ (Pythagoras' theorem)

$10^2 = BC^2 + 7^2$

$100 = BC^2 + 49$

$51 = BC^2$ (taking 49 from both sides)

$BC = \sqrt{51} = 7.141\ldots$

Length of BC = 7.14 cm   correct to 3 s.f.

> **Tip** Start each question by writing Pythagoras' theorem for the triangle.

1 Find BC.

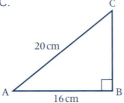

2 Find LM.

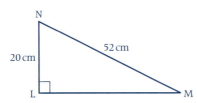

3 Find PQ.

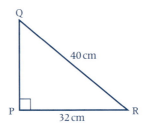

4 Find YZ.

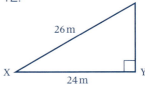

Give your answers to questions **5** to **14** correct to 3 s.f.

5 Find BC.

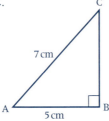

6 Find RQ.

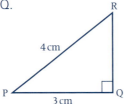

7 Find AB.

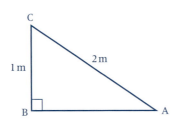

8 Find AB.

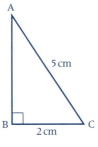

9 Find EF.

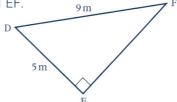

10 Find BC.

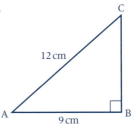

11 Find XY.

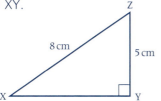

12 Find QR.

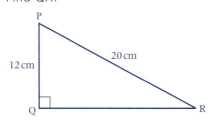

13 Find XY.

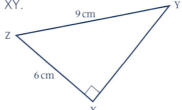

14 Find PQ.

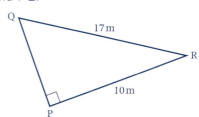

Mixed examples

*Exercise **21f***

In each case find the length of the missing side. If any answers are not exact give them correct to 3 s.f.

If you notice a 3, 4, 5 triangle or a 5, 12, 13 triangle, you can use it to get the answer quickly.

1 Find AC.

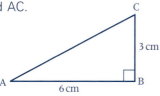

2 Find LM.

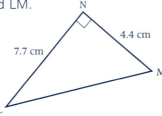

3 Find AB.

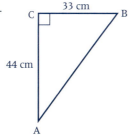

4 Find PR.

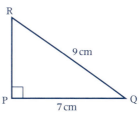

5 Find DF.

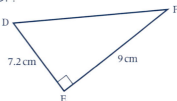

6 Find YZ.

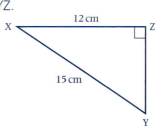

7 In $\triangle ABC$, $\hat{B} = 90°$, $AB = 2\,cm$, $AC = 4\,cm$. Find BC.

8 In $\triangle ABC$, $\hat{B} = 90°$, $AB = 1.25\,m$, $CA = 8.25\,m$. Find BC.

9 In $\triangle PQR$, $\hat{Q} = 90°$, $PQ = 65\,cm$, $QR = 60\,cm$. Find PR.

10 One number in a Pythagorean triple is 25. Find the other two.

11 In $\triangle ABC$, $\hat{C} = 90°$, $AB = 17.5\,cm$, $AC = 16.8\,cm$. Use the Pythagorean triple you found in question **10** to find BC.

12 In $\triangle DEF$, $\hat{D} = 90°$, $DE = 124\,cm$, $DF = 234\,cm$. Find EF.

13 In $\triangle ABC$, $\hat{C} = 90°$, $AC = 3.2\,cm$, $AB = 9.81\,cm$. Find BC.

14 In $\triangle XYZ$, $\hat{Y} = 90°$, $XY = 1.5\,cm$, $YZ = 2\,cm$. Find XZ.

15 In $\triangle PQR$, $\hat{P} = 90°$, $PQ = 5.1\,m$, $QR = 8.5\,m$. Find PR.

16 In $\triangle ABC$, $\hat{C} = 90°$, $AB = 92\,cm$, $BC = 21\,cm$. Find AC.

17 In $\triangle XYZ$, $\hat{X} = 90°$, $XY = 3.21\,m$, $XZ = 1.43\,m$. Find YZ.

Pythagoras' theorem using areas

The area of a square is found by squaring the length of its side, so we can represent the squares of numbers by areas of squares.

This gives us a version of Pythagoras' theorem, using areas:

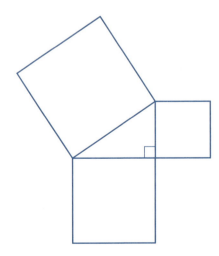

In a right-angled triangle, the area of the square on the hypotenuse is equal to the sum of the areas of the squares on the other two sides.

Practical work

Perigal's dissection

On squared paper, and using 1 cm to 1 unit, copy the left-hand diagram. Make sure that you draw an accurate square on the hypotenuse either by counting the squares or by using a protractor and a ruler. D is the centre of the square on AB. Draw a vector $\overrightarrow{DE}$ so that $\overrightarrow{DE} = \frac{1}{2}\overrightarrow{AC}$, i.e. DE must be parallel to AC.

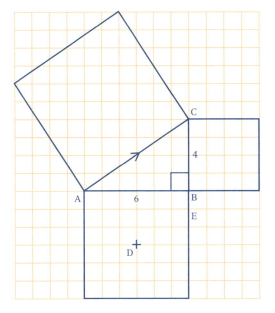

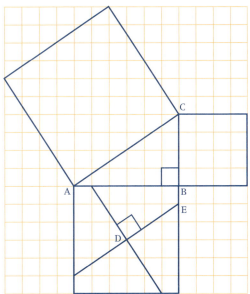

Now complete the drawing as in the right-hand diagram. Make sure that the angles at D are right angles.

Cut out the smallest square and the four pieces from the middle-sized square. These five pieces can be fitted exactly, like a jigsaw, into the outline of the biggest square.

Finding lengths in an isosceles triangle

An isosceles triangle can be split into two right-angled triangles and this can sometimes help when finding missing lengths, as it did when finding angles.

Exercise 21g   In △ABC, AB = BC = 12 cm and AC = 8 cm.
Find the height of the triangle.

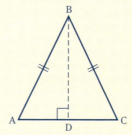 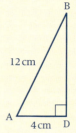

Join B to D, the mid point of AC. Then we draw one of the right-angled triangles

$$AB^2 = AD^2 + BC^2 \quad \text{(Pythagoras' theorem)}$$

$$12^2 = 4^2 + BD^2$$

$$144 = 16 + BD^2$$

$$128 = BD^2 \quad \text{(taking 16 from both sides)}$$

$$BD = \sqrt{128}.$$

$$BD = 11.31\ldots = 11.3 \text{ (correct to 3 s.f.)}$$

∴   length of BD is 11.3 cm

So the height of the triangle is 11.3 cm correct to 3 s.f.

Give your answers correct to 3 s.f.

1

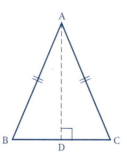

AB = AC = 16 cm. BC = 20 cm. Find the height of the triangle.

2

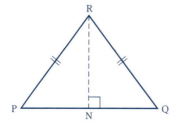

PQ = 12 cm, PR = RQ. The height of the triangle is 8 cm. Find PR.

3

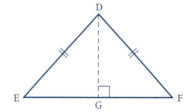

DE = DF = 20 cm. The height of the triangle is 13.2 cm. Find EG and hence EF.

4   In △ABC, AB = BC = 5.2 cm and AC = 6 cm. Find the height of the triangle.

5   In △PQR, PQ = QR = 9 cm and the height of the triangle is 7 cm. Find the length of PR.

Finding the distance of a chord from the centre of a circle

AB is a chord of a circle with centre O. OA and OB are radii and so are equal. Hence triangle OAB is isosceles and we can divide it through the middle into two right-angled triangles.

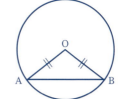

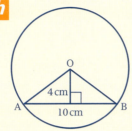

A chord AB of a circle with centre O is 10 cm long. The chord is 4 cm from O. Find the radius of the circle.

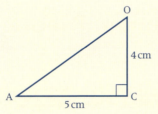

Draw one of the triangles, then use Pythagoras' Theorem on this triangle. Label the third vertex C.

The distance from the centre is the perpendicular distance so OC = 4 cm. From symmetry AC = 5 cm.

$$OA^2 = AC^2 + OC^2 \qquad \text{(Pythagoras' theorem)}$$

$$= 5^2 + 4^2$$

$$= 25 + 16$$

$$= 41$$

$$OA = \sqrt{41} = 6.403\ldots$$

$$OA = 6.40 \text{ correct to 3 s.f.}$$

The radius of the circle is 6.40 cm correct to 3 s.f.

Give your answers correct to 3 s.f.

1

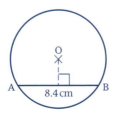

A circle with centre O has a radius of 5 cm. AB = 8.4 cm. Find the distance of the chord from the centre of the circle.

2

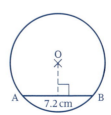

O is the centre of the circle and AB is a chord of length 7.2 cm. The distance of the chord from O is 3 cm. Find the radius of the circle.

<u>**3**</u> In a circle with centre O, a chord AB is of length 7 cm. The radius of the circle is 11 cm. Find the distance of the chord from O.

<u>**4**</u> In a circle with centre O and radius 17 cm, a chord AB is of length 10.4 cm. Find the distance of the chord from O.

<u>**5**</u> In a circle with centre P and radius 7.6 cm, a chord QR is 4.2 cm from P. Find the length of the chord.

 ## Puzzle

Molly has two 10 c stamps and two 5 c stamps.

She wants to stick then on an envelope as a block of four as shown.

How many different arrangements are possible?

Problems using pythagoras' theorem

Exercise 21i   A man starts from A and walks 4 km due north to B, then 6 km due west to C. Find how far C is from A.

Draw the triangle and then use Pythagoras' theorem.

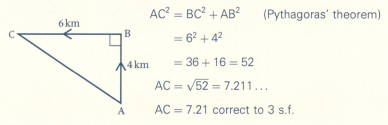

$$AC^2 = BC^2 + AB^2 \quad \text{(Pythagoras' theorem)}$$
$$= 6^2 + 4^2$$
$$= 36 + 16 = 52$$
$$AC = \sqrt{52} = 7.211\ldots$$
$$AC = 7.21 \text{ correct to 3 s.f.}$$

So the distance of C from A is 7.21 km, correct to 3 s.f.

Give your answers correct to 3 s.f.

<u>**1**</u> A ladder 3 m long is leaning against a wall. Its foot is 1.5 m from the foot of the wall. How far up the wall does the ladder reach?

<u>**2**</u> ABCD is a rhombus. AC = 10 cm and BD = 12 cm. Find the length of a side of the rhombus.

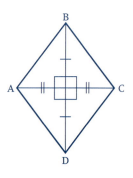

3 Find the length of a diagonal of a square of side 10 cm.

4 A hockey pitch measures 55 m by 90 m. Find the length of a diagonal of the pitch.

5 A wire stay 11 m long is attached to a telegraph pole at a point A, 8 m up from the ground. The other end of the stay is fixed to a point B, on the ground. How far is B from the foot of the telegraph pole?

6 In the kite ABCD, $\hat{A} = \hat{C} = 90°$. DC = 41 cm and BC = 62 cm. Find the length of the diagonal BD.

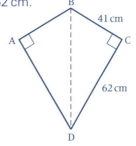

7 A diagonal of a football pitch is 130 m long and the long side measures 100 m. Find the length of the short side of the pitch.

8 The diagram shows the side view of a coal bunker. Find the length of the slant edge.

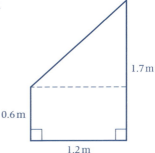

9 The slant height of a cone is 15 cm and the base radius is 5 cm. Find the height of the cone.

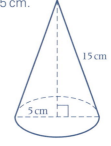

10 A man starts from A and walks 6.5 km due south to B; then he walks due east to C. He is then 9 km from A. How far is C from B?

11 A is the point (3, 1) and B is the point (7, 9). Find the length of AB.

> **Tip** First draw a diagram with AB the hypotenuse of a right-angled triangle. Mark the lengths of the other two sides.

12 A ship sails 32 nautical miles due north then 22 nautical miles due east. How far is it from its starting point?

13 A pole 4.5 m high stands on level ground. It is supported in a vertical position by two wires attached to its top and to points on opposite sides of the pole each 3.2 m from the foot of the pole. How long is each wire?

14 The diagonal AC of a rectangle ABCD is 0.67 m long and side AB is 0.32 m long. How long is side BC?

15 Find the length of the diagonal of a square of side 15 cm.

16 ABCD is a kite and AC is its line of symmetry. $\hat{B} = \hat{D} = 90°$, AB = 36 cm and BC = 16 cm. Find AC

17

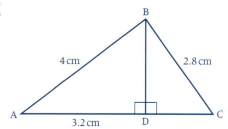

In the figure, $A\hat{D}B = 90°$
AB = 4 cm, AD = 3.2 cm and BC = 2.8 cm.

Find   **a** BD   **b** AC.

Is $A\hat{B}C$ a right angle?
Give a reason for your answer.

18

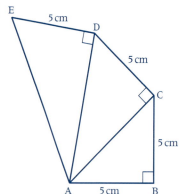

a *Construct* the figure in the diagram, starting with △ABC then adding △ADC and △ADE.

b Measure AC, AD and EA.

c *Calculate* AC, AD and AE and check the accuracy of your drawing.

> **Tip** You need a sharp pencil for this question.

19 Construct a right-angled triangle, choosing whole numbers of centimetres for the lengths of the two shorter sides, such that the hypotenuse will be $\sqrt{65}$ cm long. Check the accuracy of your drawing by measuring the hypotenuse and by calculating $\sqrt{65}$.

20 Find the length of the hypotenuse of a right-angled triangle whose sides are x cm, $(x + 1)$ cm and $(x + 3)$ cm.

21

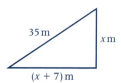

The lengths of the three edges of a triangular flower bed are given on the diagram. The length of the longest side is 35 m and the triangle is right-angled. Find the lengths of the sides containing the right angle.

22 N is the midpoint of the base BC of a triangle ABC. If AN $= x$ cm, BC $= (2x + 14)$ cm, AC $= (x + 8)$ cm and AB $=$ AC, form an equation in x and solve it. Hence find the length of the base and height of the triangle ABC.

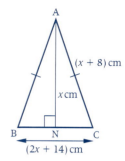

23 The sides of a square of side x cm are all extended by 4 cm so that their extremities, when joined, form another square. The area of this square is five times the area of the original square. Form an equation in x and solve it to find the length of a side of the original square.

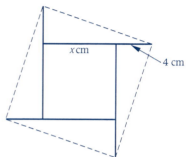

The converse of Pythagoras' theorem

If we are given three sides of a triangle, we can tell whether or not the triangle contains a right angle. Bear in mind that, *if* there is a right angle the longest side will be the hypotenuse.

Exercise **21j**   Are the following triangles right-angled?

a   △ABC: AB $=$ 17 cm, BC $=$ 8 cm, CA $=$ 15 cm

b   △PQR: PQ $=$ 15 cm, PR $=$ 7 cm, RQ $=$ 12 cm

a

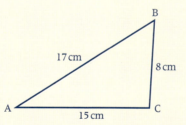

$AB^2 = 17^2 = 289$

$AC^2 + BC^2 = 15^2 + 8^2$

$= 225 + 64$

$= 289$

∴ $AC^2 + BC^2 = AB^2$

∴   by Pythagoras' theorem, $\widehat{C} = 90°$.

b

$PQ^2 = 15^2 = 225$

$PR^2 + RQ^2 = 12^2 + 7^2$

$= 144 + 49$

$= 193$

∴ $PR^2 + RQ^2 \neq PQ^2$

∴   the triangle is not right-angled.

Are the following triangles right-angled?

> **Tip** Start by drawing a triangle and marking the sides. Then find the square of the longest side and compare this value with the sum of the squares of the other two sides.

1 Triangle ABC: AB = 48 cm, BC = 64 cm and CA = 80 cm.

2 Triangle PQR: PQ = 2.1 cm, QR = 2.8 cm and RP = 3.5 cm.

3 Triangle LMN: LM = 6 cm, MN = 7.2 cm and NL = 9 cm.

4 Triangle ABC: AB = 9.2 cm, BC = 6.3 cm and CA = 4.6 cm.

5 Triangle DEF: DE = 6.4 cm, EF = 12 cm and DF = 13.6 cm.

6 Triangle XYZ: XY = 32 cm, YZ = 40 cm and ZX = 48 cm.

Exercise **21k**

Find the missing lengths in the following triangles:

1

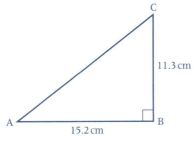

2

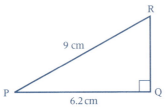

3

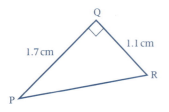

4

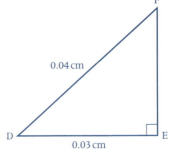

5

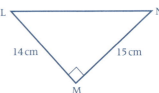

6

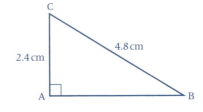

7

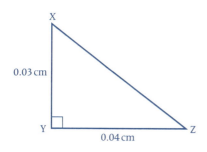

8

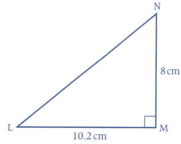

9 In triangle ABC, $\widehat{A} = 90°$, AB = 3.2 cm and BC = 4.8 cm. Find AC.

 Tip Start by drawing a diagram and marking the sides.

10 In triangle PQR, $\widehat{Q} = 90°$, PQ = 56 cm and QR = 32 cm. Find PR.

11 In triangle ABC, AB = 1 cm, BC = 2.4 cm and CA = 2.6 cm. Is $\widehat{B}$ a right angle?

12 In triangle DEF, $\widehat{F} = 90°$, DF = 2.8 cm and DE = 4.2 cm. Find EF.

13 In triangle XYZ, $\widehat{Y} = 90°$, XY = 17 cm and YZ = 20 cm. Find XZ.

14 In triangle LMN, NL = 25 cm, LM = 24 cm and MN = 7 cm. Is the triangle right-angled? If it is, which angle is 90°?

Do you think that a person could be his own worst enemy?

A great mathematician named Evariste Galois (1811–1832) was considered such a person. He had a short unhappy life filled with hate and conceit. He hated school and his teachers whom he considered to be very stupid. The teachers thought that he was bad, stupid and strange. He studied mathematics and found that he was a genius in the subject. At the age of seventeen he wrote some ideas and sent them to the French Academy. While awaiting the reply on his work from the academy his father killed himself. This caused him to hate even more. He was expelled from university for inciting a riot for the French Revolution. He fell in love but his girlfriend left him. This caused him to hate even more. He was killed in a duel at the age of twenty years. The ideas he wrote down the night before his death were finally understood around 1900, many years after his death.

IN THIS CHAPTER...

you have seen that:

- Pythagoras' theorem states that, in a right-angled triangle, the square of the hypotenuse is equal to the sum of the squares of the other two sides,

  i.e. in this triangle
  $$AC^2 = AB^2 + BC^2$$

  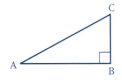

- some special triplets of numbers like 3, 4, 5 and 5, 12, 13, or multiples of these, give a right-angled triangle whatever unit of measurement is used

- if, in a triangle ABC, $AC^2 = AB^2 + BC^2$, then the triangle contains a right angle and AC is the hypotenuse, but if AC is the longest side and $AC^2 \neq AB^2 + BC^2$, then the triangle is not right-angled

TRIGONOMETRY

AT THE END OF THIS CHAPTER...

you should be able to:

1   Define sine, cosine and tangent of an angle in a right-angled triangle.

2   Use tables or a calculator to find the sine, cosine or tangent of a given angle.

3   Use tables or a calculator to find an angle given its sine, cosine or tangent.

4   Calculate the size of an angle in a right-angled triangle, given the lengths of the sides of the triangle.

5   Calculate the length of a side of a right-angled triangle, given one side and another angle.

6   Draw diagrams to show angles of elevation or depression.

7   Use trigonometric ratios to solve problems on angles of elevation and depression.

8   Use Pythagoras' Theorem and trigonometry to solve problems in three dimensions.

Did you know that the word 'Trigonometry' first appears in the English translation in 1614 of a book written by Bartholomeo Pitiscus (1561–1613) and published in 1595? The full title in the English translation is 'Trigonometry: or The Doctrine of Triangles'.

BEFORE YOU START

you need to know:
✓ how to work with decimals and fractions
✓ how to solve equations
✓ the properties of isosceles triangles
✓ the properties of similar triangles
✓ the meaning of three-figure bearings
✓ Pythagoras' theorem

KEY WORDS   adjacent side, angle of depression, angle of elevation, cosine of an angle, cuboid, hypotenuse, opposite side, pentagon, pyramid, right-angled triangle, sine of an angle, tangent of an angle, wedge

Tangent of an angle

In Book 2, we worked with right-angled triangles. This chapter is either a reminder of the work you did then or, if it is new to you, an introduction.

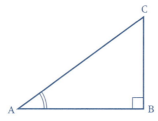

In triangle ABC, AC is the *hypotenuse*, opposite to the right angle.

BC is the *opposite side* to angle A.

AB is the *adjacent side* (or neighbouring side) to angle A.

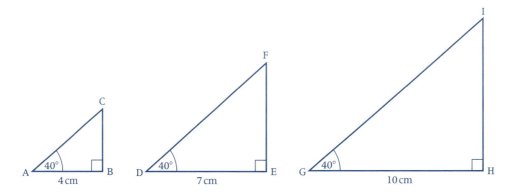

The three triangles above are similar, so their sides are proportional.

In particular, the opposite side and the adjacent side are in the same ratio in all three triangles and also in any other right-angled triangle containing an angle of 40°.

This ratio is called the *tangent* of 40° or, in shortened form, tan 40°. Its size is stored, together with the tangents of other angles, in natural tangent tables and in some calculators.

$$\tan \widehat{A} = \frac{\text{opposite side}}{\text{adjacent side}}$$

Exercise 22a

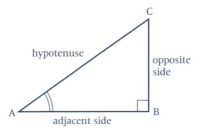

In questions **1** to **6**, copy the diagram. Identify the hypotenuse and the sides opposite and adjacent to the marked angle:

1

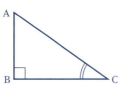

4

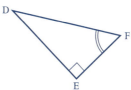

2

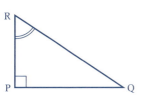

5

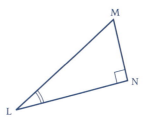

3

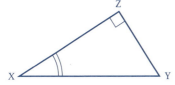

6

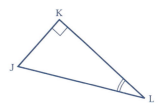

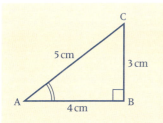

In $\triangle ABC$, $\widehat{B} = 90°$, $AB = 4\,cm$, $BC = 3\,cm$ and $AC = 5\,cm$.

Write down $\tan \widehat{A}$ as a fraction and as a decimal.

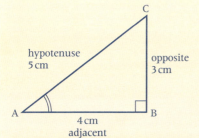

(First identify the sides and mark them on the diagram.)

$$\tan \widehat{A} = \frac{\text{opposite}}{\text{adjacent}}$$

$$= \frac{3}{4}$$

$$= 0.75$$

In each of the following questions write down the tangent of the marked angle as a fraction and as a decimal (correct to four decimal places where necessary):

Tip   Mark the sides with respect to the angle you need to find the tangent of.

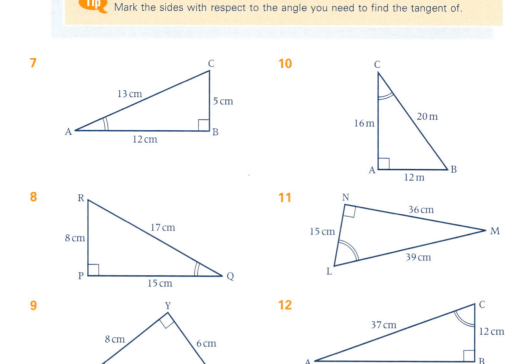

7

8

9

10

11

12

Using a calculator

To find the tangent of an angle, enter the size of the angle, then press the 'tan' button. Write the answer correct to four decimal places.

$$\tan 42.4° = 0.9131$$

To find an angle given its tangent, enter the value of the tangent and then press the inverse button followed by the tangent button. Write down the size of the angle correct to one decimal place.

If these instructions do not work consult the manual for your calculator.

Exercise 22b

Find the tangents of the following angles:

1 62°

2 14°

3 30.5°

4 16.8°

5 4.6°

6 72°

7 78.4°

8 45°

9 30°

10 48.2°

11 3°

12 29.4°

Find the angles whose tangents are given in questions **13** to **24**:

13 0.179	**16** 0.4326	**19** 0.9213	**22** 2.683
14 0.356	**17** 1.362	**20** 0.8	**23** 0.924
15 1.43	**18** 0.632	**21** 0.3214	**24** 0.0024

Finding an angle

Exercise 22c In triangle ABC, B̂ = 90°, AB = 12 cm, BC = 9 cm and AC = 15 cm. Find Â.

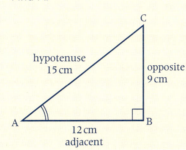

$$\tan \hat{A} = \frac{opp}{adj} = \frac{9}{12}$$
$$= 0.75$$
$$\hat{A} = 36.9°$$

Use the information given on the diagrams to find Â:

Tip Draw the triangle, mark the angle required, then label the sides with respect to this angle.

1
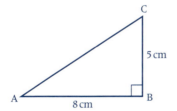

3
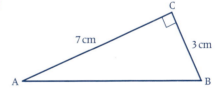

2
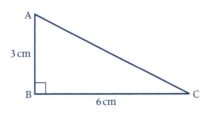

4
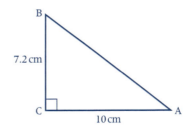

5 In triangle PQR, P̂ = 90°, QP = 6 cm and PR = 10 cm. Find R̂.

6 In triangle XYZ, Ŷ = 90°, XY = 4 cm and YZ = 5 cm. Find X̂.

7 In triangle LMN, L̂ = 90°, LM = 7.2 cm and LN = 6.4 cm. Find N̂.

8 In triangle DEF, D̂ = 90°, DE = 210 cm and DF = 231 cm. Find Ê.

9 In △ABC, Ĉ = 90°, AC = 3.2 m and BC = 4.7 m. Find B̂.

Finding a side

In triangle ABC, $\widehat{B} = 90°$, $AB = 4\,cm$ and $\widehat{A} = 32°$. Find BC.

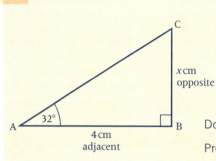

$$\frac{x}{4} = \frac{\text{opp}}{\text{adj}} = \tan 32°$$

$$\frac{x}{4} = 0.6248\ldots$$

$$\cancel{4} \times \frac{x}{\cancel{4}} = 4 \times 0.6248\ldots$$

Do not clear the display on your calculator.

Press

$$x = 2.4994\ldots$$

$$\therefore \quad BC = 2.50\,cm \text{ (to 3 s.f.)}$$

Use the information given in the diagram to find the required side:

> **Tip** When you use a calculator write down the first four figures in the display for intermediate steps. Do not clear the display, use the entry for the next step in the calculation.

1 Find RQ.

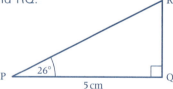

4 Find PR.

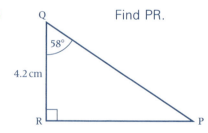

2 Find BC.

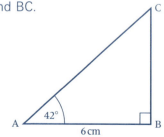

5 Find AB.

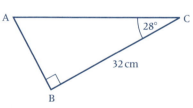

3 Find YZ.

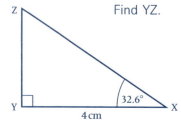

6 Find LN.

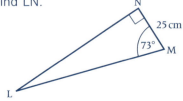

7 In triangle ABC, $\widehat{B} = 90°$, $\widehat{A} = 32°$ and AB = 9 cm. Find BC.

8 In triangle DEF, $\widehat{D} = 90°$, $\widehat{E} = 48°$ and DE = 20 cm. Find DF.

9 In triangle PQR, $\widehat{R} = 90°$, $\widehat{Q} = 10°$ and RQ = 16 cm. Find PR.

10 In triangle XYZ, $\widehat{Z} = 90°$, $\widehat{Y} = 67°$ and ZY = 3.2 cm. Find XZ.

In $\triangle$ABC, $\widehat{B} = 90°$, $\widehat{A} = 24°$ and BC = 6 cm. Find AB.

(It is easier to find AB if it is on top of the tangent ratio, i.e. if AB is the opposite side. AB is opposite to $\widehat{C}$, so find $\widehat{C}$ first.)

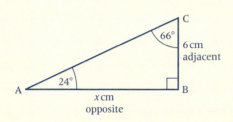

$\widehat{C} = 66°$ ($\angle$s of a triangle)

$$\frac{x}{6} = \frac{opp}{adj} = \tan 66°$$

$$\frac{x}{6} = 2.246$$

$$\cancel{6} \times \frac{x}{\cancel{6}} = 6 \times 2.246$$

$$x = 13.476$$

$\therefore$ AB is 13.5 cm   (to 3 s.f.)

Use the information given in the diagram to find the required side. It may be necessary to find the third angle of the triangle first.

11 Find AB.

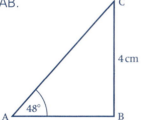

12 Find PQ.

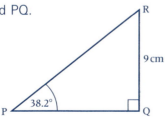

13 Find NL.

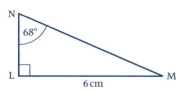

14 Find ZY.

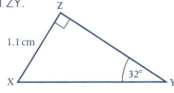

15 Find DE.

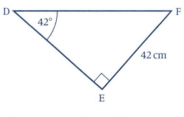

16 Find AC.

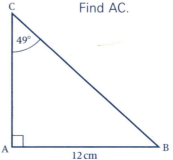

17 In triangle PQR, $\hat{P} = 90°$, $\hat{Q} = 52°$ and PR = 6 cm. Find QP.

18 In triangle ABC, $\hat{A} = 90°$, $\hat{B} = 31°$ and AC = 220 cm. Find AB.

19 In triangle XYZ, $\hat{Z} = 90°$, $\hat{X} = 67°$ and YZ = 2.3 cm. Find XZ.

20 In triangle LMN, $\hat{L} = 90°$, $\hat{M} = 9°$ and LN = 11 m. Find LM.

Using the hypotenuse

So far, we have used only the opposite and adjacent sides. If we wish to use the hypotenuse we need different ratios.

The sine of an angle

For an angle in a right-angled triangle, the name given to the ratio $\dfrac{\text{opposite side}}{\text{hypotenuse}}$ is the *sine* of the angle.

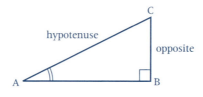

In triangle ABC $\qquad \dfrac{BC}{AC} = \sin \hat{A}$

The use of sines is similar to the use of tangents.

Exercise 22e

Find the sines of the following angles:

1 62.4°

2 70°

3 14.3°

4 9°

5 15.2°

6 37.5°

7 59.6°

8 30°

9 82°

10 27.8°

11 15.8°

12 87.2°

> **Tip** Press 62.4 sin (or sin 62.4 if this does not work).

Find the angles whose sines are given:

13 0.271

14 0.442

15 0.524

16 0.909

17 0.6664

18 0.3720

19 0.614

20 0.7283

21 0.1232

> **Tip** Press 0.271 sin⁻¹ (or sin⁻¹ 0.271 if this does not work).

Exercise 22f

In triangle ABC, $\widehat{B} = 90°$, $BC = 3\,cm$ and $AC = 7\,cm$. Find $\widehat{A}$.

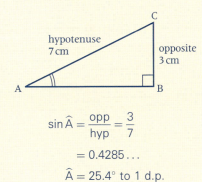

$$\sin \widehat{A} = \frac{opp}{hyp} = \frac{3}{7}$$

$$= 0.4285\ldots$$

$$\widehat{A} = 25.4° \text{ to 1 d.p.}$$

Use the information given in the diagram to find the marked angle:

Tip Draw the triangle and label the sides with respect to the required angle.

1

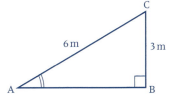

2

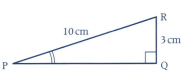

3

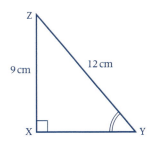

4

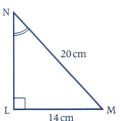

5

6

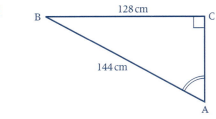

7 In triangle ABC, $\widehat{C} = 90°$, $BC = 7\,cm$ and $AB = 10\,cm$. Find $\widehat{A}$.

8 In triangle PQR, $\widehat{Q} = 90°$, $PQ = 30\,cm$ and $PR = 45\,cm$. Find $\widehat{R}$.

9 In triangle LMN, $\widehat{M} = 90°$, $MN = 3.2\,cm$ and $LN = 8\,cm$. Find $\widehat{L}$.

10 In triangle DEF, $\widehat{E} = 90°$, $EF = 36\,cm$ and $DF = 108\,cm$. Find $\widehat{D}$.

In triangle PQR, $\widehat{P} = 90°$, $\widehat{Q} = 32.4°$ and RQ = 4 cm. Find PR.

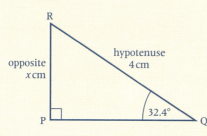

$$\frac{x}{4} = \frac{\text{opp}}{\text{hyp}} = \sin 32.4°$$

$$\require{cancel}\cancel{4} \times \frac{x}{\cancel{4}} = 4 \times 0.5358\ldots$$

$$x = 2.1433\ldots$$

$$\therefore \quad PR = 2.14 \text{ cm (to 3 s.f.)}$$

Use the information given in the diagram to find the required length:

11 Find AC.

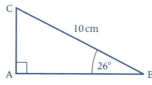

12 Find XY.

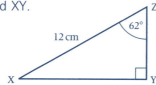

13 Find EF.

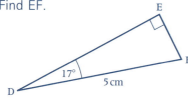

14 Find PR.

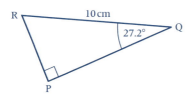

15 Find BC.

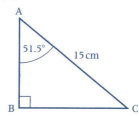

16 Find PR.

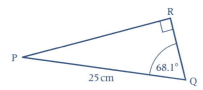

17 In triangle ABC, $\widehat{A} = 90°$, BC = 11 cm and $\widehat{C} = 35°$. Find AB.

18 In triangle PQR, $\widehat{P} = 90°$, QR = 120 m and $\widehat{Q} = 10.5°$. Find PR.

19 In triangle XYZ, $\widehat{X} = 90°$, YZ = 3.6 cm and $\widehat{Y} = 68°$. Find XZ.

20 In triangle DEF, $\widehat{F} = 90°$, DE = 48 m and $\widehat{D} = 72°$. Find EF.

The cosine of an angle

For an angle in a right-angled triangle, the name given to the ratio $\dfrac{\text{adjacent side}}{\text{hypotenuse}}$ is the *cosine* of the angle.

In triangle ABC $\quad \dfrac{AB}{AC} = \cos \widehat{A}$

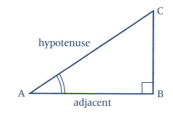

Exercise 22g

Find the cosines of the following angles:

1 32°   **3** 82°   **5** 60°   **7** 52.1°

2 41.8°   **4** 47.8°   **6** 15.6°   **8** 49°

Find the angles whose cosines are given:

9 0.347   **11** 0.719   **13** 0.6281   **15** 0.865

10 0.936   **12** 0.349   **14** 0.3149   **16** 0.014

Exercise 22h

In triangle ABC, $\hat{B} = 90°$, AC = 20 cm and AB = 15 cm. Find $\hat{A}$.

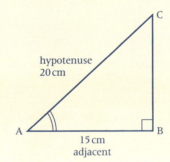

$$\cos\hat{A} = \frac{adj}{hyp} = \frac{15}{20}$$

$$= 0.75$$

$$\hat{A} = 41.40\ldots°$$

$$= 41.4° \text{ to 1 d.p.}$$

Find the marked angles in the following triangles:

1

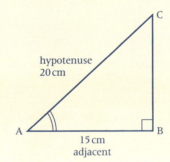

2

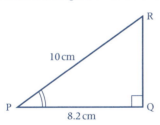

3

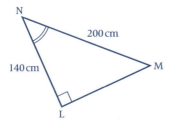

4

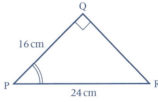

5

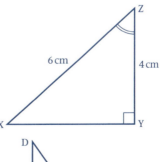

6

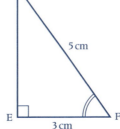

327

7 In triangle ABC, $\hat{B} = 90°$, AB = 3.2 cm and AC = 5 cm. Find $\hat{A}$.

8 In triangle PQR, $\hat{P} = 90°$, QR = 12 cm and PQ = 4.8 cm. Find $\hat{Q}$.

9 In triangle LMN, $\hat{L} = 90°$, MN = 20 cm and ML = 3 cm. Find $\hat{M}$.

10 In triangle DEF, $\hat{F} = 90°$, DE = 18 cm and DF = 16.2 cm. Find $\hat{D}$.

11 In triangle XYZ, $\hat{Z} = 90°$, XY = 14 m and YZ = 11.6 m. Find $\hat{Y}$.

In triangle XYZ, $\hat{X} = 90°$, $\hat{Y} = 27°$ and ZY = 3.2 cm. Find XY.

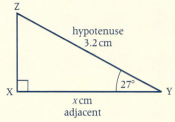

$$\frac{x}{3.2} = \frac{\text{adj}}{\text{hyp}} = \cos 27°$$

$$\frac{x}{3.2} = 0.8910$$

$$\cancel{3.2}^{1} \times \frac{x}{\cancel{3.2}_{1}} = 0.8910 \times 3.2$$

$$x = 2.8512$$

∴ XY = 2.85 cm   (to 3 s.f.)

Use the information given in the diagrams to find the required lengths:

12 Find AB.

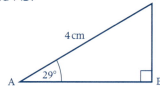

15 Find MN.

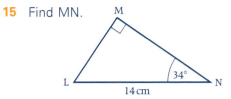

13 Find XY.

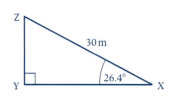

16 Find PQ.

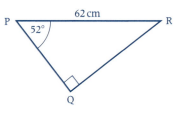

14 Find ED.

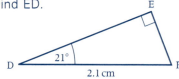

17 Find YZ.

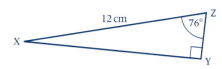

18 In triangle PQR, $\hat{Q} = 90°$, $\hat{P} = 31°$ and PR = 20 cm. Find PQ.

19 In triangle LMN, $\hat{N} = 90°$, $\hat{L} = 42°$ and LM = 3 cm. Find LN.

20 In triangle DEF, $\hat{D} = 90°$, $\hat{E} = 68°$ and EF = 11 cm. Find DE.

21 In triangle XYZ, $\hat{Z} = 90°$, $\hat{Y} = 15°$ and YX = 14 cm. Find ZY.

Summary

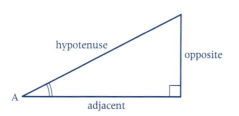

$$\text{Sin } \widehat{A} = \frac{\textbf{O}\text{pposite}}{\textbf{H}\text{ypotenuse}} \quad (\,\text{SOH}\,)$$

$$\text{Cos } \widehat{A} = \frac{\textbf{A}\text{djacent}}{\textbf{H}\text{ypotenuse}} \quad (\,\text{CAH}\,)$$

$$\text{Tan } \widehat{A} = \frac{\textbf{O}\text{pposite}}{\textbf{A}\text{djacent}} \quad (\,\text{TOA}\,)$$

Some people remember by using the word 'SOHCAHTOA' or a sentence like 'Some Old Hangars Can Almost Hold Two Old Aeroplanes'.

Sines, cosines and tangents

Exercise 22i

In questions **1** to **8**, find the marked angles.

> **Tip** Remember to label the given sides with respect to the angle required first and then decide which ratio you will have to use.

1

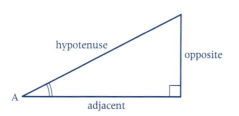

4

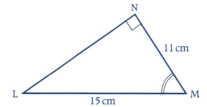

2

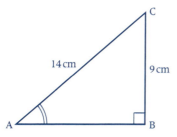

5

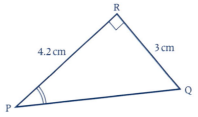

3

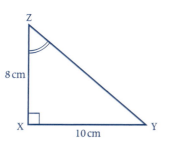

6

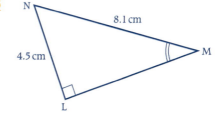

7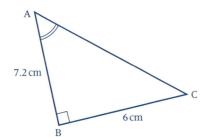

9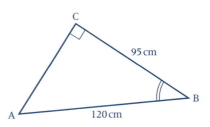

8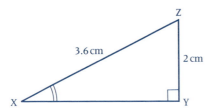

10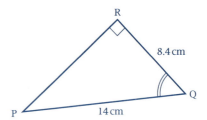

11 In triangle ABC, $\hat{B} = 90°$, AC = 60 cm and BC = 22 cm. Find $\hat{C}$.

12 In triangle PQR, $\hat{R} = 90°$, PQ = 24 cm and QR = 6 cm. Find $\hat{P}$.

13 In triangle ABC, $\hat{B} = 90°$, AC = 1.5 cm and BC = 0.82 cm. Find $\hat{C}$.

14 In triangle PQR, $\hat{R} = 90°$, RQ = 8 cm and RP = 6.2 cm. Find $\hat{Q}$.

15 In triangle DEF, $\hat{F} = 90°$, DF = 16.2 cm and EF = 19.8 cm. Find $\hat{E}$.

16 In triangle XYZ, $\hat{X} = 90°$, YZ = 1.6 m and XY = 1.32 m. Find $\hat{Z}$.

17 In triangle DEF, $\hat{E} = 90°$, DE = 1.9 m and EF = 2.1 m. Find $\hat{F}$.

18 In triangle GHI, $\hat{H} = 90°$, GI = 52 cm and IH = 21 cm. Find $\hat{I}$.

Use the information given in the diagram to find the required length:

19 Find BC.

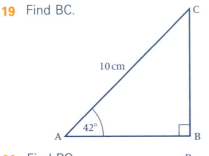

21 Find ZY.

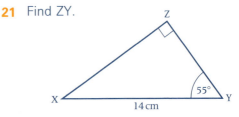

20 Find PQ.

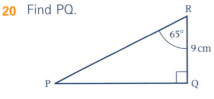

22 Find AB.

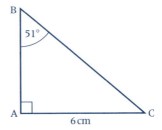

23 Find MN.

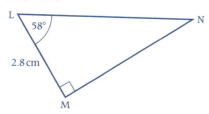

24 Find DE.

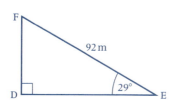

25 Find YZ.

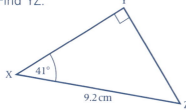

26 Find AB.

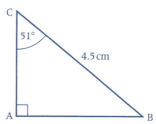

27 Find BC.

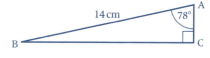

28 Find PQ.

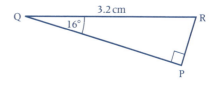

29 In triangle ABC, $\hat{C} = 90°$, $\hat{A} = 78°$ and AC = 24 cm. Find BC.

30 In triangle PQR, $\hat{P} = 90°$, $\hat{Q} = 36°$ and QR = 3.2 cm. Find PQ.

31 In triangle XYZ, $\hat{X} = 90°$, $\hat{Y} = 36°$ and YZ = 17 cm. Find XZ.

32 In triangle DEF, $\hat{F} = 90°$, $\hat{E} = 51°$ and DF = 9.2 cm. Find EF.

33 In triangle LMN, $\hat{M} = 90°$, $\hat{N} = 25°$ and LN = 16 cm. Find MN.

34 In triangle LMN, $\hat{L} = 90°$, $\hat{M} = 56.2°$ and LN = 32 cm. Find ML.

35 In triangle ABC, $\hat{C} = 90°$, $\hat{B} = 72.8°$ and AB = 78 cm. Find AC.

36 In triangle PQR, $\hat{R} = 90°$, $\hat{P} = 31.2°$ and PQ = 117 cm. Find QR.

 Puzzle

As part of an aerobatics display, six aeroplanes fly at the same speed away from each other in a fan of 60° to each other.

This is what it looks like from the ground.

What does it look like to the pilot of aircraft A when he looks back?

Finding the hypotenuse

Up to now, when finding the length of a side, we have been able to form an equation in which our unknown length is on the top of the fraction. If we wish to find the hypotenuse, this is not possible and the equation we form takes slightly longer to solve.

In triangle ABC, $\hat{B} = 90°$, $AB = 8\,cm$ and $\hat{C} = 62°$. Find AC.

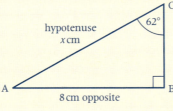

$$\frac{8}{x} = \frac{opp}{hyp} = \sin 62°$$

$$\frac{8}{x} = \sin 62°$$

Multiply both sides by x

$$\not{x} \times \frac{8}{\not{x}} = \sin 62° \times x$$

Divide both sides by $\sin 62°$

$$\frac{8}{\sin 62°} = x$$

$$x = 9.060\ldots$$

$$\therefore \ AC = 9.06\,cm \ \ (to \ 3\,s.f.)$$

Use the information given in the diagram to find the hypotenuse:

1

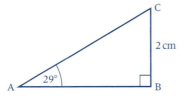

4

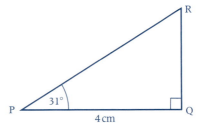

2

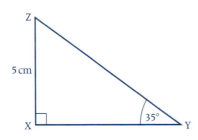

5

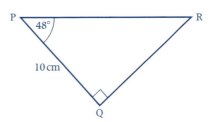

3

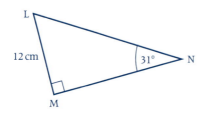

6

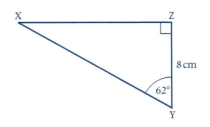

7 In triangle ABC, $\hat{B} = 90°$, $\hat{A} = 43°$ and BC $= 3$ cm. Find AC.

8 In triangle PQR, $\hat{P} = 90°$, $\hat{Q} = 28°$ and PR $= 7$ cm. Find QR.

9 In triangle LMN, $\hat{L} = 90°$, $\hat{M} = 14°$ and LN $= 8$ cm. Find MN.

10 In triangle XYZ, $\hat{Z} = 90°$, $\hat{Y} = 62°$ and ZY $= 20$ cm. Find XY.

Angles of elevation and depression

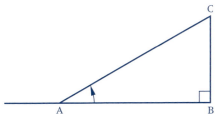

Imagine that you are at A, a point on level ground some distance from a cliff BC. You look horizontally at B, then *elevate* your line of view until you are looking at C.

$\hat{BAC}$ is the *angle of elevation* of C from A.

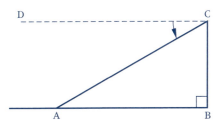

Now imagine you are on top of the cliff, at C. You look out horizontally to D, then *depress* your line of view until you are looking at A.

$\hat{DCA}$ is the *angle of depression* of A from C.

Problems

Exercise 22k   A flagpole stands on level ground. From a point on the ground 30 m away from its foot, the angle of elevation of the top of the pole is 22°. Find the height of the pole.

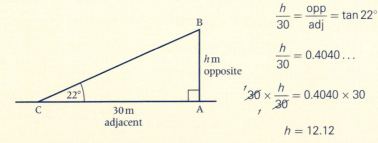

$$\frac{h}{30} = \frac{\text{opp}}{\text{adj}} = \tan 22°$$

$$\frac{h}{30} = 0.4040\ldots$$

$$\cancel{30} \times \frac{h}{\cancel{30}} = 0.4040 \times 30$$

$$h = 12.12$$

The pole is 12.1 m high   (to 3 s.f.)

> **Tip** Draw a diagram. Mark the sides and angles given and required. Label the sides with respect to the angles. Then you can see which ratio you need to use.

1 In triangle ABC, AC = CB = 10 m and $\widehat{A}$ = 64°. Find the height of the triangle.

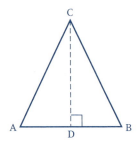

2 From a point on level ground 40 m from the base of a pine tree, the angle of elevation of the top of the tree is 50°. Find the height of the tree.

3 The angle of elevation of the top of a church tower, from a point on level ground 500 m away, is 16°. Find the height of the tower.

4 A is the point (2, 0), B is (8, 0) and C is (8, 5). Calculate the angle between AC and the x-axis.

5 ABCD is a rectangle with AB = 26 cm and BC = 48 cm. Find the angle between the diagonal AC and side AB.

6 A is the point (1, 2), B(3, 2) and C(1, 5). Find $A\widehat{B}C$.

7 ABCD is a rhombus of side 15 cm. The diagonal AC is of length 20 cm. Find the angle between AC and the side CD.

8 A boat C is 200 m from the foot B of a vertical cliff, which is 40 m high. What is the angle of depression of the boat from the top of the cliff?

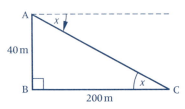

9

In the figure, AB = 10 cm, $\widehat{A}$ = 32° and $A\widehat{B}C = B\widehat{D}C$ = 90°. Copy the figure and then mark in the sizes of the remaining angles. Find

  a BD

  b BC

10 Triangle ABC is an equilateral triangle of side 6 cm.

Find

  a its height

  b its area.

11 A lamp post stands on level ground. From a point which is 10 m from its foot, the angle of elevation of the top is 25°. How high is the lamp post?

12 ABCDE is a regular pentagon of side 10 cm.

Find

  a AÔB and OÂB

  b OF

  c the area of triangle AOB and hence find the area of the pentagon.

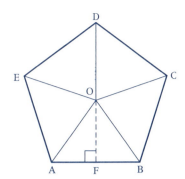

Three-dimensional problems

We can use Pythagoras' theorem and the trigonometry of right-angled triangles to find lengths and angles of various solids.

Cuboids

Exercise 22I

1

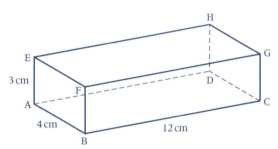

The solid above is a cuboid. Each face is a rectangle.

  a Copy the diagram, noticing that, in the drawing, some faces look like parallelograms.

Which edges are equal in length to EA?

Which edges are equal in length to AB?

Which edges are equal in length to BC?

How many right angles are there?

> **Tip** Remember that a vertical line and a horizontal line that meet always do so at 90°.

  b Join B to E on your diagram and draw triangle ABE (notice that EÂB = 90° and should now be drawn as a right angle).

Find EB by using Pythagoras' theorem, and EB̂A by using its tangent.

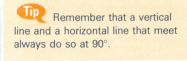

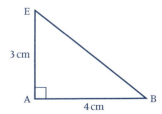

  c Join F and C on your drawing of the cuboid. Draw triangle FBC, marking in all the information you know about it. Find FC and FĈB.

2 a Copy the picture of the cuboid again. Join A and C. Draw triangle ABC (notice $A\hat{B}C = 90°$) and find AC.

b Join E and C. What is the size of $E\hat{A}C$? Draw triangle EAC and mark in the sizes of any sides and angles you know. Find EC and $E\hat{C}A$.

3 Draw a picture of a cuboid like the first one but such that AB = 5 cm, EA = 2 cm and BC = 8 cm.

a Draw the appropriate triangle and find FC.

b Find AF and $F\hat{A}B$.

c Find EG and the angle between FG and EG.

When dealing with solids like cuboids and the other types in this chapter, remember always to make a separate drawing of the triangle you are using.

Exercise **22m**

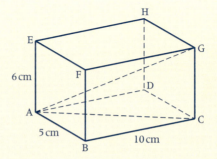

In the cuboid, EA = 6 cm, AB = 5 cm and BC = 10 cm.
Find   **a** AG   **b** $C\hat{A}G$.

(To find AG, we use triangle AGC so we need to calculate AC^2 first.)

Draw the right-angled triangles you are going to use and mark in any measurements.

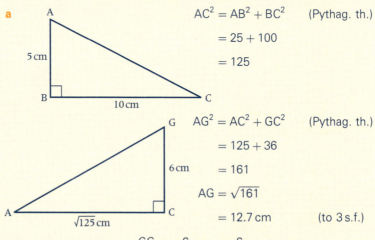

a
$$AC^2 = AB^2 + BC^2 \quad \text{(Pythag. th.)}$$
$$= 25 + 100$$
$$= 125$$

$$AG^2 = AC^2 + GC^2 \quad \text{(Pythag. th.)}$$
$$= 125 + 36$$
$$= 161$$
$$AG = \sqrt{161}$$
$$= 12.7 \text{ cm} \quad \text{(to 3 s.f.)}$$

b In $\triangle CAG$, $\tan C\hat{A}G = \dfrac{GC}{AC} = \dfrac{6}{\sqrt{125}} = \dfrac{6}{11.18\ldots} = 0.5366\ldots$

So $C\hat{A}G = 28.2°$ (to 3 s.f.)

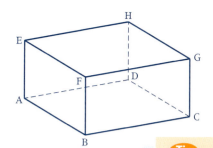

1 In the cuboid above, AB = 8 cm, BC = 12 cm and CG = 5 cm. Find

    a AC   **b** EC   **c** $E\hat{C}A$

> **Tip** Draw triangle ABC first and mark the sides. Use Pythagoras' Theorem to find AC. Now draw triangle AEC and mark the sides. Use this triangle to find EC and $E\hat{C}A$ by using the appropriate trig ratios.

2 In the cuboid above, EF = 3 cm, EA = 2 cm and AD = 6 cm. Find

    a EB   **b** the angle between AB and EB   **c** EG

3 In the cuboid above, HG = 6 cm, GC = 8 cm and BC = 12 cm. Find

    a HC   **b** HB   **c** $H\hat{B}C$

4 In the cuboid above, AD = 11 cm, AB = 10 cm and EA = 12 cm. Find

    a AC   **b** EC   **c** BH   **d** $G\hat{B}C$ **e** $E\hat{G}H$

5 In the cuboid above, BC = 20 cm, CG = 8 cm and DC = 12 cm. Find EC.

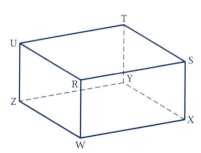

6 In the cuboid above, RS = 12 cm, SX = 7 cm and XY = 9 cm. Find

    a WY   **b** WT   **c** $T\hat{W}Y$

7 In the cuboid above, TX = 10 cm, $T\hat{X}Y = 45°$ and UT = 12 cm. Find

    a XY

    b TY

    c the volume of the cuboid.

8 In the cuboid above, SX = 6 cm, WX = 9 cm and XY = 4 cm. Find

    a $S\hat{W}X$ **b** $S\hat{Y}X$ **c** $S\hat{Z}X$

9 In the cuboid above, WX = 11 cm, XY = 5 cm and WT = 14 cm. Find

    a TY   **b** the surface area of the cuboid.

Investigation

An open rectangular box is *a* units long, *b* units wide and *c* units deep.

The length of the longest straight stick that will fit into this box, without projecting outside it is *d* units.

Find a relationship between d^2 and *a*, *b* and *c*.

Investigate the possible values of *a*, *b*, *c* and d if all four lengths are whole numbers.

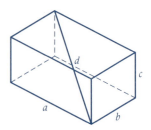

Wedges

1

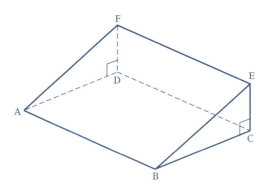

The solid above is a *wedge*. Two faces are right-angled triangles and three faces are rectangles.

a Copy the diagram. (Notice that, in the drawing, each rectangular face looks like a parallelogram.)

AB = 10 cm, BC = 6 cm and CE = 4 cm.
Which edges are equal to AB?
Which edges are equal to BC?
Which edges are equal to EC?
How many right angles are there?

b Draw triangle EBC. Find $E\hat{B}C$ and BE.

c Join A to C. Draw triangle ABC. Find AC and $C\hat{A}B$. Is AC at right angles to EC?

d Join A to E. Draw triangle AEC. Find AE. What other length (not yet drawn in) is equal to AE?

2 Draw a wedge similar to the wedge in question 1 but such that AB = BE = 8 m and $E\hat{B}C = 22°$. Find

    a EC   **b** BC   **c** AC   **d** $E\hat{A}C$

3 Draw a wedge similar to the wedge in question 1 but such that EF = 40 m, EC = 10 m and BÊC = 70°. Find

   a BC   **b** AC   **c** BE   **d** AE   **e** EÂC   **f** EÂF

4 Draw a wedge similar to the wedge in question 1, but such that BC = 4.4 m, AB = 6.8 m and EC = 2 m. Find

   a EB̂C   **b** EÂC

5 Draw a wedge similar to the wedge in question 1, but such that AE = 7 cm, EÂC = 22° and AD = 3 cm. Find

   a EC   **b** BE   **c** AB

Pyramids

Exercise **22p**

1

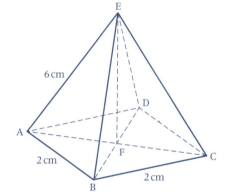

> **Tip** Before you calculate the length of a side or the size of an angle draw the appropriate triangle (or other shape) and mark as many lengths and angles as you can.

The solid above is called a square pyramid because its base is a square. Its vertex, E, is directly above the centre, F, of its base.

   a Copy the diagram. (Start by drawing the base as a parallelogram. Draw the diagonals of the base, to cut at F, and then mark E directly above F.) Name as many right angles as you can find. Which edges are equal to AE?

   b Draw the base ABCD as a square. Find AC and FC.

   c Draw triangle EFC. Find EF and EĈF.

2 Draw a square pyramid similar to the pyramid in question 1, but such that AB = 4 cm and EF = 5 cm.

   a Find AC and AF.

   b Find AE and EÂF.

   c G is the midpoint of AB. Draw triangle EFG and find EG and EĜF.

3 The base of the given pyramid is a rectangle and E is directly above A. AB = 4 cm, EA = BC = 3 cm. Find

   a $E\hat{B}A$ and $E\hat{D}A$

   b AC

   c EC.

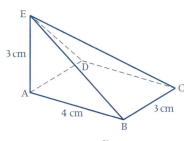

4 The base of this pyramid is a rectangle. PQ = 8 cm and QR = 3 cm. Its vertex, X, is directly above the centre, Y, of the base. XY = 6 cm. Find

   a PR   b PY

   c $X\hat{P}Y$ d XR

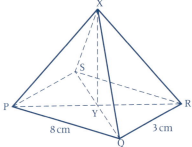

Mixed questions

Exercise 22q

1 EA is a pole standing at the corner of a rectangular plot of level ground. AD = 20 m and DC = 12 m. The angle of elevation of the top of the pole from D is 20°.

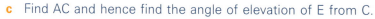

   a Find EA.

   b Find the angle of elevation of E from B (i.e. $A\hat{B}E$).

   c Find AC and hence find the angle of elevation of E from C.

2 ABCD is the base of a cube of side 4 cm. A′, B′, C′ and D′ are the four vertices of the cube directly above A, B, C and D respectively.

   a Find AC, CD′ and AD′. What sort of triangle is △ACD′?

   b What sort of quadrilateral is ACC′A′? Find AC′. What other lengths are equal to AC′?

3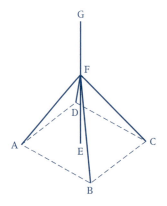

A pole, GE, is held upright by four equal guy ropes fastened to its midpoint, F. The other ends of the ropes are fastened to the four corners of a square, ABCD, on the level ground. AB = 6 m. Each rope is 6 m long.

   a Find BD and BE.

   b Find EF and the height of the pole.

   c Find the angle that each rope makes with the pole.

4 This is the net of a square pyramid with base ABCD and vertex E. AB = 5 cm. AE = 6 cm.

  a Find AC.

  b Draw a picture of the pyramid.

  c Find the height of the pyramid.

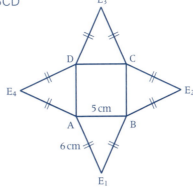

5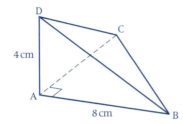

This is a triangular pyramid. The base is triangle ABC. $C\widehat{A}B = 90°$ and AB = AC = 8 cm. D is directly above A and AD = 4 cm.

Find **a** BD **b** $D\widehat{B}A$ **c** BC

         d DC **e** $D\widehat{C}A$

Puzzle

The integral triples (40, 42, 58), (24, 70, 74) and (15, 112, 113) are Pythagorean triples. Calculate the areas of triangles having these triples as the lengths of their sides.

Teacher: How many times can you subtract 7 from 83, and what is left over?

Student: I can subtract it as often as I want to, and it leaves 76 every time.

IN THIS CHAPTER...

you have seen that:

● in a right-angled triangle ABC,

$$\tan C = \frac{\text{opp}}{\text{adj}}, \quad \sin C = \frac{\text{opp}}{\text{hyp}}, \quad \cos C = \frac{\text{adj}}{\text{hyp}}$$

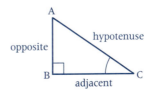

THREE-FIGURE BEARINGS

AT THE END OF THIS CHAPTER...

you should be able to:

1   Give the bearings of one point relative to another using three figures.

2   Draw diagrams to show the bearing of a point relative to another point.

3   Solve problems involving distances and bearings.

Several words used in mathematics with precise definitions have very different meanings in other contexts, e.g. table, factor, matrix and gradient.

The word 'bearing' is another example. Apart from the meaning it has in this chapter, it has more than twelve other meanings including 'a coat of arms' and 'a device reducing friction between rotating parts of a machine'.

How many different meanings can you find?

BEFORE YOU START

you need to know:
- ✓ that the sum of the angles in a triangle is 180°
- ✓ the properties of angles formed by a transversal and parallel lines
- ✓ Pythagoras' theorem and how to use it
- ✓ how to use trigonometry to find angles in right-angled triangles

KEY WORDS

anticlockwise, bearing, clockwise, interior angles, Pythagoras' theorem, three-figure bearing, trigonometry

If we wish to describe a direction in which one point lies relative to another, we need some sort of reference line. On paper, we can use axes. On the surface of the earth we use a line pointing north.

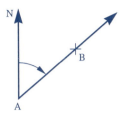

We start by standing at the first point A looking north, then turn clockwise until we are looking at the second point B.

The angle we have turned through gives the *bearing* of B from A.

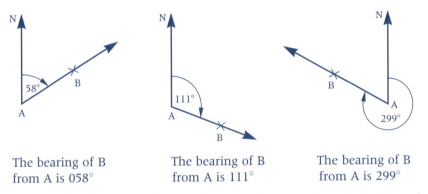

The bearing of B
from A is 058°

The bearing of B
from A is 111°

The bearing of B
from A is 299°

Notice that we write 0 in front of the two-figure number to make it into a three-figure number.

A bearing gives a *line*. If a point is on a given bearing, it lies on that line.

Exercise 23a   Mark a point P, then mark point Q, so that the bearing of Q from P is 212°

Imagine that you are standing at P looking north. Now turn clockwise through 212°. You are now looking towards Q.

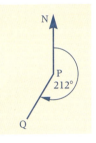

In questions **1** to **9**, mark point A, then mark point B so that the bearing of B from A is:

1 042°	**4** 082°	**7** 108°
2 140°	**5** 222°	**8** 355°
3 320°	**6** 008°	**9** 092°

Use the information in the diagram to find the bearing of Q from P.

You are standing at P looking north. The angle you must turn through in a clockwise direction to look towards Q is the reflex angle at P.

The reflex angle at P = $360° - 72° = 288°$

The bearing of Q from P is 288°.

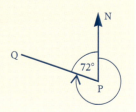

Use the information given in each diagram to find the bearing of B from A:

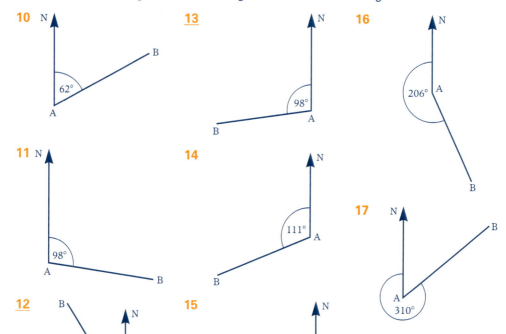

10

11

12

13

14

15

16

17

A man starts from point A and walks 4 km on a bearing of 036° to point B. Draw a diagram to represent this information.

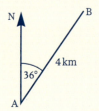

Tip Start by marking A. Next draw a north line from A. Then mark the angle turning clockwise around A by 36°.

Draw diagrams to represent the information given in questions **18** to **27**:

18 A car is driven 30 km on a bearing of 168°. It starts at point P and finishes at point Q.

19 A boy starts from C and cycles for 12 km, on a bearing of 213°, to D.

20 A town X is 260 km from town Y. The bearing of X from Y is 282°.

21 A man walks from A on a bearing of 027° to B. B is 11 km from A.

A car starts from A and is driven 50 km to B on a bearing of 072°. It is then driven 30 km to C on a bearing of 326°. Draw a diagram to represent this information.

Stand at A and face north. Turn through 72° clockwise and mark a point B at a suitable distance from A in this direction – say about 5 cm.

At B face north and turn clockwise through 326°. Mark a point C a suitable distance from B in this direction – say 3 cm.

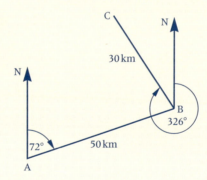

22 A ship sails 10 km from A, on a bearing of 100°, to point B. It then sails 12 km to C on a bearing of 070°.

23 Ship X is 5 km from ship Y on a bearing of 080°. Ship Z is 6 km from Y on a bearing of 300°.

24 A man walks 7 km from A, on a bearing of 285°, to B. Then he walks 3 km due east to C.

25 From town P, town Q is 45 km away on a bearing of 098°. Town R is 60 km from P on a bearing of 003°.

26 ABC is a triangular field. B is 100 m from A on a bearing of 127° and C is 130 m from B on a bearing of 330°.

27 Ship L is 10 km due east of ship M. The bearing of ship N from L is 343° and the bearing of N from M is 044°.

Finding bearings

If we are given the bearing of P from Q, we can draw a diagram in order to find the bearing of Q from P.

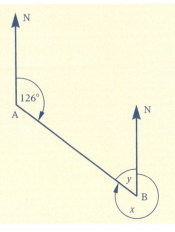

The bearing of B from A is 126°. What is the bearing of A from B?

(Draw a line pointing north at B. The bearing of A from B is x.)

$y = 180° - 126°$ (interior angles)

$\quad = 54°$

$x = 360° - 54°$ (angles at a point)

$\quad = 306°$

The bearing of A from B is 306°.

1 The bearing of P from Q is 060°. What is the bearing of Q from P?

> **Tip** Start by drawing lines pointing north at your chosen positions for P and Q.

2 The bearing of C from D is 292°. What is the bearing of D from C?

3 The bearing of Y from X is 162°. What is the bearing of X from Y?

4 The bearing of A from B is 212°. What is the bearing of B from A?

5 The bearing of D from C is 352°. What is the bearing of C from D?

6 The bearing of Z from Y is 125°. What is the bearing of Y from Z?

Angle calculations

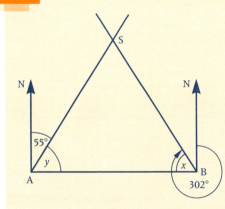

The bearing of a ship S from a lighthouse A is 055°. A second lighthouse B is due east of A. The bearing of S from B is 302°. Find $A\hat{S}B$.

Mark $A\hat{B}S = x$ and $B\hat{A}S = y$ then
$x = 302° - 270° = 32°$

and $y = 90° - 55° = 35°$

$A\hat{S}B = 180° - 32° - 35°$

(angles of a triangle)

$\quad = 113°$

Tip You will need to draw a line pointing north at each of the points referred to.

1 A car starts from C and is driven for 10 km on a bearing of 278°, to D. The car is then driven for 15 km due north to E. Find ED̂C.

2 The bearing of a town P from a town Q is 031°. The bearing of a town R from a town Q is 300°. Find PQ̂R.

3 A man starts from A and walks 5 km to B on a bearing of 112°. He then walks 3 km to C on a bearing of 260°. Find AB̂C.

4 A yacht starts from L and sails 12 km due east to M. It then sails 9 km on a bearing of 142° to K. Find LM̂K.

5 Point B is 60 km from A, on a bearing of 062°. C is 60 km from B on a bearing of 182°. Find AB̂C and BĈA.

6 The bearing of a ship X from a lighthouse Y is 101°. Ship Z is due west of X. The bearing of Z from X is 230°. Find the angles of triangle XYZ.

7 A ship sails 10 km from point A, on a bearing of 005°, to B. Then it sails another 10 km from B, on a bearing of 095°, to C. Find CÂB.

8 From a church tower P, the bearing of a bridge Q is 340° and the bearing of a crossroads R is 111°. Find QP̂R.

Problems

These problems use Pythagoras' theorem and trigonometry.

Exercise 23d P is 12 km due east of Q. R is 8 km due north of Q. Find

a the distance of R from P   **b** the bearing of P from R.

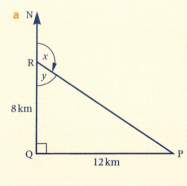

First draw a diagram and on it mark the details you are given. Mark the angle you want to find x. You will need to find PR̂Q so mark it y. Then in triangle PRQ,

$$RP^2 = RQ^2 + PQ^2 \quad (\text{Pythag. th.})$$
$$= 64 + 144 = 208$$
$$RP = \sqrt{208}$$
$$= 14.42\ldots$$

R is 14.4 km from P (to 3 s.f.)

b $\tan y = \dfrac{\text{opp}}{\text{adj}} = \dfrac{12}{8} = 1.5$

$y = 56.3°$ (correct to 1 d.p.)

$x = 180° - 56.3°$ (angles on a straight line)

$= 123.7°$

The bearing of P from R is 124° (to the nearest degree).

Give bearings correct to the nearest degree:

1 From a point P, a man walks 9 km north to Q, then 5 km east to R. What is the bearing of R from P?

2 Three towns P, Q and R, lie in such positions that the bearing of P from R is 027°, the bearing of Q from R is 297° and PR = QR = 21 km. Find $P\hat{R}Q$. How far is P from Q?

3 A ship sails 4 km on a bearing of 032° then 5 km on a bearing of 122°. How far is it from its starting point?

4 The bearing of a ship A from a ship B is 324°.
Ship C is 8 km due north of B and is due east of A.

    a How far is C from A?

    b What is the bearing of B from A?

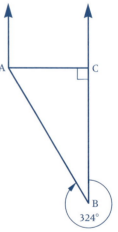

5 A is 20 km due west of B. The bearing of C from A is 044° and the bearing of C from B is 314°. Find

    a the angles of triangle ABC   **b** the distance BC.

6 P is 16 km due north of Q. The bearing of R from P is 152°. The bearing of R from Q is 062°. Find

    a RQ   **b** the bearing of P from R.

7 X is 25 km due west of Y. The bearing of Z from X is 052°. $X\hat{Y}Z = 90°$. Find the distances of Z from X and Y.

8 L is 5 km from M on a bearing of 241°. K is 7 km from M on a bearing of 331°. Find

    a $M\hat{L}K$

    b the bearing of L from K

    c the bearing of K from L.

9 A ship leaves a point P and travels on a bearing of 042° to a point Q 15 km from P. Here the ship changes course and travels 30 km on a bearing of 126° to a point R.

    a Draw a sketch of the ship's journey, showing clearly north, the three-figure bearings and the points P, Q and R.

    b Calculate

        i the distance PR, in kilometres, correct to 2 decimal places

        ii the bearing of R from P

        iii the bearing of P from R.

10 Starting from A, a ship sails 20 km on a bearing of 225° to a point B. From B it sails 24 km on a bearing of 330° to a point C.

    a Draw a sketch to show this information.

    b Calculate the distance and bearing of C from A.

11 ABC is a triangular field. From the corner A , B is 250 metres on a bearing of 064°. From B, C is 310 metres on a bearing of 218°. Find the distance and bearing of A from C.

12 From Kingston, Ocho Rios is 55 km on a bearing of 324° and Port Morant is 44 km on a bearing of 103°.

    a Draw a sketch to show the positions of the three places, marking clearly due north, and the distances and bearings given.

    b Calculate the distance and bearing of Ocho Rios from Port Morant.

Puzzle

There are three different routes from Ian's home to the nearest post box and two different routes from the post box to school. How many different ways are there for Ian to go to school if he must pass the post box on the way?

IN THIS CHAPTER...

you have seen that:

- the bearing of B from A is the angle turned through if you stand at A looking north, then turn clockwise until you are looking at B

- if this angle is less than 100° put a 0 in front of it to make it a three-figure bearing

- a bearing and a distance fix the position of one point relative to another.

AT THE END OF THIS CHAPTER...

you should be able to:

1   Find the mean, mode and median of a set of raw data.

2   Find the mean, mode and median from a grouped frequency table.

3   Identify the modal group.

MATHS IS OUT THERE

Did you know that three famous mathematicians, Blaise Pascal (1623–1662), Pierre de Fermat (1601–1665) and Christian Huygens (1629–1695) laid the basis of probability theory? They studied games of chance, in particular cards and dice. Between them they came up with what they called 'the laws of random events'.

BEFORE YOU START

you need to know:
✓ how to draw and label bar charts
✓ how to add, subtract and multiply fractions
✓ how to find simple probabilities

KEY WORDS

arithmetic average, bar chart, central tendency, data, frequency table, mean, median, modal group, mode, probability, raw data

In Book 2 we saw that, when we have a set of numbers, there are three different measures that we can use that attempt to give a 'typical' member that is representative of the set. These measures are the mean, the mode and the median and they are called measures of central tendency.

To start we will revise finding the mean, mode and median from a list and from a frequency table.

Mean

The mean of a list of n values is the sum of the values divided by n.

The mean of the list $4, $7, $4, $6 is $\dfrac{\$4 + \$7 + \$4 + \$6}{4} = \dfrac{\$21}{4} = \$5.25$

Mode

The mode of a list of values is the value that occurs most often.

The mode of the list $4, $7, $4, $6 is $4 as no other value occurs more than once.

When the values in a list are all different, there is no mode.

If there are two or more values that equally occur most often, there are two or more modes.

Median

When a list of values is arranged in order of size, the median is the middle value.

For this list of 5 values: 4 cm, 5 cm, 5 cm, 6 cm, 6.5 cm,   the median is 5 cm.

↑
middle value

When there is an even number of values, there are two middle values and the median is the mean of these two values.

For the list $4, $4, $6, $7,   the median is the mean of $4 and $6, i.e. $5.

two middle values

In general, the median of a list of n values is the $\left(\frac{n+1}{2}\right)$th value when they have been arranged in order of size.

Exercise 24a

1 a Find the mean, mode and median of these five marks: 24, 35, 44, 28, 34.

   b Find the mean, mode and median of these six lengths:

   2 m, 3 m, 5 m, 4.5 m, 2.5 m, 6 m.

2 The ages, in years, of the children in a swimming club are

9, 10, 8, 10, 11 ,8, 12, 9, 10, 11, 10, 12.

    a Find the mean, mode and median age.

    b One of these children is chosen at random.

       What is the probability that the child is older than the mean age?

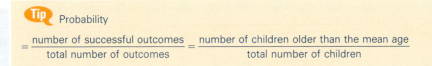

> **Tip** Probability
>
> $$= \frac{\text{number of successful outcomes}}{\text{total number of outcomes}} = \frac{\text{number of children older than the mean age}}{\text{total number of children}}$$

3 The buses that passed the school gate in four hours were counted.

The mean number of buses was 3 each hour.

How many buses were counted?

> **Tip** The mean number of buses an hour = (number counted) ÷ (number of hours)
> So the number counted = mean number per hour × number of hours.

4 Twenty-eight students took a maths test.
The average mark was 15.

> **Tip** You get this by assuming that every pupil scored the average mark.

    a What is the sum of the marks for all these students?

    b Carl was away on the day of the test and took it later. His mark was 24.

       i Will Carl's mark increase or decrease the average mark for the test?

         Explain your answer.

> **Tip** If Carl's mark is more than the old average, adding it to the total will raise the average; if it is less than the average, adding it to the total will lower the average.

       ii Find the new mean mark when Carl's mark is added into the total. Give your answer correct to 1 decimal place.

> **Tip** You need to divide the new total by the increased number of marks.

5 The lengths of six leaves were

4.9 cm, 5.2 cm, 5.6 cm, 5.2 cm, 5.7 cm and 5.2 cm.

    a Find the mean. mode and median length of these leaves.

The lengths of the next four leaves were 5.5 cm, 4.7 cm, 5.0 cm and 4.4 cm.

    b Find the mean length of these four leaves.

    c Find the mean length of all ten leaves.

    d One of these ten leaves is chosen at random. What is the probability that its length is greater than the mean length?

6 The same practical test was given to all the students on an electronics course.

> **Tip** You need to find the total marks scored by all the students. Then divide this total by the number of students.

The mean mark for the 18 day students was 55.

The mean mark for the 12 evening students was 50.

Work out the mean mark for the whole group.

7 Kit has four numbered cards. The mean of the numbers is 4.

He picks another card. The mean of his five numbers is 5.

What number is on the new card?

8 Greg recorded the number of people entering a store each hour for 24 hours.

This stem-and-leaf diagram shows the results.

Number of people   3|45 means 345

2	12	32	45	56	78	81		
3	15	45	67	89	92	96		
4	08	17	34	44	45	67	88	92
5	11	22	36	55				

    a Find the mean, median and mode.

    b One hour out the 24 hours is chosen at random. What is the probability that the number of people entering the store in that hour is greater than the mean?

9 This pie chart shows the number of bedrooms in the homes of a group of students.

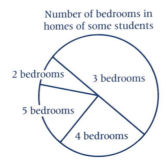

Number of bedrooms in homes of some students

Write down the modal number of bedrooms.

Finding measures of central tendency from a frequency table

It is easy to identify the mode from a frequency table; it is the item with the highest frequency.

This table shows the marks, out of 5, obtained in a maths test.

The modal mark is 3 because this mark has the highest frequency.

We can also find the median mark. There are 30 marks, so the median is the average of the 15th and 16th mark. Counting the frequencies, we can see that the first mark is 0, the 2nd mark is 1, the 3rd to 10th marks are all 2, the 11th to the 21st mark are all 3. Therefore the 15th and 16th mark are both 3, so the median is 3.

Mark	Frequency
0	1
1	1
2	8
3	11
4	5
5	4
Total: 30	

To find the mean, we need to add up all 30 marks. Rather than list them, it is easier to find the totals for each mark, then add these. This is done by adding a column to the table:

Therefore the mean mark is $90 \div 30 = 3$.

Mark, x	Frequency, f	fx
0	1	0
1	1	1
2	8	16
3	11	33
4	5	20
5	4	20
Total: 30	Total: 90	

Exercise 24b

1 Some students gathered this information about themselves.

Number of children in each family	Frequency
1	8
2	12
3	4
4	2

Tip Add a column like the one in the text. Add up the numbers in this column. Now divide by the total number of families.

a Find the mean number of children per family.

b Write down the mode and median number of children.

Finding measures of central tendency from a frequency table

2 Joshua tossed three coins several times and recorded the number of heads that showed at each toss.

His results are shown in the table.

  a Find the mean number of heads per toss.

  b Write down the mode and the median.

Number of heads obtained when three coins are tossed	Frequency
0	9
1	7
2	16
3	3

3 This table shows the number of defective items found by a technician in each sample of 12 tested.

  a How many defective items were found?

  b How many items were tested?

  c Find the mean number of defective items per sample.

Number of defective items	Frequency
0	25
1	9
2	4
3	2

  d One sample is chosen at random.

What is the probability that the number of defective items is less than the mean number?

4 A sample of students were asked to count the number of quarters that they had with them.

The distribution of these coins is shown in the bar chart.

  a Find the size of the sample.

  b Find the mode and the median

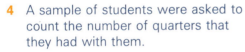 **c** Work out the total value of the coins.

  d The total sum of money was shared out equally among the students.
How much did each student get?

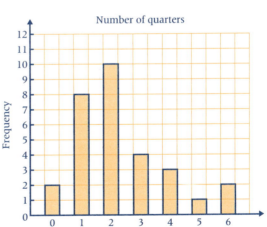

Number of quarters

Frequency

Number of coins

> **Tip** 8 students each had 1 quarter: the total value of these coins is $8 \times 25c = \$2.00$
> 10 students each had 2 quarters: the total value of these coins is $10 \times 50c = \$5.00$, and so on.

Grouped frequency tables

When we need to organise data that has a large number of values, such as examination marks, we group them.

For an examination marked out of 40, the marks can be recorded in groups of ten marks.

These groups go in the first column.

Marks	Frequency
1–10	4
11–20	8	← This tells us that 8 students got a mark between 11 and 20
21–30	25
31–40	13

↑
Adding the numbers in this column gives the total frequency, i.e. the total number of marks

When you need to group data, use between three and ten groups.

Make the groups the same size when you can (you may have to make one group slightly larger than the others).

Exercise 24c

1 This table was used to record the scores of teams in each round of quiz.

Score	Frequency
0–10	22
11–20	28
21–30	24
	Total:

a Write down the number of scores recorded.

b How many times was more than 20 scored?

2 A general knowledge quiz was marked out of 30.

These are the scores of a class of students who took the quiz.

```
12   13   18   25   23   17   23   27   17   15
12   19   29   30   24    2    7    6   22   11
 5   11   15   26   18   17    9   26
```

Organise these scores in a grouped frequency table.

Use groups 1–10, 11–20, 21–30.

3 **a** This list gives the number of letters in each word from a paragraph in a book.

$$6 \quad 4 \quad 6 \quad 3 \quad 7 \quad 4 \quad 6 \quad 4 \quad 1 \quad 3 \quad 5 \quad 3 \quad 4$$

$$4 \quad 10 \quad 5 \quad 6 \quad 4 \quad 3 \quad 6 \quad 7 \quad 8 \quad 4 \quad 5 \quad 3 \quad 1$$

$$9 \quad 5 \quad 6 \quad 3 \quad 7 \quad 1 \quad 12 \quad 1 \quad 1 \quad 9 \quad 4 \quad 1 \quad 15$$

Organise these numbers in a frequency table. Use groups 1–5, 6–10, 11–15

b How many words are in the paragraph?

4 This list gives the weights of some books. The weights are in grams to the nearest gram.

$$45 \quad 52 \quad 71 \quad 95 \quad 51 \quad 38 \quad 44 \quad 62 \quad 80 \quad 72$$

$$81 \quad 39 \quad 48 \quad 60 \quad 66 \quad 75 \quad 36 \quad 63 \quad 92 \quad 108$$

$$25 \quad 115 \quad 95 \quad 91 \quad 56 \quad 72 \quad 61 \quad 97 \quad 110 \quad 88$$

a Copy and complete this frequency table.

Weight, grams	Tally	Frequency
21–50		
51–80		
81–110		
111-150		

Tip Do not go through the list looking for all the weights between 21 and 50 grams; work down the rows putting a tally mark in the appropriate row. Then add the tally marks.

b How many books weighed less than 81 g?

5 The table shows the times, to the nearest minute, that Emma had to wait for a bus each morning.

Time, minutes	Frequency
0–4	7
5–9	9
10–14	3
15–19	1

a One morning Emma waited $8\frac{1}{2}$ minutes.

Write down the group where this time is recorded.

b Write down the number of mornings on which Emma recorded the time she waited for a bus.

6 This table shows the weights, to the nearest kilogram, of a sample of women.

Weight, kg	Frequency
41–60	36
61–80	108
81–100	52
101–120	14

Tip Proportion means describe the share that weighs over 100 kg are of the whole sample.

a Write down the number of women whose weight is in the group 61 to 80 kg.

b Calculate the number of women in this sample.

c i How many of the sample of women weighed over 100 kg, measured to the nearest kilogram?

ii What proportion of the sample is this? Give your answer as a fraction.

7 This table is used to record the masses. in grams to the nearest gram, of some oranges.

Mass, grams	Tally	Frequency
200–209	卌 I	
210–219	卌 卌 II	
220–229	卌 卌 IIII	
230–239	卌	

Two more oranges with masses 208 grams and 231 grams are included.

a Add tally marks for the two extra oranges.

b Complete the table.

c How many oranges are there altogether?

Finding the mode from a grouped frequency table

For grouped data, the group with the largest frequency is called the *modal group*.

This table shows ambulance response times to emergency calls.

Response time (minutes)	5–9	10–14	15–19	20–24	25–29
Frequency	3	6	27	18	6

Because the times have been placed into groups, we have lost some of the detail; we do not know how many times an ambulance took 20 minutes to respond. It is possible that 20 minutes is the mode, but we do not know. We do know that the largest number of response times is in the group 15 to 19 minutes. For this distribution, 15–19 minutes is the modal group.

Finding the mean from a grouped frequency table

The pupils in one class were asked to count the number of items in their pockets and the following frequency table was drawn up:

Number of items	0–4	5–9	10–14	15–19	20–24
Frequency	6	11	6	4	3

We can see that six pupils had between 0 and 4 items. We do not know the exact number of items that each of these pupils had, so we cannot find the exact total number of items that all six pupils had in their pockets.

However, if we *assume* that the average number of items in that group is halfway between 0 and 4, i.e. 2, then we can find an approximate total, i.e. $6 \times 2 = 12$ items.

Using the halfway value in the same way for the other groups, we can find (approximately) the total number of items in the pockets of all 30 pupils:

Number of items	Frequency f	Halfway values x	fx
0–4	6	2	12
5–9	11	7	77
10–14	6	12	72
15–19	4	17	68
20–24	3	22	66
	Total: 30		Total: 295

Tip Check your two totals before you use them to find the mean.

Therefore, the mean number of items is $\frac{295}{30} = 9.8$ (to 1 d.p.)

Remember that this calculation is based on the assumption that the average of each group is the halfway value of that group. This means that the mean value is an estimate.

Exercise 24d

1   Fifty boxes of peaches were examined and the number of bad peaches in each box was recorded with the following result:

No. of bad peaches per box	0–4	5–9	10–14	15–19
Frequency	34	11	4	1

Find the mean number of bad peaches per box.

Tip Make a table like the one above the exercise.

2 Twenty tomato seeds were planted in a seed tray. Four weeks later the heights of the resulting plants were measured and the following frequency table was made:

Height (in cm to the nearest cm)	1–3	4–6	7–9	10–12
Frequency	2	5	10	3

Find the mean height of the seedlings.

3 The table shows the result of a survey amongst 100 pupils on the amount of money each of them spent in the school canteen on one particular day:

Amount (cents)	0–24	25–49	50–74	75–99
Frequency	26	15	38	21

Find the mean amount of money spent.

4 The diagram shows the result of an examination of 20 boxes of screws:

Make a frequency table and find the mean number of defective screws per box.

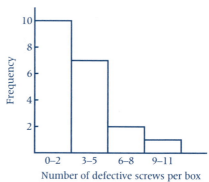

5 The table shows the distribution of heights of 50 adult females, measured to the nearest cm:

Height (cm)	145–149	150–154	155–159	160–164	165–169	170–174
Frequency	1	3	21	18	5	2

Find the mean height.

6 Ninety pupils in one year sat a mathematics examination and the following frequency table was drawn up from the results:

Mark	0–14	15–29	30–44	45–59	60–74	75–89	90–100
Frequency	7	5	6	34	22	12	4

Find the mean mark.

7 The table shows the results of a survey amongst 70 pupils on the amount of money that each of them brought to school on a particular morning:

Amount (cents)	0–49	50–99	100–149	150–199
Frequency	11	35	15	9

Find the mean amount of money.

8 Use the information gathered from a project of your own to find the mean

    a from the raw data

    b from a grouped frequency distribution made from the raw data.
Suggestions for projects follow in the next exercise.

Projects

Exercise 24e

From a chosen project, collect the information (raw data), decide on the groups it can be divided into and make a frequency table. Illustrate your results with a histogram and find the mean value from the frequency table.

Suggestions for class projects

1 Heights of pupils in the class.

2 Weights of pupils in the class.

3 Costs of journeys to school this morning.

4 Times of journeys to school (in minutes).

5 Numbers of brothers and sisters per pupil.

6 Numbers of items in school bags.

7 Ages of pupils (in months).

8 Results of an experiment in science (where every pupil did the same experiment).

9 The prices of a 500 g packet of cornflakes (or any other item) from as many shops as possible.

Suggestions for individual projects

10 Choose a page of text from a book and record the number of letters per word (about 100 words is enough).

11 Use the same page of text and record the number of words per sentence.

12 Choose a completely different type of book from that used in question 10 and repeat questions 10 and 11.

13 Count the number of people per car passing a particular place in the evening rush hour.

14 Repeat question 13 at a different time of day.

15 From as many brochures as possible, find the cost of a two-week holiday in August to a particular place, say Miami.

16 Use either a daily newspaper, or your school weather station if you have one, to record the daily temperature in your area for a period of 30 days.

17 Use the class register to find the number of absences for each pupil in your class last term.

Investigation

Are games always fair?

Definition: A game is **fair** if each player has a fifty-fifty chance of winning, or if the probability of winning is $\frac{1}{2}$.

Try the following games:

Version 1
This is played by two people.
One person is called 'Even' and the other is called 'Odd'. Each person writes down one number from 0 through 9, and hides the number.
Next show the numbers. If the sum of the two numbers is even, then Even wins. If the sum is odd, then Odd wins.
Play the game several times, and record wins by tally marks in the chart.
Is the game fair? Argue your case.

	Total wins
Even	
Odd	

Version 2
Repeat the game above, using the numbers 1 through 9 this time.
Is this version fair? Who has the advantage?

Definition: The probability of winning is the number of favourable outcomes divided by the total number of possible outcomes.

Use the table on the right to find the **probability** that Odd will win.

From the table we have 81 possible outcomes.
How many of these have an odd sum?
What is the probability that Odd will win?

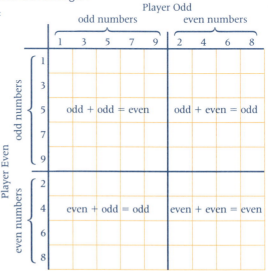

ADDITION TABLE

MATHS IS OUT THERE

Winston Churchill is alleged to have said:

"I gather, young man, that you wish to be a Member of Parliament. The first lesson that you must learn is, when I call for statistics about the rate of infant mortality, what I want is proof that fewer babies died when I was Prime Minister than when anyone else was Prime Minister. That is a political statistic."

IN THIS CHAPTER...

you have seen that:

- the mean of a set of numbers is the sum of the numbers divided by the number of numbers

- an estimate of the mean of grouped data can be calculated by assuming that the average of each group is the halfway value for that group

- the mode of a set of numbers is the number that occurs most often. If the numbers are grouped, the group with the highest frequency is called the modal group

- the median of a list of n values is the $\left(\dfrac{n+1}{2}\right)$ value when the values are in order of size

25

SETS

AT THE END OF THIS CHAPTER...

you should be able to:

1   Find the complement of a set.

2   Solve harder problems using Venn diagrams.

MATHS IS OUT THERE

One of the best selling mathematics books ever published was *Mathematics for the Million*.

It was written by Lancelot Hogben, an English zoologist and geneticist who held academic posts in the UK, Canada and South Africa. He applied mathematical principles to genetics and was concerned with the way statistical methods could be used in the biological and behavioural sciences. This book was published in 1933.

BEFORE YOU START

you need to know:
- ✓ the meaning of: equal sets, empty or null set, finite and infinite sets, intersection and union of two sets, subset, universal set
- ✓ the meaning of the symbols $\in$, $\notin$, $\subset$, $\subseteq$, $\cup$, $\cap$, $\varnothing$ and U
- ✓ how to draw a Venn diagram
- ✓ the meaning of $\mathbb{N}$, $\mathbb{Z}$, $\mathbb{Q}$ and $\mathbb{R}$.

KEY WORDS

complement, element, empty set, equal set, finite set, infinite set, intersection of sets, member, null set, proper subset, set, subset, union of sets, universal set, Venn diagram, the symbols $\in$, $\notin$, $\subset$, $\subseteq$, $\cup$, $\cap$, $\varnothing$ and U

Reminders

In previous books we introduced set notation and used sets to solve simple problems. We saw that

- a set is a collection of things having something in common
- things that belong to a set are called members or elements
- the symbol $\in$ means 'is a member of' and $\notin$ means 'is not a member of'
- in a finite set all the members can be written down
- the number of members in set A is written $n(A)$
- two sets are equal if they contain exactly the same elements or members
- a set with no members is called an empty set or null set. It is denoted by $\{\,\}$ or $\varnothing$
- a set that contains all the elements of the sets under consideration (and possibly some more) is called a universal set. It is denoted by U or $\mathscr{E}$
- if all the members of a set B are also members of a set A, then B is a subset of A. If B does not contain all the members of A, B is called a proper subset of A. This is written $B \subset A$. If B contains all the members of A we write $B \subseteq A$.

Notation

We can describe a set in words, e.g. the set of whole numbers that are less than 20.

We can also describe this set by listing the elements enclosed in curly brackets,

e.g. $\{1, 2, 3, 4, 5, 6, 7, 8, 9, 10, 11, 12, 13, 14, 15, 16, 17, 18, 19\}$.

A shorter way of describing this set is to write $\{x : 1 \leqslant x < 20, x \in \mathbb{N}\}$, where the colon (:) means 'such that'.

$\{x : 1 \leqslant x < 20, x \in \mathbb{N}\}$ reads 'the values of x such that x is greater than or equal to 1 and less than 20 and x is a member of the set of natural numbers.'

Exercise 25a

1 Are the following sets finite or infinite?

   a {even numbers}   b {vowels}

   c {number of students in Trinidad}   d {prime numbers less than 30}

2 a Find the number of elements in each set

   i A = {consonants}   ii B = {players in a cricket team}

   b C = {prime numbers between 10 and 20}

   i Find $n(C)$.

   ii Write down the subset of C whose elements are odd numbers. Is this a proper subset of C?

   iii Write down the subset of C whose elements are even numbers. What special name is given to this set?

3 $U = $ {integers bigger than 10 but smaller than 30}
   $A = $ {prime numbers}, $B = $ {integers exactly divisible by 2 and by 3},
   $C = $ {factors of 18}.

   List the sets A, B and C.

4 List the members of the set $\{x : x \leqslant 10, x \in \mathbb{N}\}$

5 a List any ten members of the set $\{x : x = 2n, n \in \mathbb{N}\}$

   b Is this set finite or infinite?

6 List the members of the set $\{x : -2 < x < 3, x \in \mathbb{Z}\}$

7 List the members of the set $A = \{(x, y): y = 3x, -1 \leqslant x \leqslant 1, x \in \mathbb{Z}\}$

8 Find $n(A)$ where $A = \{x : x = \sqrt{m}, 0 < m < 20, x \in \mathbb{Z}\}$

9 $U = \{x : 1 \leqslant x < 15, x \in \mathbb{N}\}$

   $P = $ {multiples of 3}, $Q = $ {even numbers}, $R = $ {multiples of 5}.

   List the sets P, Q and R.

10 $U = \{x : -6 \leqslant x < 6, x \in \mathbb{Z}\}$

   $A = $ {even numbers}, $B = $ {negative numbers}, $C = $ {positive prime numbers}.

   List the sets A, B and C.

Complement of a set

If $U = $ {pupils in my school}
and $A = $ {pupils who represent the school at games}
then the *complement of A* is the set of all the members of U that are *not* members of A.

In this case, the complement of A is
{pupils in my school who do not represent the school at games}

The complement of A is denoted by A'

Similarly if $U = $ {the whole numbers from 1 to 10 inclusive}

and $A = $ {odd numbers} $= \{1, 3, 5, 7, 9\}$

the complement of A, i.e. A', is $\{2, 4, 6, 8, 10\} = $ {even numbers}

Exercise **25b**

Give the complement of P where

    $P = $ {Thursday, Friday}   if $U = $ {days of the week}
    $P' = $ {Monday, Tuesday, Wednesday, Saturday, Sunday}

Give the complement of each of the following sets.

1 $A = \{5, 15, 25\}$ if $U = \{5, 10, 15, 20, 25\}$

2 $B = \{7, 8, 9, 10\}$ if $U = \{5, 6, 7, 8, 9, 10, 11\}$

3 $V = \{a, e, i, o, u\}$ if $U = \{\text{letters of the alphabet}\}$

4 $P = \{\text{the consonants}\}$ if $U = \{\text{letters of the alphabet}\}$

5 $A = \{\text{Monday, Wednesday, Friday}\}$ if $U = \{\text{days of the week}\}$

6 $X = \{\text{children}\}$ if $U = \{\text{human beings}\}$

7 $M = \{\text{British motor cars}\}$ if $U = \{\text{motor cars}\}$

8 $S = \{\text{male tennis players}\}$ if $U = \{\text{tennis players}\}$

9 $C = \{\text{Jamaican towns}\}$ if $U = \{\text{Caribbean towns}\}$

10 $D = \{\text{squares}\}$ if $U = \{\text{quadrilaterals}\}$

11 $E = \{\text{adults over 80 years old}\}$ if $U = \{\text{adults}\}$

12 $F = \{\text{male doctors}\}$ if $U = \{\text{doctors}\}$

> If $A = \{\text{men}\}$ and $A' = \{\text{women}\}$ what is U?
>
> $U = \{\text{adults}\}$

13 If $A = \{\text{homes with television sets}\}$ and $A' = \{\text{homes without television sets}\}$ what is U?

14 If $A = \{\text{vowels}\}$ and $A' = \{\text{consonants}\}$ what is U?

15 If $X = \{a, b, c, d, e\}$ and $X' = \{f, g, h, i, j\}$ what is U?

16

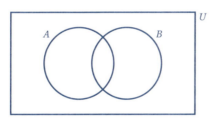

a Copy the Venn diagram and shade the region representing A'.

b Copy the Venn diagram and shade the region representing B'.

Harder problems

Exercise 25c   Given $U = \{1, 2, 3, 4, 5, 6, 7, 8, 9, 10\}$

$A = \{\text{odd numbers}\} = \{1, 3, 5, 7, 9\}$

$B = \{\text{multiples of 3}\} = \{3, 6, 9\}$

Show these sets on a Venn diagram.

Use your diagram to list the following sets:

a A' b B' c $A \cup B$ d the complement of $A \cup B$.

(Each of the numbers in the given sets is placed in the correct position on the following Venn diagram.)

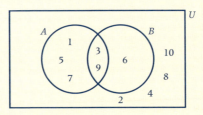

From the diagram

a $A' = \{2, 4, 6, 8, 10\}$ (A' is the set of all members not members of A.)

b $B' = \{1, 2, 4, 5, 7, 8, 10\}$

c $A \cup B = \{1, 3, 5, 6, 7, 9\}$
(the set of members in A or in B. Don't count any member twice.)

d $(A \cup B)' = \{2, 4, 8, 10\}$

e $A' \cap B' = \{2, 4, 8, 10\}$ (the set of members that are not in A and also not in B.)

Notice that $(A \cup B)' = A' \cap B'$

---

1 $U = \{1, 2, 3, 4, 5\}$ $A = \{1, 2, 3, 5\}$ $B = \{2, 4\}$

Show the sets on a Venn diagram and use it to find

   a A' **b** B' **c** $A \cup B$ **d** $(A \cup B)'$ **e** $A' \cap B'$

2 $U = \{\text{whole numbers less than 17}\}$ $P = \{\text{multiples of 3}\}$
     $Q = \{\text{even numbers}\}$

Show these on a Venn diagram and use it to find

   a P' **b** Q' **c** $P \cup Q$ **d** $(P \cup Q)'$ **e** $P' \cap Q'$

3 $U = \{1, 2, 3, 4, 5, 6, 7, 8, 9, 10, 11, 12\}$

    $A = \{\text{multiples of 4}\}$ $B = \{\text{even numbers}\}$

Show these sets on a Venn diagram and use this diagram to list the sets

   a A' **b** B' **c** $A \cup B$ **d** $(A \cup B)'$ **e** $A' \cap B'$

4 $U = \{\text{whole numbers from 10 to 25}\}$

    $P = \{\text{multiples of 4}\}$ $Q = \{\text{multiples of 5}\}$

Show these sets on a Venn diagram and use this diagram to list the sets

   a P' **b** Q' **c** $P \cup Q$ **d** $(P \cup Q)'$ **e** $P' \cap Q'$

5 $U = \{\text{different letters in the word GENERAL}\}$

    $A = \{\text{different letters in the word ANGEL}\}$

    $B = \{\text{different letters in the word LEAN}\}$

Show these sets on a Venn diagram and use this diagram to list the sets

   a A' **b** B' **c** $A \cap B$

   d $A \cup B$ **e** $(A \cap B)'$ **f** $A' \cap B'$

6 $U = \{p, q, r, s, t, u, v, w\}$

$X = \{r, s, t, w\}$ $Y = \{q, s, t, u, v\}$

Show U, X and Y on a Venn diagram entering all the members. Hence list the sets

  a X' **b** Y' **c** $X' \cap Y'$ **d** $X \cup Y'$ **e** $(X \cup Y)'$

Which two sets are equal?

7 $U = \{1, 2, 3, 4, 5, 6, 7, 8, 9, 10, 11, 12\}$

$X = \{\text{factors of } 12\}$ $Y = \{\text{even numbers}\}$

Show U, X and Y on a Venn diagram entering all the members. Hence list the sets

  a X' **b** Y' **c** $X' \cap Y'$
  d $X' \cup Y'$ **e** $X \cup Y$ **f** $(X \cup Y)'$

Which two sets are equal?

8 Draw the diagram given opposite
six times. Use shading to illustrate
each of the following sets.

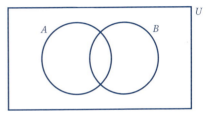

  a A' **b** B' **c** $A' \cap B'$
  d $A' \cup B'$ **e** $A \cup B$ **f** $(A' \cup B)'$

9 $U = \{\text{different letters in the word MATHEMATICS}\}$

$A = \{\text{different letters in the word ATTIC}\}$

$B = \{\text{different letters in the word TASTE}\}$

Show U, A and B on a Venn diagram entering all the elements. Hence list the sets

  a A' **b** B' **c** $A \cup B$
  d $(A \cup B)'$ **e** $A' \cup B'$ **f** $A' \cap B'$

10 $U = \{\text{pupils in my class}\}$

$P = \{\text{those with compasses}\}$

$Q = \{\text{those with protractors}\}$

Describe

  a P' **b** Q' **c** $P' \cap Q'$ **d** $(P \cup Q)'$ **e** $P \cup Q$

11 $U = \{x : 1 \leqslant x < 12, x \in \mathbb{N}\}$

$A = \{\text{multiples of } 3\}$, $B = \{\text{even numbers}\}$.

List the sets

  a A' **b** B' **c** $A' \cap B'$ **d** $(A \cup B)'$

12 $U = \{x : -6 \leqslant x < 8, x \in \mathbb{Z}\}$

$P = \{\text{odd numbers}\}$, $Q = \{\text{prime numbers}\}$.

List the sets

  a P' **b** Q' **c** $P' \cap Q'$ **d** $P' \cup Q'$ **e** $(P \cup Q)'$

13 $U = \{x : 10 \leqslant x < 25, x \in \mathbb{N}\}$

$A = \{$multiples of 4$\}$, $B = \{$multiples of 3$\}$.

List the sets

a A' **b** B' **c** $A' \cap B'$ **d** $(A \cup B)'$.

Number of members

Exercise **25d**   Illustrate on a Venn diagram the sets A and B if

$A = \{$Sunday, Monday, Tuesday, Wednesday$\}$
$B = \{$Wednesday, Thursday, Friday, Saturday, Sunday$\}$

Use your diagram to find

a $n(A)$ **b** $n(B)$ **c** $(A \cup B)$ **d** $n(A \cap B)$

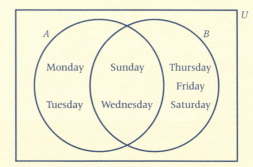

Counting the number of elements in the various regions gives

a $n(A) = 4$ **b** $n(B) = 5$
c $n(A \cup B) = 7$ **d** $n(A \cap B) = 2$

In questions **1** to **4** count the number of elements in the various regions to find

a $n(A)$ **b** $n(B)$ **c** $n(A \cup B)$ **d** $n(A \cap B)$

1
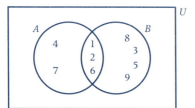

3
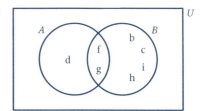

2
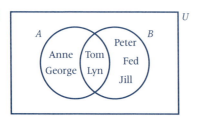

4
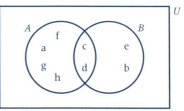

In the remaining questions in this exercise, the numbers in the various regions of the Venn diagrams show the *number of elements in*, or *members of*, the set in that region.

In questions **5** to **8** use the the information given in the Venn diagrams to find

a $n(X)$ **b** $n(Y)$ **c** $n(X \cup Y)$ **d** $n(X \cap Y)$

5

U
X   *Y*
5   4   3

This means that there are 5 members in set *A* but not in set *B*

This means that there are 4 members in set *A* and also in set *B*

7

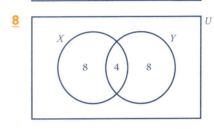

8

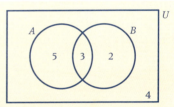

6

U
X   *Y*
5   0   2

Use the information given in the Venn diagram to find

$n(A)$, $n(B)$, $n(A')$, $n(B')$, $n(A \cup B)$, $n(A \cap B)$, $n(A' \cup B')$, $n[(A \cap B)']$ for the given sets.

The numbers in the regions show the number of elements in that region.

U
A   *B*
5   3   2
4

$n(A) = 8$ (the number of elements in set *A*, i.e. $5 + 3$)

$n(B) = 5$ (the number of elements within the *B* circle, i.e. $3 + 2$)

$n(A') = 6$ (the number of elements not in set *A*)

$n(B') = 9$ (the number of elements not in set *B*)

$n(A \cup B) = 10$ (the sum of the numbers in either *A* or *B*, i.e. $5 + 3 + 2$)

$n(A \cap B) = 3$ (the number in both *A* and *B*)

$n(A' \cup B') = 11$ (the number not in set *A* ($2 + 4$) plus the number not in set *B* and not already accounted for (5))

$n[(A \cap B)'] = 11$ (the number not in both A and B)

Use the information given in the following Venn diagrams to find $n(A)$, $n(B)$, $n(A')$, $n(B')$, $n(A \cup B)$, $n(A \cap B)$, $n(A' \cup B')$ and $n[(A \cap B)']$ for each of the given pairs of sets.

9

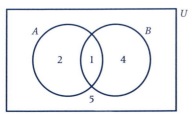

10

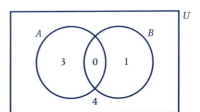

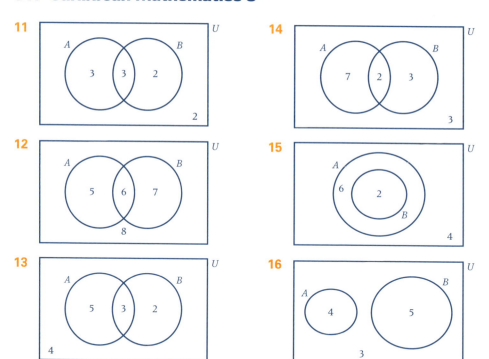

11 A 3 3 2 B U 2

14 A 7 2 3 B U 3

12 A 5 6 7 B U 8

15 A 6 2 B U 4

13 A 5 3 2 B U 4

16 A 4 B 5 U 3

Problems

Exercise 25e

The Venn diagram shows how many pupils in a class have cellphones (C) and video recorders (V).

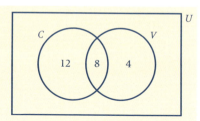

C 12 8 4 V U

How many pupils have:

a both a cell phone and a video recorder

b a cell phone

c a cell phone and/or a video recorder

d a video recorder but not a cell phone?

a The number of pupils with both is 8. (The number in both circles.)

b The number of pupils with a cell phone is $12 + 8$, i.e. 20. (The sum of the numbers in circle C.)

c The number of pupils with at least one of the two is $12 + 8 + 4$, i.e. 24.

(The sum of the numbers in either circle.)

d 4 pupils have a video recorder but not a cell phone. (The number in V but not in C.)

1 The Venn diagram shows the number of boys in a class who play soccer (*S*) and who play cricket (*C*).

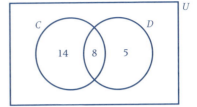

How many boys play

   a both games

   b only cricket

   c soccer

   d exactly one of these games?

2 The students in a form were asked if they did any cooking (*C*) or dressmaking (*D*) at home. Their replies are shown in the Venn diagram.

If all the students in the form took part in at least one of these activities, how many students

   a are there in the form   **b** did only cooking

   c did both   **d** did exactly one of these activities?

3 In a group of 24 children, each had a dog or a cat or both. If 18 kept a dog and 5 of these also kept a cat, show this information on a Venn diagram and hence find the number of children who kept

   a a cat   **b** only a dog   **c** just one of these as a pet.

4 A group of 50 television addicts were asked if they watched sport programmes and nature programmes. The replies revealed that 21 watched both sport and nature programmes but 9 watched nature programmes only. All 50 people watched either sport or nature programmes or both. Show this information on a Venn diagram and use it to find the numbers of viewers who

   a watched Sport

   b did not watch Nature programmes

   c watched either Sport or Nature programmes but not both.

5 In a youth club 35 teenagers said that they went to football matches, discos or both. Of the 22 who said they went to football matches, 12 said they also went to discos. Show this information on a Venn diagram. How many went to football matches or discos, but not to both?

6 There are 28 pupils in a form, all of whom take history or geography or both. If 14 take history, 5 of whom also take geography, show this information on a Venn diagram and hence find the number of pupils who take

   a geography

   b history but not geography

   c just one of these subjects.

7 In a squad of 35 cricketers 20 said that they could bat and 8 said that they could bat and bowl. Show this information on a Venn diagram. How many more were willing to bowl than to bat?

373

The Venn diagram shows how many houses in a street have new windows (W) and how many have new front doors (D).

How many houses

a are there in the street

b do not have a new front door

c have either a new front door or new windows but not both?

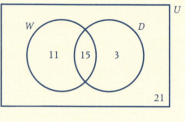

a Number of houses in the street is $11 + 15 + 3 + 21$, i.e. 50 (The sum of all the numbers in U.)

b Number of houses without a new front door is $11 + 21$, i.e. 32

(The sum of the numbers outside W.)

c The numbers with either a new front door or new windows but not both is $11 + 3$, i.e. 14.

(The number in either W or D but not in both.)

8 The Venn diagram shows how many pupils in a class passed the English examination (E) and how many passed the mathematics examination (M).

How many pupils

a passed in only one examination

b did not pass in English

c passed in at least one examination?

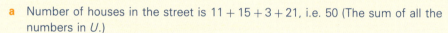

9 The Venn diagram shows how many pupils in a class kept goldfish (G), budgerigars (B) or both.

How many pupils

a were there in the class

b did not have a budgerigar

c had at least one of these pets?

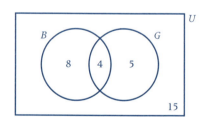

10 The passengers on a coach were questioned about the newspapers and weekly magazines they bought. 3 bought both a daily newspaper and a weekly magazine. 15 bought a daily newspaper. 8 bought a weekly magazine. 8 did not buy either a daily paper or a weekly magazine. Show this information on a Venn diagram.

a How many passengers were there on the coach?

b How many passengers bought a daily newspaper, a weekly magazine or both?

11 One evening all 78 members of a Youth Club were asked whether they liked swimming (S) and/or dancing (D). It was found that 34 liked swimming, 41 liked dancing and 19 liked both. Show this information on a Venn diagram. How many were

    a swimmers but not dancers

    b dancers or swimmers but not both

    c neither dancers nor swimmers?

12 During April, 36 cars were taken to a Testing Station for an MOT certificate. The results showed that 8 had defective brakes and lights, 10 had defective brakes, and 13 had defective lights. How many cars

    a failed the test

    b passed the test

    c had exactly one defect?

13 The 32 pupils in a class were asked whether they studied French or art or both. It was found that 8 studied both, 13 studied French and 6 did not study either subject How many pupils studied

    a art but not French

    b French or art but not both?

 ## Puzzle

Every map or shape divided into regions can be coloured so that no two touching edges of any regions are the same colour using not more than four different colours. Make a copy of the diagram given below and colour it so that no two touching edges of any regions are the same colour.

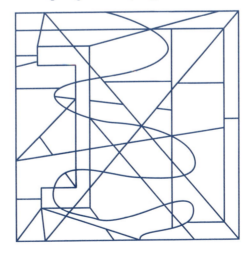

If you are successful, try using three different colours. Is it possible?

The following exercise extends the ideas that we have used so far, to three intersecting sets.

In a certain group of pupils, some are in one or more of the school swimming, debating and trampoline teams. The Venn diagram shows these numbers where

$S = \{$those in the swimming teams$\}$

$D = \{$those in debating teams$\}$

and $T = \{$those in trampoline teams$\}$

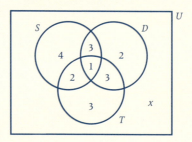

How many take part in

a   debating teams only

b   at least one type of team

c   exactly two types of teams?

If there are 24 in the group find

d   the value of x

e   the number who do not belong to a debating team

a   2 take part in debating teams and nothing else.

b   The number taking part in at least one type of team is $4 + 2 + 3 + 3 + 2 + 3 + 1$, i.e. 18.

c   Taking part in exactly two types of teams are $3 + 3 + 2$ i.e. 8 pupils.

d   Since $18 + x = 24$, 6 pupils do not belong to any team.

e   The number who do not belong to any debating team is $24 - (3 + 2 + 3 + 1)$, i.e. 15 pupils.

1   The Venn diagram shows the number of students taking geography (G), history (H) and accounts (A) in a class of 43. Every student takes at least one of these subjects

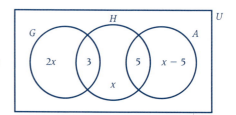

a   Write down an expression, in terms of x, for the number of students who take history.

b Write down an expression, in terms of x, which shows the information given in the Venn diagram.

c Work out the number of students who take geography only.

d Work out the number of students who take accounts.

2 Given that $U = \{p, q, r, s, t, u, v\}$

$$A = \{p, q, s, t\}, B = \{q, r, s, u, v\} \text{ and } C = \{s, t, u\}$$

a Draw a Venn diagram showing the sets A, B, C and U.

b List the members of the set represented by $A \cup (B \cap C)$.

c Write down the value of $n(A \cup (B \cap C))'$.

3 The universal set U contains the sets A, B and C such that

$$B \subset A, C \subset A, B \cap C = \varnothing$$

Draw a Venn diagram to show the relation between the sets A, B and C.

4 The universal set U contains the sets A, B and C such that $P \cap Q = \varnothing$, $P \cap R \neq \varnothing$, $Q \cap R \neq \varnothing$.

Draw a Venn diagram to show the relation between the sets P, Q and R.

5 The universal set U contains the sets P, Q and R such that

$$P \not\subset Q, P \cap Q, \neq \varnothing, R \subset Q', R \subset P.$$

Draw a Venn diagram to show the relation between the sets P, Q and R.

6 The universal set U contains the sets A, B and C such that

$$B \subset A, A \cap C \neq \varnothing, B \cap C = \varnothing$$

Draw a Venn diagram to show the relation between the sets A, B and C.

7 $U = \{1, 2, 3, 4, 5, ..., 10\}; A = \{1, 2, 3, 4\}, B = \{1, 2, 5, 6\}$ and $C = \{2, 4, 6, 8, 10\}$.

a List the members of $A \cup B$ and $B \cup C$. Hence show that $(A \cup B) \cup C = A \cup (B \cup C)$.

b List the members of $A \cap B$ and $B \cap C$. Hence show that $(A \cap B) \cap C = A \cap (B \cap C)$.

c Draw a Venn diagram to show the relation between sets A, B and C.

8 The Venn diagram shows the relation between those who run (set R), those who hurdle (set H) and those who throw the javelin (set J).

Copy the diagram and shade in different colours $J \cup H$ and $R \cap (J \cup H)$. Copy the diagram again and shade the area representing the union of $(R \cap H)$ and $(R \cap J)$. On comparing the diagrams you should find that $R \cap (H \cup J) = (R \cap H) \cup (R \cap J)$.

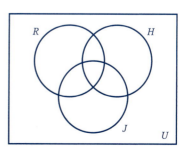

9 The universal set is U where $U = \{$positive integers from 1 to 24 inclusive$\}$. X, Y and Z are subsets of U and are defined as follows: $X = \{$multiples of 4$\}$. $Y = \{$multiples of 5$\}$, and $Z = \{$prime numbers$\}$. Draw a Venn diagram to illustrate this information and use it to list the elements of the sets.

    a $X \cup Y$ **b** $X' \cap Z$

    c $X' \cap Y' \cap Z'$ **d** $(X \cup Y) \cap (Y \cup Z)$

10 Given $U = \{$letters of the alphabet$\}$, $A = \{$letters used to make the word ALGEBRA$\}$, $B = \{$letters used to make the word ARITHMETIC$\}$, write down

    a the elements in the set $(A \cup B) \cap C$ **b** $n(B \cup C)$

11 If $U = \{$quadrilaterals$\}$, $A = \{$parallelograms$\}$, $B = \{$rectangles$\}$ and $C = \{$squares$\}$, draw a Venn diagram to illustrate the connection between the sets.

12 If $U = \{$triangles$\}$, $X = \{$right-angled triangles$\}$, $Y = \{$equilateral triangles$\}$ and $Z = \{$isosceles triangles$\}$, draw a Venn diagram to show the relationship between the sets.

Describe the elements of

    a $X \cap Z$ **b** $X' \cap Y$ **c** $X \cup Y$

13 Draw a Venn diagram to show three sets A, B and C in a universal set U. Enter numbers in the correct parts of your diagram using the following information.

$n(A \cap B \cap C) = 2$, $n(A \cap B) = 7$, $n(B \cap C) = 6$, $n(A \cap C) = 8$, $n(A) = 16$, $n(B) = 20$, $n(C) = 19$ and $n(U) = 50$.

Use your diagram to find

    a $n(A' \cap C')$ **b** $n(A \cup B')$ **c** $n(A' \cap B' \cap C')$

14

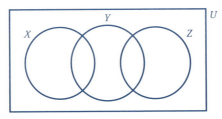

The Venn diagram shows three sets X, Y and Z contained within a universal set U. Enter numbers in the correct regions of your diagram using the following information.

$n(X \cap Y) = 3$, $n(Y \cap Z) = 4$, $n(X) = 8$, $n(Y) = 18$, $n(Z) = 10$, and $n(U) = 35$.

Use your diagram to find

    a $n(X \cap Y \cap Z)$ **b** $n(X' \cup Y)$ **c** $n(X' \cap Z')$

15 Forty travellers were questioned about the various methods of transport they had used the previous day. Every one of them had used at least one of the methods shown in the Venn diagram.

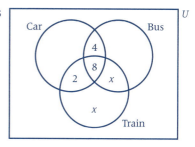

Of those questioned 8 had used all three methods of transport, 4 had travelled by bus and car only, and 2 by car and train only. The number (x) who had travelled by train only was equal to the number who had travelled by bus and train only.

If 20 travellers had used a train and 33 had used a bus find

a the value of x

b the number who travelled by bus only

c the number who used exactly two methods of transport

d the number who travelled by car only.

16 Some tests were held in the fourth form in physics (P), chemistry (C) and biology (B). Forty-five pupils took part and the recorded passes were as follows: physics 24, chemistry 25, biology 30, physics and chemistry but not biology 3, physics and biology but not chemistry 6, biology and chemistry but not physics 7. If 10 pupils passed in all three and 2 pupils failed in all three draw a Venn diagram to illustrate this information entering the correct number in each region.

Use your diagram to find the number of pupils who

a passed in chemistry only

b passed in more than one of these subjects

c passed in exactly one of these subjects

d did not pass in either biology or chemistry.

17 In a particular sixth form all the pupils belong to one or more of the groups studying arts subjects (A), science subjects (B) and craft subjects (C) as indicated in the Venn diagram.

Which of the following statements are true?

a some students do not study an arts subject

b some students study subjects of all three types

c all students studying a craft subject also study an arts subject

d all students studying an arts subject also study a craft subject

e every student studying an arts subject also studies a science subject

f every student studying a craft subject also studies a science subject.

The famous French mathematician René Descartes (1596–1650) believed that the whole of human knowledge should be founded on the belief *cogito ergo sum* – I think, therefore I am. He also believed that the entire material universe could be explained in terms of mathematical physics. If you go on to study mathematics for a few more years you will come across his name again in the branch called Cartesian or coordinate geometry.

IN THIS CHAPTER...

you have seen that:

- the complement of a set *A* is the set of all members that are not members of *A*

- many problems involving details about two or three sets of data can be solved using Venn diagrams.

26 OPERATIONS, RELATIONS AND FUNCTIONS

Do you know how dependent we all are on notation and convention?

Ask any mathematician to solve the equation $ax = b$ and you will get $x = \dfrac{b}{a}$. If you get $a = \dfrac{x}{b}$, you would probably say that it is wrong. But is it?

Without realising it, we use the convention that letters near the end of the alphabet represent unknowns and letters near the beginning represent known quantities.

BEFORE YOU START

you need to know:
- ✓ the language and notation used for sets, including the meaning of $\mathbb{N}$, $\mathbb{Z}$, $\mathbb{Q}$, $\mathbb{R}$
- ✓ the meaning of intersection and union of sets
- ✓ how to add and multiply matrices
- ✓ how to sketch a graph of a straight line from its equation
- ✓ how to draw the graph of a curve from its equation

KEY WORDS

associative, closure, commutative, distributive, domain, function, identity, image, inverse, operation, range, relation

Operations

We know that numbers can be combined in different ways.

For example, we can combine 4 and 5 by addition, i.e $4 + 5 = 9$,
by subtraction, i.e $4 - 5 = -1$,
by multiplication, i.e. $4 \times 5 = 20$,
by division, i.e. $4 \div 5 = 0.8$.

We also know that we can combine matrices, vectors and sets in different ways, e.g. $\{1, 2, 3\} \cap \{2, 4, 6\} = \{2\}$ and $\{1, 2, 3\} \cup \{2, 4, 6\} = \{1, 2, 3, 4, 6\}$

The rule for combining two elements is called an *operation*.

We know the rules for the operations of addition, subtraction, multiplication and division on numbers, and we can define other operations.

For example, if a and b are members of the set of real numbers, we can define an operation such as combine a and b to give $\dfrac{ab}{a^2 + b^2}$.

This operation is written more briefly using a symbol, e.g. $a * b$ means $\dfrac{ab}{a^2 + b^2}$.

(Any symbol can be used as long as it does not have a commonly known meaning.)

Using $a * b = \dfrac{ab}{a^2 + b^2}$, we can find $2 * 3$ by substituting 2 for a and 3 for b in $\dfrac{ab}{a^2 + b^2}$,

i.e. $2 * 3 = \dfrac{2 \times 3}{2^2 + 3^2} = \dfrac{6}{13}$

Exercise 26a

1 Given that $a \, \triangle \, b$ means $a + 2b$, find

    a $3 \triangle 2$ **b** $-1 \triangle 6$ **c** $\frac{1}{2} \triangle \frac{1}{3}$

2 Given that $p * q$ means $\dfrac{p}{p + q}$, find

    a $5 * 4$ **b** $-3 * 2$ **c** $\frac{2}{3} * \frac{1}{4}$

3 If $y \dagger y = x^2 - y^2$, find

    a $2 \dagger 5$ **b** $-2 \dagger -1$ **c** $\frac{1}{2} \dagger \frac{1}{4}$

4 If $x \, \star \, y = \dfrac{xy}{x - y}$, find

    a $-3 \star 2$ **b** $5 \star -4$ **c** $\frac{1}{2} \star \frac{3}{4}$

> Given that $a \, \square \, b = a^2 b$, find $(2 \, \square \, 3) \, \square \, 5$.
>
> First find $2 \, \square \, 3$: $\ 2 \, \square \, 3 = 2^2 \times 3 = 12$
>
> Then $(2 \, \square \, 3) \, \square \, 5 = 12 \, \square \, 5 = 12^2 \times 5 = 720$

5 Given that $x * y = x + 2y$, find

    a $2 * 4$ **b** $(2 * 4) * 3$ **c** $2 * (4 * 3)$

6 Given that $p \blacktriangle q = p^2 q$, find

   a $3 \blacktriangle -2$ **b** $(4 \blacktriangle 3) \blacktriangle -1$ **c** $4 \blacktriangle (3 \blacktriangle -1)$

7 If $x \oplus y = \dfrac{x}{y}$, find

   a $4 \oplus 2$ **b** $-2 \oplus (3 \oplus 2)$ **c** $(-2 \oplus 3) \oplus 2$

8 If $a \bullet b = a(a - b)$, find

   a $2 \bullet 6$ **b** $3 \bullet (4 \bullet 3)$ **c** $(2 \bullet 3) \bullet 4$

9 Given that $a \ddagger b = \dfrac{1}{a} + \dfrac{1}{b} =$ and that $a \square b = a^2 - b$, find

   a $4 \square (2 \ddagger 3)$ **b** $4 \ddagger (2 \square 3)$ **c** $a \ddagger (b \square c)$

10 If $x * y = 2xy$ and $x \# y = x + 2y$, find

   a $(3 * 2) \# 2$ **b** $(3 \# 2) * 2$ **c** $(p \# q) * r$

Properties of operations

Commutative

We know that we can add numbers in any order, e.g. $8 + 2 = 2 + 8$.

We say that the operation of addition on $\mathbb{R}$ is *commutative*.

However, we also know that $8 - 2$ is not the same as $2 - 8$, so subtraction is not commutative.

> Any operation is commutative when the elements can be interchanged and still give the same result.

Associative

We also know that, to find $2 + 4 + 8$, we can either find $2 + 4$ first or $4 + 8$ first, i.e. $(2 + 4) + 8 = 2 + (4 + 8)$. This shows that under addition on three numbers, we can add any two numbers first. We say that the operation of addition on $\mathbb{R}$ is *associative*.

However, $(2 - 4) - 8 = -2 - 8 = -10$ and $2 - (4 - 8) = 2 - (-4) = 2 + 4 = 6$,

i.e $(2 - 4) - 8$ is *not* the same as $2 - (4 - 8)$, so subtraction is *not* associative.

> An operation * is associative when three elements can be combined such that $(a * b) * c = a * (b * c)$.

Distributive

We can combine elements using two or more operations, for example, $2 \times (3 + 4)$.

We know that $2 \times (3 + 4) = 2 \times 7 = 14$ and that $2 \times 3 + 2 \times 4 = 6 + 8 = 14$, i.e. $2 \times (3 + 4) = 2 \times 3 + 2 \times 4$. This shows that we can 'distribute' the operation of multiplication over addition and we say that multiplication is *distributive* over addition on $\mathbb{R}$.

However, addition is not distributive over multiplication, e.g. $2 + (3 \times 4) = 14$, but $2 + 3 \times 2 + 4 = 12$.

We use the fact that multiplication is distributive over addition whenever we factorise or expand brackets, e.g. $a(b + c) = ab + ac$ and $3x + 6y = 3(x + 2y)$.

An operation * is distributive over an operation $\diamond$ when $a * (b \diamond c) = a * b \diamond a * c$

Exercise **26b**

1 a Show that the operation of subtraction on $\mathbb{Z}$ is not commutative.

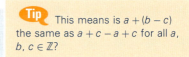

Tip This means is $a + (b - c)$ the same as $a + c - a + c$ for all a, $b, c \in \mathbb{Z}$?

b Is addition distributive over subtraction on $\mathbb{Z}$?

2 a Show that division is not commutative on the set of real numbers.

b Is multiplication on $\mathbb{R}$ commutative?

c Is multiplication on $\mathbb{R}$ associative?

3 a Is division on $\mathbb{R}$ distributive across multiplication?

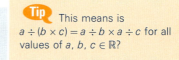

Tip This means is $a \div (b \times c) = a \div b \times a \div c$ for all values of $a, b, c \in \mathbb{R}$?

b Is multiplication on $\mathbb{R}$ distributive across division?

4 Given $\mathbf{A} = \begin{pmatrix} 1 & 5 \\ 3 & 2 \end{pmatrix}$ and $\mathbf{B} = \begin{pmatrix} 2 & 4 \\ 3 & 6 \end{pmatrix}$

a Find **i** $\mathbf{A} + \mathbf{B}$ **ii** $\mathbf{B} + \mathbf{A}$

Is addition of two-by-two matrices commutative?

b Find **i** $\mathbf{AB}$ **ii** $\mathbf{BA}$

Is multiplication of two-by-two matrices commutative?

5 $A = \{1, 2, 3, 4\}$, $B = \{2, 4, 6, 8\}$ and $C = \{3, 5, 7\}$

a Find $A \cup (B \cap C)$.

b Find $A \cup B \cap A \cup C$.

c Is the union of sets distributive across the intersection of sets?

Given that $a \square b = a^2 b$, $a, b \in \mathbb{R}$, is the operation $\square$ commutative?

The operation $\square$ is commutative if $a \square b = b \square a$ for all values of a and b.

$a \square b = a^2 b$ and $b \square a = b^2 a$;

when $a = 2$ and $b = 3$, $a^2 b = 4 \times 3 = 12$ but $b^2 a = 9 \times 2 = 18$

12 is not equal to 18, so the operation $\square$ is not commutative.

6 If $x * y = 2xy$ for $x, y \in \mathbb{N}$, show that

    a $x * y = y * x$

    b $(x * y) * z = x * (y * z)$

    c $x * (y + z) = x * y + x * z$

    d Describe the properties that **a**, **b** and **c** show about the operation $*$.

7 Given that $p \# q = \dfrac{p}{q}$, determine whether the operation $\#$ is

    a commutative   **b** associative.

8 If $a \bullet b = a(a - b)$, determine whether the operation $\bullet$ is

    a commutative   **b** associative.

Properties of sets under an operation

Closure

If x and y are any two integers, i.e. $x, y \in \mathbb{Z}$, then we know that $x + y$ is also an integer.

For example, $2 + 8 = 10$, $-6 + 37 = 31$, and so on.

This means that the operation of addition on the set of integers is *closed*.

But we also know that $a \div b$ is not always an integer, e.g. $3 \div 4 = 0.75$.

So the operation of division on the set of integers is not closed.

In general, if a and b are *any* two members of a set A, and if $a * b$ always gives a member of A, we say that the set is closed under the operation $*$.

Identity element

For any integer x, $x + 0 = 0 + x = x$, i.e. if we add 0 to any integer, we do not change the value of that integer. The integer 0 is called the *identity element* for the set of integers under addition.

In general, if a is a member of the set A under an operation $*$, and there is another member b such that $a * b = b * a = a$, i.e. b leaves a unchanged under $*$, then b is called the identity element.

Inverse

We know that, for any integer x, there is another integer $-x$, such that $x + (-x) = 0$.

$-x$ is called the *inverse* of x under the operation of addition.

An element, a, has an inverse under an operation if there is another member of the set which when combined with a gives the identity.

Exercise **26c**

1 This question refers to ℕ, i.e. {1, 2, 3, ...}

> **Tip** This means what is the value of a such that $8 \times a = 8$.

  a What is the identity element for 8 under multiplication?

  b Explain why 8 does not have an inverse under addition.

> **Tip** Remember that, if an inverse exist, it must be a member of {1, 2, 3,...}.

  c Is ℕ closed under

    i multiplication **ii** addition

    iii division?

> **Tip** A set is closed if the operation always gives another member of the set.

2 This question refers to ℚ, i.e. the set of all rational numbers under multiplication.

> **Tip** What number, when multiplied by any other number, does not change the second number?

  a What is the identity element?

  b What is the inverse of 3?

  c Is ℚ closed?

3 $A = \begin{pmatrix} 1 & 5 \\ 3 & 2 \end{pmatrix}$

  a Find **B** such that $A + B = B + A = A$

  b Find **I** such that $AI = IA = A$

  c Show that there is a matrix **D** such that $AD = I$.

4 $B = \begin{pmatrix} 2 & 4 \\ 3 & 6 \end{pmatrix}$.

  Show that **B** does have an inverse under addition but does not have an inverse under multiplication.

If $a, b \in A = \{0, 1, 2\}$ and $a \sim b =$ the difference between a and b, i.e. the larger minus the smaller, show that A is closed under the operation $\sim$.

To show that A is closed under $\sim$, we need to show that all possible values of $a \sim b$ give either 0, 1, or 2.

$0 \sim 1 = 1 \sim 0 = 1$ and $0 \sim 2 = 2 \sim 0 = 2$

$1 \sim 2 = 2 \sim 1 = 1$

$0 \sim 0 = 1 \sim 1 = 2 \sim 2 = 0$

The results are all members of A, so the operation is closed.

5 Given that a and b are members of the set {0, 1, 2, 3} and $a * b$ means $a(a - b)$, show that the set is not closed under this operation.

6 $a, b \in A = $ {5, 10, 15} and $a \sim b = $ the difference between a and b. Show that A is not closed under $\sim$.

Relations

Consider the set $A = \{(1, 3), (2, 6), (3, 9), (4, 12)\}$.

Each element is a pair of numbers in which the second number is three times the first number.

In the set, $B = \{(\text{UK, Jamaica}), (\text{UK, Trinidad}), (\text{Jamaica, Trinidad})\}$, each element is a pair of countries where the second country is south of the first country.

Both of these sets are examples of a relation.

> A *relation* is a set of ordered pairs with a rule that connects the two objects in each pair.

We say that the first object maps to the second object.

The set of the first objects in the ordered pairs is called the *domain* of the relation.

The domain of A is {1, 2, 3, 4} and the domain of B is {UK, Jamaica}.

The set of the second objects in the ordered pairs is called the *range* of the relation.

The range of A is {3, 6, 9, 12} and the range of B is {Jamaica, Trinidad}.

Exercise **26d**

1 The second number in each pair in a relation is the cube of the first number.

The domain of the relation is {1, 2, 3, 4}.

Write down the relation as a set of ordered pairs.

2 The range of a relation is {4, 9}.
The second number in each ordered pair is the square of the first number.

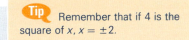

Tip Remember that if 4 is the square of x, $x = \pm 2$.

Write down the relation.

3 A relation is defined by $\{(a, b): a > b, \quad a, b \in \{1, 2, 3\}\}$

a List the set of ordered pairs in the relation.

b Write down the domain and the range of this relation.

Functions

In the relation $\{(1, 3), (2, 6), (3, 9), (4, 12)\}$, the first numbers in the ordered pair are all different.

This means that each number in the domain is paired with just one number in the range.

This relation is an example of a *function*.

> A function is a relation where each member of the domain is paired with just one member of the range.

In the relation $B = \{(\text{UK}, \text{Jamaica}), (\text{UK}, \text{Trinidad}), (\text{Jamaica}, \text{Trinidad})\}$, the first objects in the ordered pairs are all not all different because the UK is paired with two different countries. The relation B is not a function.

We can write the relation A above as $\{(x, y): y = 3x \text{ for } x \in \{1, 2, 3, 4\}\}$

As we know that this relation is a function, we can describe it as $f : x \rightarrow 3x, x \in \{1, 2, 3, 4\}$ which we read as 'the function such that x maps to $3x$ for $x = 1, 2, 3$ and 4.'

This means that x is the first number in the ordered pair and $3x$ is the second number in the ordered pair.

$3x$ is sometimes called the image of x under the function f.

We can describe this function even more briefly as $f(x) = 3x, x \in \{1, 2, 3, 4\}$.

We read $f(x) = 3x$ as 'f of x equals $3x$' and it means that the second number in each ordered pair is three times the first number.

For example, if $f(x) = x^2 - 1, x \in \{1, 2\}$
the ordered pairs are $(1, 1^2 - 1)$ and $(2, 2^2 - 1)$, i.e. $(1, 0)$ and $(2, 3)$.

Exercise **26e**

1 The second number in each pair in a relation is 5 times the first number. The domain of the relation is $\{-1, 0, 1, 2\}$.

   List the ordered pairs in the relation and determine whether this relation is a function.

2 A relation is defined by $\{(x, y): x < y, x, y \in \{1, 4, 6\}\}$.

   List the ordered pairs in this relation and determine whether or not it is a function.

3 A relation in defined by $\{(x, y): y = x^2 + 1, x \in \{0, 2, 4\}\}$.

   List the ordered pairs in the relation and determine whether it is a function.

4 A function is given by $f : x \rightarrow 5x$ for $x \in \{3, 7, 9\}$.

   List the ordered pairs for this function.

5 A function is given by $f : x \rightarrow 2x^2 - 1$ for $x \in \{0, 1, 2\}$.

   List the ordered pairs for this function.

6 A function is defined as $f(x) = x^2 - 2x$ for $x \in \{-1, 0, 2\}$.

   List the ordered pairs for this function.

7 A function is defined as $f(x) = 5x - 4$ for $x \in \{2, 4, 6\}$.

List the ordered pairs for this function.

8 Which of these ordered pairs are members of the function $f : x \to 3x - 1,\ x \in \mathbb{R}$?

a $(3, 1)$ **b** $(2, 5)$ **c** $(-1, 2)$

> **Tip** The ordered pairs that are members of f are $(x, 3x - 1)$ for all values of x. When $x = 3$, $3x - 1 = 3 \times 3 - 1 = 8$, so $(3, 1)$ is not a member of f.

9 Which of these ordered pairs are members of the function $f(x) = 2x^2,\ x \in \mathbb{R}$?

a $(2, 8)$ **b** $(2, 16)$ **c** $(-1, 2)$

10 A function f is given by $f(x) = 3x - 5,\ x \in \mathbb{R}$. Find

a $f(2)$ **b** $f(5)$ **c** $f(-1)$

> **Tip** $f(2)$ means the second number in the ordered pair when the first number is 2, i.e. $f(2) = 3(2) - 5$.

11 A function f is given by $f(x) = x^3 + x,\ x \in \mathbb{R}$. Find

a $f(1)$ **b** $f(0)$ **c** $f(-1)$

12 A function g is given by $g(x) = 2x^2 - 3x + 1,\ x \in \mathbb{R}$. Find

a $g(1)$ **b** $g(0)$ **c** $g(-1)$

Representing functions graphically

Consider the function $f(x) = 2x - 1,\ x \in \mathbb{R}$

If we let $y = f(x)$, i.e. $y = 2x - 1$, we can represent this function by drawing the graph of $y = 2x - 1$.

We cannot draw the graph for all possible values of x, but we can draw part of the graph by choosing a range of values for x.

This graph represents $f(x) = 2x - 1$ for values of x from -1 to 2.

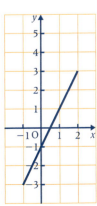

*Exercise **26f***

Use squared paper and 1 square for 1 unit on both axes.

 1 Sketch the graph that represents the function

> **Tip** Let $y = f(x)$ and use what you learnt in Chapter 9.

     a $f(x) = x + 2$ for $-3 < x < 1$

     b $f(x) = 3 - x$ for $-1 < x < 4$

2 Draw the graph that represents
$f(x) = x^2 - 1$ for $-2 \leqslant x \leqslant 2$

> **Tip** Make a table using half unit values of x.

Investigation

     i Choose any odd number n, greater than 1, and calculate:

$$\frac{n^2 - 1}{2}, \quad \frac{n^2 + 1}{2}$$

What do you notice about the values of n, $\dfrac{n^2 - 1}{2}$ and $\dfrac{n^2 + 1}{2}$?

     ii Choose any number n, greater than 1, and calculate $2n$, $n^2 - 1$ and $n^2 + 1$. What do you notice about these numbers?

Towards the end of the 19th century, mathematicians tried to formalise mathematics by attempting to define every mathematical object as a set. As part of this quest, Dirichlet and Lobachevsky independently and at almost the same time, gave the modern definition of function.

IN THIS CHAPTER...

you have seen that:

- an operation is a rule for combining two elements from a set

- an operation is commutative when the elements can be interchanged and still give the same result

- an operation is associative when three elements can be combined by combining the first two elements first or the second two elements first

- an operation $*$ is distributive over an operation $\diamond$ when
  $a * (b \diamond c) = a * b \diamond a * c$

- a set is closed under an operation when the combination of any two elements gives an element in the set

- the identity element is one that combines with any other element a and leaves it unchanged

- the inverse of an element, when combined with the element, gives the identity

- a relation is a set of ordered pairs with a rule that connects the objects in each pair

- the domain of a relation is the set of the first objects in each pair and the range is the set of the second objects in each pair

- a function is a relation where the first objects in each pair are all different

- the ordered pairs in a function are $(x, f(x))$

- a function can be represented graphically by graphing the equation $y = f(x)$

27 GEOMETRIC PROOF, ALTERNATE SEGMENT THEOREM

AT THE END OF THIS CHAPTER...

you should be able to:

1   Understand what a deductive proof is.

2   Prove the angle properties of triangles.

3   Understand how to show that a hypothesis is not true.

4   Know the properties of tangents to circles.

5   Know and be able to use the alternate segment theorem.

Do you realise that circles, in the form of wheels, have been making life easier for man for thousands of years? There were four-wheeled wagons for carrying goods in Mesopotamia four and a half thousand years ago. In Egypt, three and a half thousand year ago wooden wheels were made using very intricate joinery. A set of Vs were cut out of solid pieces of elm and put together so that each spoke was the join of two of these pieces. Wooden spokes thus became an integral part of the hub. The design gave a light, strong wheel, very suitable for Egyptian war-chariots.

BEFORE YOU START

you need to know:
- ✓ how to use trigonometry in right-angled triangles
- ✓ Pythagoras' theorem
- ✓ the facts about angles in circles
- ✓ how to show that two triangles are congruent
- ✓ how to construct a perpendicular to a line at a given point
- ✓ how to construct the bisector of an angle

KEY WORDS

hypothesis, proof, segment, tangent

Geometric proof

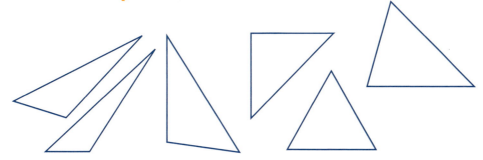

Jim has drawn about thirty different triangles of all shapes and sizes. For each one he measured the three angles and found their sum. The results he obtained varied from 178° to 181.5°. It was from demonstrations such as these in Book 1 that it was concluded that the sum of the angles of *any* triangle is 180°.

Could it be that this method is 'jumping to conclusions' and is unsatisfactory for many reasons? After all, it is impossible to draw a line, for a line has no thickness and if it did not have thickness we could not see it! Furthermore it is impossible to measure angles with absolute accuracy. The protractor Jim uses is probably capable of measuring angles at best to the nearest degree. The only conclusion that Jim can draw from his results is 'it seems likely that the angle sum of any triangle is 180°, but he must remain aware that 'proof' by examining particular cases leaves open the possibility that somewhere, as yet unfound, there lurks an exception to the rule.

Jim needs to know if the result can be *proved* to be true for every triangle. If it can, several other results follow. For example, if the angle sum of any triangle is 180°, the angle sum of the four angles in every quadrilateral must be 360° since one diagonal always divides a quadrilateral into two triangles.

This chapter shows how certain geometric properties can be proved and how other properties follow from them.

Deductive proof

Learning geometrical properties from demonstrations gives the impression that each property is isolated. However geometry can be given a logical structure where one property can be deduced from other properties. This forms the basis of deductive proof; we quote known and accepted facts and then make logical deductions from them.

For example, if we accept that

- vertically opposite angles are equal
- corresponding angles are equal,

then, using just these two facts, we can prove that
alternate angles are equal.

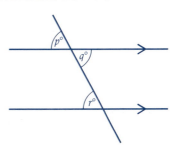

In the diagram $p° = q°$ (vertically opposite angles)

$\qquad\qquad p° = r°$ (corresponding angles)

$\Rightarrow\ q° = r°$

Therefore the alternate angles are equal.

The symbol $\Rightarrow$ means 'implies that' and indicates the logical deduction made from the two stated facts.

This proof does not involve angles of a particular size; $p°$, $q°$ and $r°$ can be any size. Hence this proves that alternate angles are *always* equal whatever their size.

As a further example of deductive proof we will prove that in *any* triangle, the sum of the interior angles *is* 180°. Note that angles on a straight line by definition, add up to 180°.

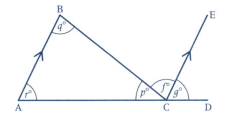

If $\triangle ABC$ is any triangle and if AC is extended to D and CE is parallel to AB then

$$p° + f° + g° = 180° \qquad \text{(angles on a straight line)} \qquad (1)$$

$$f° = q° \qquad \text{(alternate angles)} \qquad (2)$$

$$g° = r° \qquad \text{(corresponding angles)} \qquad (3)$$

$$\Rightarrow \quad p° + q° + r° = 180°$$

i.e. the sum of the interior angles of *any* triangle is 180°.

Notice how this proof uses the property proved in the first example, that is, this proof follows from the previous proof. The angle sum property of triangles can now be used to prove further properties.

These statements also lead to another useful fact about angles in triangles:

$$(2) \text{ and } (3) \quad \Rightarrow \quad f° + g° = q° + r°$$

i.e. an exterior angle of a triangle is equal to the sum of the two interior opposite angles.

Because this proof does not involve measuring angles in a particular triangle it applies to all possible triangles thus closing the loophole that there may exist a triangle whose angles do not add up to 180°.

Euclid was the first person to give a formal structure to Geometry. He started by making certain assumptions, such as 'there is only one straight line between two points'. Using only these assumptions (called axioms), he then proved some facts and used those facts to prove further facts and so on. Thus the proof of any one fact could be traced back to the axioms.

However when *you* are asked to give a geometric proof you do not have to worry about which property depends on which; you can use *any* facts that you know. One aspect of proof is that it is an argument used to convince other people of the truth of any statement, so whatever facts you use must be clearly stated.

It is a good idea to marshal your ideas before starting to write out a proof. This is most easily done by marking right angles, equal angles and equal sides etc. on the diagram.

The exercises in this chapter give practice in writing out a proof.

For the next exercise the following facts are needed;

- vertically opposite angles are equal,
- corresponding angles are equal,
- alternate angles are equal,
- interior angles add up to 180°,
- angle sum of a triangle is 180°,
- an exterior angle of a triangle is equal to the sum of the interior opposite angles,
- an isosceles triangle has two sides of the same length and the angles at the base of those sides are equal,
- an equilateral triangle has three sides of the same length and each interior angle is 60°.

Exercise 27a

In a triangle ABC the bisectors of angles B and C intersect at I.
Prove that $\widehat{BIC} = 90° + \frac{1}{2}\widehat{A}$

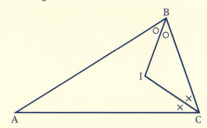

In triangle BIC

$\widehat{BIC} + \widehat{IBC} + \widehat{ICB} = 180°$ (angles in a △)

i.e.

$\widehat{BIC} + \frac{1}{2}\widehat{B} + \frac{1}{2}\widehat{C} = 180°$ (BI bisects $\widehat{B}$ and CI bisects $\widehat{C}$) (1)

But $\widehat{A} + \widehat{B} + \widehat{C} = 180°$ (angles in a △)

i.e. $\widehat{B} + \widehat{C} = 180° - \widehat{A}$

so $\frac{1}{2}\widehat{B} + \frac{1}{2}\widehat{C} = 90° - \frac{1}{2}\widehat{A}$

Substituting in (1)

$\widehat{BIC} + 90° - \frac{1}{2}\widehat{A} = 180°$

i.e. $\widehat{BIC} = 90° + \frac{1}{2}\widehat{A}$

1 Prove that $\widehat{ACD} = \widehat{ABC} + \widehat{DEC}$.

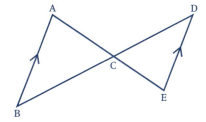

2 Prove that $\widehat{ACB} = 2\widehat{CDB}$.

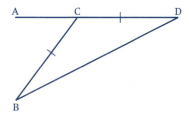

3 Prove that AD bisects $B\hat{A}C$.

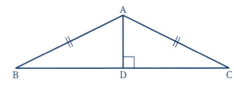

4 CE bisects $A\hat{C}D$ and CE is parallel to BA.
Prove that $\triangle ABC$ is isosceles.

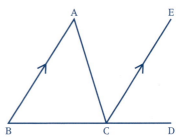

5 $A\hat{M}L = A\hat{B}C$.

Prove that $A\hat{L}M = A\hat{C}B$.

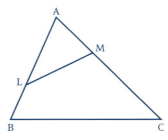

6 O is the centre of the circle.

  a Prove that $A\hat{O}X = 2A\hat{C}O$

  b Prove that $A\hat{O}B = 2A\hat{C}B$.
  Express this result in words.

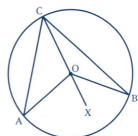

7 O is the centre of the circle. Use the second
result you obtained in question **6** to prove that
$A\hat{C}B = A\hat{D}B$.
Express this result in words.

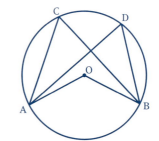

8 O is the centre of the circle and $A\hat{O}B = 180°$.
Use the results from questions **6** and **7** to
prove that $A\hat{C}B = A\hat{D}B = 90°$.
Express this result in words.

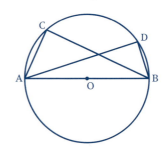

Showing that a hypothesis is false

We saw in the last section that drawing a few triangles and measuring the angles led us to say that 'it looks as though' the angles of any triangle add up to 180°. At that stage we had a *hypothesis*, which we then *proved* to be true for any triangle.

It is also important to be able to show that certain hypotheses are in fact false.

Suppose that students were asked to investigate the relationship between the number of lines drawn across a circle and the number of regions that the circle is divided into by those lines.

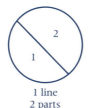

1 line
2 parts

3 lines
6 parts

2 lines
4 parts

These three drawings led John to the hypothesis that n lines drawn across a circle give $2n$ regions.

Jocelyn however, drew the lines this way, showing that 3 lines can give 7 regions, and therefore that John's hypothesis is false.

Jocelyn used a *counter example* to disprove the hypothesis.

Now consider the hypothesis 'the product of two irrational numbers is itself irrational'.

This can be shown to be untrue using this counter example: $\sqrt{2}$ and $\sqrt{8}$ are both irrational, but $\sqrt{2} \times \sqrt{8} = \sqrt{16} = 4$, which is rational.

You may like to see if you can find a counter example to disprove that all prime numbers are odd.

Not every hypothesis can be either proved or shown to be false. In mathematics there are several in this category that are well known, one being Goldbach's conjecture. That is that every even number greater than or equal to 6, can be written as the sum of two odd prime numbers. At the time of writing no one has yet proved this to be true; on the other hand no one has found a counter example.

Exercise 27b

In questions **1** to **4** see if you can find a counter example to disprove each hypothesis.

1 The square root of a positive number is always smaller than the number.

2 If the side of a square is x cm long, the number of units of area of the square is always different from the number of units of length in the perimeter.

3 The diagonals of a parallelogram never cut at right angles.

4 The sum of any two angles in a triangle is always greater than the third angle.

Questions **5** and **6** give 'proofs' that are obviously invalid since they lead to untruths. Find, in each case, the flaw in the argument.

5 It is a fact that

$$4 - 10 = 9 - 15$$

Adding $\frac{25}{4}$ to each side gives

$$4 - 10 + \frac{25}{4} = 9 - 15 + \frac{25}{4}$$

Factorising

$$\left(2 - \frac{5}{2}\right)\left(2 - \frac{5}{2}\right) = \left(3 - \frac{5}{2}\right)\left(3 - \frac{5}{2}\right)$$

i.e.

$$\left(2 - \frac{5}{2}\right)^2 = \left(3 - \frac{5}{2}\right)^2$$

Take the square root of each side

$$2 - \frac{5}{2} = 3 - \frac{5}{2}$$

Add $\frac{5}{2}$ to each side

$$2 = 3 \quad \text{which is nonsense.}$$

6 Let

$$x = y$$

and obviously

$$x^2 - xy = x^2 - xy$$

Now

$$x = y \text{ so } xy = y^2$$

i.e. line 2 can be rewritten

$$x^2 - xy = x^2 - y^2$$

Factorise

$$x(x - y) = (x - y)(x + y)$$

Divide both sides by $(x - y)$

$$x = x + y$$

but $x = y$, so

$$x = x + x$$

i.e.

$$x = 2x$$

i.e.

$$1 = 2 \quad \text{which is nonsense.}$$

 **Puzzle**

Which shaded area, if any, is the largest?

Secants and tangents

A straight line which cuts a circle at two distinct points is called a *secant*.

PQ is a secant and AB is a *chord*.

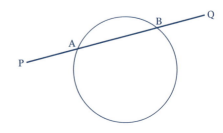

Imagine that this line is pivoted at P. As PQ rotates about P, we get successive positions of the points A and B. As PQ moves towards the edge of the circle, the points A and B move closer together, until eventually they coincide.

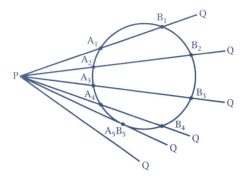

When PQ is in this position it is called a *tangent* to the circle and we say that PQ touches the circle. (When PQ is rotated beyond this position it loses contact with the circle and is no longer a tangent.)

We therefore define a tangent to a circle as a straight line which touches the circle.

The point at which the tangent touches the circle is called the point of contact.

PT is a tangent to the circle.
T is the point of contact.

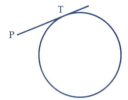

The *length of a tangent* from a point P outside the circle is the distance between P and the point of contact. In the diagram the length of the tangent from P to the circle is the length PT.

1 The diagram shows a disc, of radius 20 cm, rolling along horizontal ground. Describe the path along which O moves as the disc rolls.

At any one instant,

a how many points on the disc are in contact with the ground

b how far is O from the ground

c how would you describe the line joining O to the ground and what angle does it make with the ground?

2 Copy the diagram and draw any line(s) of symmetry.

3 Copy the diagram and draw any line(s) of symmetry.

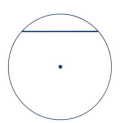

4 a Show that the chord AB is perpendicular to the radius ON which bisects AB. (Join OA and OB.)

b Now imagine that the chord AB slides down the radius ON. When the points A and B coincide with N, what has the line through A and B become? What angle does this line make with ON?

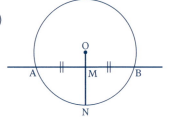

First tangent property

The investigational work in the last exercise suggests that

a tangent to a circle is perpendicular to the radius drawn from the point of contact.

The general proof of this property is an interesting exercise in logic. We start by assuming that the property is *not* true and end up by contradicting ourselves. (This is called 'proof by contradiction'.)

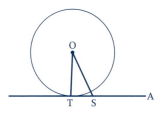

TA is a tangent to the circle and OT is the radius from the point of contact.

If we *assume* that $O\hat{T}A$ is *not* 90° then it is possible to draw OS so that OS *is* perpendicular to the tangent, i.e. $O\hat{S}T = 90°$.

Therefore △OST has a right angle at S.

Hence OT is the hypotenuse of △OST

i.e.   OT > OS

∴   S is inside the circle, as OT is a radius.

∴   the line through T and S must cut the circle again.

But this is impossible, as the line through T and S is a tangent.

Hence the assumption that $O\hat{T}A \neq 90°$ is wrong, i.e. $O\hat{T}A$ *is* 90°

Exercise 27d

Some of the questions in this exercise require the use of trigonometry.

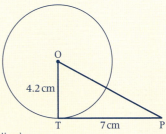

The tangent from a point P to a circle of radius 4.2 cm is 7 cm long. Find the distance of P from the centre of the circle.

$O\hat{T}P = 90°$ (tangent perpendicular to radius)

$OP^2 = OT^2 + TP^2$ (Pythagoras' theorem)

      $= (4.2)^2 + 7^2$

      $= 17.64 + 49 = 66.64$

∴ $OP = 8.163\ldots$

The distance of P from O is 8.16 cm, correct to 3 s.f.

In questions **1** to **8**, O is the centre of the circle and AB is a tangent to the circle, touching it at A.

1

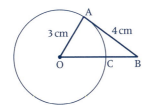

Find OB and CB.

2

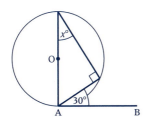

Find the marked angle.

3

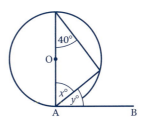

Find the marked angles.

4

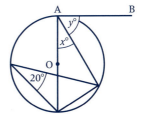

Find the marked angles

5

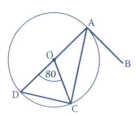

Find **a** DÂC **b** BÂC

6

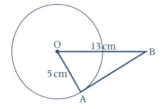

Find AB and OB̂A.

7

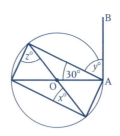

Find the marked angles.

8

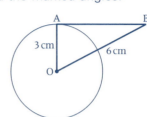

Find AB̂O.

9

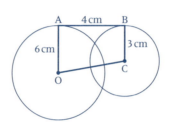

AB is a tangent to the circles with centres O and C, touching them at A and B respectively. Find OC.

10

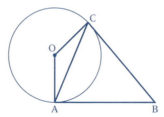

AB and BC are tangents to the circle centre O touching it at A and C. Show that △ABC is isosceles.

11

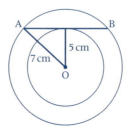

AB is a chord of the larger circle and a tangent to the smaller circle. If O is the centre of both circles, find the length of AB.

AD is the diameter of a circle and AB is a tangent to the circle at A. BD meets the circle again at E and DE = EB. Prove that $E\hat{A}B = 45°$.

AD is a diameter

$\therefore$ $D\hat{E}A = 90°$ (angle in semicircle)

In $\triangle AED$ and $\triangle AEB$

     DE = EB   (given)

AE is common

     $D\hat{E}A = A\hat{E}B$ (both 90°)

$\therefore$ $\triangle s$ $\begin{matrix} AED \\ AEB \end{matrix}$ are congruent   (SAS)

$\therefore$ $D\hat{A}E = E\hat{A}B$

But $D\hat{A}B = 90°$ (angle between tangent and radius)

$\therefore$ $D\hat{A}E = 45°$

12 AB is the diameter of a circle and D is a point on the circumference of the circle. A circle is drawn on AD as diameter. Prove that BD is a tangent to this circle.

13 P is a point outside a circle with centre O. Tangents from P to the circle touch the circle at R and S. Prove that $\triangle ROP$ and $\triangle SOP$ are congruent. Hence show that tangents from P to the circle are equal in length.

14 A circle centre A is drawn to cut a circle, centre B, at points C and D such that $A\hat{C}B = 90°$. Prove that AC is a tangent to the circle centre B.

15 AOB is a diameter of a circle, centre O. AD is a tangent to the circle at A and DB meets the circle again at C. Prove that $D\hat{A}C = A\hat{B}C$.

16 AOB is a diameter of a circle centre O. AP is a tangent to the circle at A. A chord AC is drawn so that C and P are on the same side of AB. Prove that $C\hat{A}P = A\hat{B}C$.

Second tangent property

The tangent property proved in question **13** of the last exercise can be quoted, i.e.

the two tangents drawn from an external point to a circle are the same length.

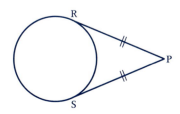

Exercise 27e

1

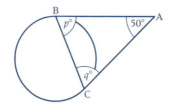

Find the marked angles.

2

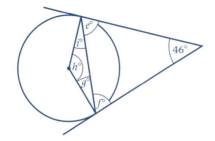

Find the marked angles.

3

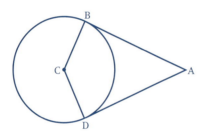

a If $\hat{BCD} = 130°$, find $\hat{BAD}$.

b What type of quadrilateral is ABCD and why?

4

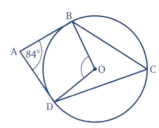

Find **a** $\hat{BOD}$ **b** $\hat{BCD}$

5

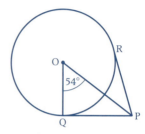

Find **a** $\hat{OPQ}$ **b** $\hat{OPR}$

6

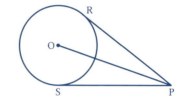

$PR = 8\,cm$ and $OP = 10\,cm$.

Calculate

a the radius of the circle

b the angle between the tangents.

7

The diagram shows the cross-section through the centre of a ball placed in a hollow cone. The vertical angle of the cone is $60°$ and the diameter of the ball is $8\,cm$. Find the depth of the vertex of the cone below the centre of the ball.

8 A second ball, of diameter $20\,cm$, is now placed in the cone described in question **7**. Will it touch the first ball?

9

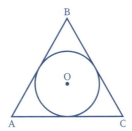

The circle, centre O, is inscribed in the equilateral triangle ABC (i.e. it is the circle that fits inside the triangle and touches all three sides). The sides of the triangle are each 20 cm long. Calculate the radius of the circle.

10 A circle of radius 4 cm is circumscribed by an equilateral triangle (i.e. the triangle is outside the circle and its sides touch the circle). Write down the angles between the sides of the triangle and the lines joining the centre of the circle to the vertices of the triangle. Hence calculate the lengths of the sides of the triangle.

11 A circle of radius 4 cm is circumscribed by an isosceles right-angled triangle. Find the lengths of the sides of the triangle.

12 ABCD is a quadrilateral circumscribing a circle. If AC goes through the centre of the circle, prove that ABCD is a kite.

13 Construct a circle, centre O and radius 4 cm. Mark a point A on the circumference. Construct $O\hat{A}B = 90°$ and hence draw the tangent AB. Mark any two points D and C on the circumference. Join A to D and A to C. Measure $C\hat{A}B$ and $A\hat{D}C$. How do they compare?

14 **a** Construct △ABC with $\hat{A} = 90°$, AB = 24 cm and AC = 10 cm.

   b Find by construction the centre of the circle which touches all three sides of △ABC.

   c Find by construction the radius of this circle and hence draw the circle (which is called the inscribed circle of △ABC).

 Puzzle

Linford met two friends. One friend had three bars of chocolate while the other had five similar bars. They decided to divide the eight bars equally between the three of them. Linford thanked his friends but insisted on paying 80 c for his share – It was all he had with him.

How should the two friends divide the 80 c between them?

Third tangent property – alternate segment theorem

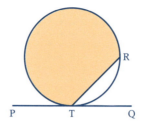

PQ is a tangent to the circle and TR is a chord. The major segment (which is shaded) is called the alternate (or other) segment with respect to the angle $R\hat{T}Q$. Similarly the minor (unshaded) segment is alternate to the angle $P\hat{T}R$.

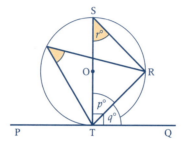

If TS is a diameter then

$$S\hat{R}T = 90° \qquad \text{(angle in semi-circle)}$$

$$S\hat{T}Q = 90° \qquad \text{(angle between tangent and radius)}$$

Now $p° + q° = 90°$

and $p° + r° = 90°$ (angles of $\triangle$)

$\Rightarrow$ $q° = r°$

But $r°$ is equal to any angle subtended by the chord TR, i.e.

the angle between a tangent and a chord drawn from the point of contact is equal to any angle in the alternate segment.

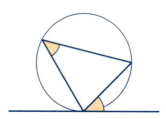

The result is known as the *alternate segment theorem* and can be quoted.

Exercise 27f

In questions **1** to **3**, copy the diagram and shade the alternate segment with respect to the angle marked $x°$.

1

2

3

Find the size of the angles marked $x°$ and $y°$ in the diagram.

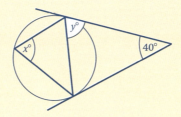

$y° = 70°$ (base angle of isosceles triangle)

$x° = y°$ (alternate segment theorem)

$\therefore x° = 70°$

Find the sizes of the angles marked by the letters.

4
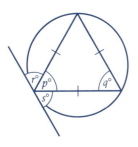

6
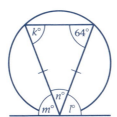

8
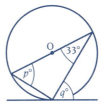

5

7
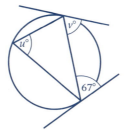

9
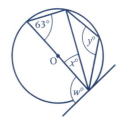

10

12

14

11

13

15

16

Find **a** DB̂C **c** BĈD **e** BD̂A
     b CD̂B **d** BÂD **f** BÊD

18

Find **a** OP̂S **b** ST̂P **c** SQ̂P

17

Find **a** RP̂S **b** PŜR **c** PR̂S **d** PŜU

19

Find **a** BĈD **b** AD̂E **c** AD̂B

20 AE is a tangent at E to the circle centre O.

a Find, in terms of x **i** AÊC

ii AB̂D

iii BÂE

b What can you deduce about

i the lines AB and ED **ii** the line BE?

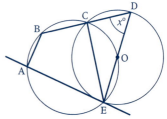

21 EDF is a tangent and BC is parallel to AD.

 a Find   **i** DB̂C

         ii BÂD

         iii BĜC

 b What kind of triangle is △BCG?

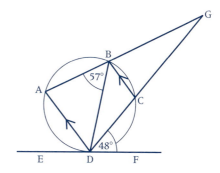

22 DE is a tangent at D to a circle centre O.

Find   **a** BD̂C   **c** BÂD

       b BĈD   **d** AD̂B

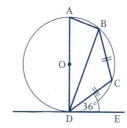

23 Find   **a** ED̂B   **b** ED̂F   **c** BÂE

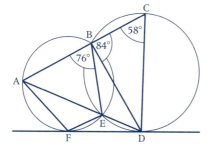

24 a Find the value of x.

 b Hence find the size of

     i CF̂D   **ii** BÂD   **iii** BD̂C

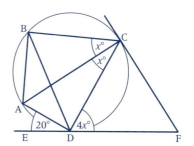

The remaining questions use the circle facts to deduce other results.

25 AB is a diameter and O is the centre of the circle.
AC = BD and E, F are the midpoints
of AC, BD. Prove that △AEO and △BFO
are congruent. Hence prove that

 a EOF is a straight line

 b AC and DB are parallel.

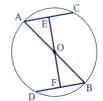

26 AB is a diameter and O is the centre
of the larger circle. AO is a
diameter of the smaller circle.
Prove that CO is parallel to DB.

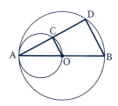

27 AE is a diameter of the circle and AD is perpendicular to BC. Prove that △AEC and △ABD are equiangular.

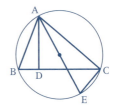

28

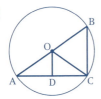

AOB is a diameter of the circle and OD bisects AÔC.
Prove that OD is parallel to BC.

Investigation

The diagram shows two circles with fixed, but different, radii.

Investigate the difference between the shaded area marked A and the shaded area marked B.

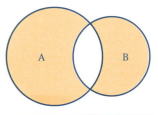

Did you know that you can divide a circle into four equal areas using three lines, of equal length, none of which intersect?

IN THIS CHAPTER...

you have seen that:

- it is possible to give a proof that geometric statements are always true

- a tangent to a circle is perpendicular to the radius at the point of contact

- two tangents drawn from a point to a circle are the same length

- the angle between a tangent and a chord is equal to the angle in the alternate segment of the circle

REVIEW TEST 3
CHAPTERS 20–27

In questions **1** to **10**, choose the letter for the correct answer.

1 442 acres of a plantation include 120 acres of vegetables. The rest is divided between citrus and bananas in the ratio 5 : 2. How many acres are planted with bananas?

   A 48   **B** 60   **C** 92   **D** 161

2 Which of the three sets of numbers are measures of the sides of a right angled triangle?

   i {6, 8, 10}   **ii** {5, 12, 13}   **iii** {2, 3, 4}

   A i and ii only   **B** ii and iii only   **C** i and iii only   **D** i, ii and iii

3 Which of the following cannot be the value of the sine of an angle?

   i 0.235   **ii** 2.156   **iii** 1.32

   A ii only   **B** iii only   **C** i only   **D** ii and iii only

4 From the top of a building 20 metres high the angle of depression of a point is 50 degrees. The distance of the point from the foot of the building is

   A $\dfrac{20}{\tan 50°}$ **B** $20\tan 50°$ **C** $20\cos 50°$ **D** $\dfrac{20}{\cos 50°}$

5 If $x * y = \dfrac{x}{x+y}$, $3 * -1 =$

   A -1 **B** $\frac{3}{4}$ **C** $1\frac{1}{2}$ **D** 4

6

Scores	40	50	60	70	80	90
Frequency	6	5	5	10	3	2

In the above distribution the median score is

   A 60   **B** 65   **C** 70   **D** 80

7 A function f is given by $f(x) = 3x - x^2$. $f(2) =$

   A 0   **B** 1   **C** 2   **D** 4

8 In triangle ABC, AB = 3 cm, AC = 4 cm and BC = 5 cm. Sin A$\hat{B}$C =

   A $\frac{3}{5}$ **B** $\frac{4}{3}$ **C** $\frac{4}{5}$ **D** $\frac{5}{4}$

9 For a set of scores, the median score is 65.

Which statement is true?
   A 65 scores are at or below the median.
   B There are 130 scores in the set.
   C Half the scores are at or above 65.
   D The middle mark is 50.

10 The bearing of P from Q is 30°. What is the bearing of Q from P?

    A 60°   **B** 150°   **C** 210°   **D** 330°

11 $A = \{1, 2, 3\}$, $B = \{2, 4, 6\}$ and $U = \{1, 2, 3, 4, 5, 6, 7\}$.

  a Draw a Venn diagram to illustrate the sets A, B and U.

  b Shade the part of the diagram that represents $A \cup B'$

12 ABC is a triangle whose base $BC = 17.5$ cm. The point X on BC is such that $BX = 10.5$ cm, $AX = 8$ cm and $A\hat{X}B = 60°$. Calculate **a** AB   **b** AC   **c** $B\hat{A}C$

13

Runs scored	Frequency
1–10	4
11–20	11
21–30	6
31–40	5
41–50	2

The table above shows the distribution of runs scored by a batsman in 28 innings.

  a Draw a bar chart for the data.

  b Calculate the mean score.

  c Which interval contains the median?

14 a John, James and Mary invest $2000, $2500 and $3500 respectively in a business. If the year's profits of $16 000 are divided proportionally, how much does each receive?

  b A relation is given by $\{(x, y), x \geqslant y, \ x, y \in \{1, 2, 3\}\}$. State with a reason whether the relation is a function.

15 Given that $a * b = b(a - b)$ and $a \oplus b = \dfrac{a - b}{b}$ find

  a $a * (b \oplus c)$ **b** $(a * b) \oplus c$

16

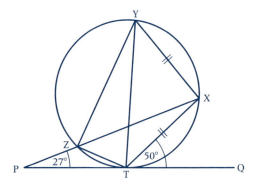

In the diagram, PTQ is a tangent and PZX is a straight line. Calculate

  a the size of angle TXY   **b** the size of angle XYZ.

MULTIPLE CHOICE QUESTIONS

1 The value of $6 + 12 \div 3$ is

    A 4   **B** 6   **C** 10   **D** 12

2 $0.35 \qquad \frac{9}{25}$

    The correct symbol to place between these two numbers is

    A $<$ **B** $\leqslant$ **C** $=$ **D** $>$

3 The value of $\left(\frac{1}{6}\right)^{-1}$ is

    A $-\frac{1}{6}$ **B** 0   **C** $\frac{1}{6}$ **D** 6

4 For the prime numbers between 1 and 30 the largest exceeds the smallest by

    A 17   **B** 21   **C** 26   **D** 27

5 3^2 exceeds $\left(\frac{1}{2}\right)^{-2}$ by

    A 2   **B** 5   **C** $5\frac{3}{4}$ **D** $8\frac{3}{4}$

6 The value of $4 - 3 - 5(-2)$ is

    A -9 **B** -6 **C** 4   **D** 11

7 The value of x that satisfies the equation $x - 3(x - 2) = 5$ is

    A $-\frac{1}{2}$ **B** $\frac{1}{2}$ **C** 2   **D** $5\frac{1}{2}$

8 Given that $a = \dfrac{3}{b - c}$ the value of a when $b = \frac{1}{2}$ and $c = \frac{1}{3}$ is

    A $\frac{1}{2}$ **B** 3   **C** 9   **D** 18

9 Given that $P = \dfrac{3Q}{R}$ the expression for R is

    A $\dfrac{3Q}{P}$ **B** $\dfrac{P}{3Q}$ **C** $3Q - P$ **D** $3QP$

10 The solution of the simultaneous equations

    $5x - y = 7$

    $3x + 2y = -1$ is

    A $x = 1, y = 2$ **B** $x = 2, y = 3$ **C** $x = -1, y = 2$ **D** $x = 1, y = -2$

11 The inverse of the matrix $\begin{pmatrix} 5 & 7 \\ 2 & 3 \end{pmatrix}$

    A $\begin{pmatrix} -3 & 7 \\ 2 & -5 \end{pmatrix}$ **B** $\begin{pmatrix} 3 & -7 \\ -2 & 5 \end{pmatrix}$ **C** $\begin{pmatrix} -5 & 2 \\ 7 & -3 \end{pmatrix}$ **D** $\begin{pmatrix} 5 & -2 \\ -7 & 3 \end{pmatrix}$

12 The equation of the straight line with gradient $\frac{1}{2}$ that passes through the point $(2, -3)$ is

    A $x + 2y + 8 = 0$ **B** $x + 2y = 8$ **C** $x - 2y = 8$ **D** $2x - y = 7$

13 For the equation $y = x^2 - x + 5$ the value of y when $x = -2$ is

    A 3   **B** 5   **C** 7   **D** 11

14 The expansion of $(3 - 4x)^2$ is

    A $9 - 12x + 16x^2$ **B** $9 - 12x + 4x^2$ **C** $9 - 24x + 16x^2$ **D** $9 - 16x^2$

15 If n is a positive integer, then $n + 7$ is always

    A an integer that is positive   **B** an integer that is even

    C an integer that is odd   **D** an integer that is prime

16 $\dfrac{2a^3}{3} \times \dfrac{3a^{-3}}{2} =$

    A 0   **B** 1   **C** a **D** $\frac{4}{9}a^6$

17 If $a * b$ denotes $a^b \div b^a$, then $1 * 2$ is

    A $\frac{12}{21}$ **B** $\frac{1}{2}$ **C** 1   **D** 2

18 An item of jewellery appreciates by 10% each year. If its original cost was $2000, then its value at the end of two years will be

    A $2020   **B** $2200   **C** $2400   **D** $2420

19 30% of 30 exceeds 20% of 20 by

    A 1   **B** 4   **C** 5   **D** 10

20 If M and N are both 2×2 matrices and $M + N = M$ then N is

    A $\begin{pmatrix} 1 & 1 \\ 1 & 1 \end{pmatrix}$ **B** $\begin{pmatrix} 0 & 0 \\ 0 & 0 \end{pmatrix}$ **C** $\begin{pmatrix} 1 & 0 \\ 0 & 1 \end{pmatrix}$ **D** $\begin{pmatrix} 0 & 1 \\ 1 & 0 \end{pmatrix}$

21 $(1 \quad -1)\begin{pmatrix} -1 \\ 1 \end{pmatrix} =$

    A (-1) **B** (-2) **C** $\begin{pmatrix} 1 & -1 \\ -1 & 1 \end{pmatrix}$ **D** $\begin{pmatrix} -1 & 1 \\ 1 & -1 \end{pmatrix}$

22

The most likely equation suggested for this straight line is

    A $y = x$ **B** $y = -x$ **C** $y = 2x + 1$ **D** $y = -2x + 1$

23 Which of the following points does not lie on the line with equation $y = 2x - 3$?

    A (2, 1)   **B** (−1, −5)   **C** (0, −3)   **D** (−3, 0)

24 The straight line $2y - 5x + 1 = 0$ has a gradient of

    A 5   **B** $\frac{5}{2}$ **C** $\frac{2}{5}$ **D** −5

25 The point (11, −3) lies on the line $3x + 4y = 3n$. The value of n is

    A 21   **B** 15   **C** 8   **D** 7

26 A car uses 6 litres of fuel at $2 per litre, when a man drives to work from Monday to Friday. If he travels by bus the cost is $1.75 per day. By travelling to work by bus instead of driving, the man would save

    A $1.50   **B** $3.25   **C** $3.50   **D** $10.25

27 A circle of diameter 14 cm is inscribed in a square of side 14 cm. If $\pi = \frac{22}{7}$, the fraction that the area of the unshaded part of the diagram is of the area of the square is

    A $\frac{1}{5}$ **B** $\frac{3}{14}$ **C** $\frac{3}{11}$ **D** $\frac{3}{10}$

28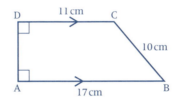

The area of trapezium ABCD, in cm², is

    A 112   **B** 136   **C** 140   **D** 168

29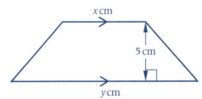

In this trapezium, the parallel sides are x cm and y cm. If the perpendicular distance between the parallel sides is 5 cm and the area of the trapezium is 30 cm² then $x + y =$

    A 6 cm   **B** 12 cm   **C** 15 cm   **D** 25 cm

30 An aircraft leaves P at 09:15 hrs and arrives at Q, a distance of 1125 km away after travelling at an average speed of 450 kmh⁻¹. The time of arrival is

    A 11:00 hrs   **B** 11:25 hrs   **C** 11:45 hrs   **D** 12:00 hrs

31 A man walks at the rate of 2 ms⁻¹ for 10 s and 1 ms⁻¹ for 20 s. His average speed, in ms⁻¹ is

    A 3   **B** 2   **C** $1\frac{1}{2}$ **D** $1\frac{1}{3}$

32 $3(2x - 1) - 2(3x + 1) =$

   A $12x - 1$ **B** $12x - 5$ **C** 0 **D** -5

33 If $2x^2 + 6x + 1 = 0$, then $x =$

   A $\dfrac{-6 \pm \sqrt{28}}{4}$ **B** $\dfrac{3 \pm \sqrt{7}}{2}$ **C** $\dfrac{-3 \pm \sqrt{7}}{2}$ **D** $\dfrac{6 \pm \sqrt{44}}{2}$

34 $\dfrac{6x - 12y}{-3} =$

   A $-2x + 4y$ **B** $2x - 4y$ **C** $-2x - 4y$ **D** $2x + 2y$

35 If $\dfrac{1}{a} + \dfrac{2}{b} = \dfrac{3}{c}$ then $c =$

   A $\dfrac{2a + b}{3ab}$ **B** $\dfrac{2a + b}{ab}$ **C** $\dfrac{ab}{2a + b}$ **D** $\dfrac{3ab}{2a + b}$

36 The interior angle of a regular polygon is twice its exterior angle.
The number of sides of the polygon is

   A 4   **B** 5   **C** 6   **D** 8

37 In a regular polygon, each exterior angle is $36°$.
The number of sides of the polygon is

   A 12   **B** 10   **C** 8   **D** 5

38 The diagram shows a sector of a circle of radius 7 cm.
Using $\pi = \frac{22}{7}$, the perimeter of the figure, in cm, is

   A 47   **B** 44   **C** 40   **D** 33

39 In the quadrilateral ABCD, $\widehat{A} = 120°$ and $\widehat{C} = 60°$.
The quadrilateral is best described as

   A a trapezium   **B** a rhombus

   C a cyclic quadrilateral   **D** a kite

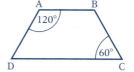

40 The largest circle that may be inscribed in a rectangle of sides 8 cm by 12 cm,
will be of area, in cm^2,

   A 8π **B** 10π **C** 12π **D** 16π

41 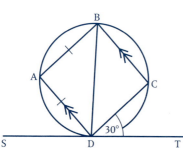 ABCD is a cyclic quadrilateral with AD parallel
to BC and AB = AD. SDT is the tangent to the
circle at D and $\angle CDT = 30°$. The size of $\angle BAD$ is

   A $90°$ **B** $120°$ **C** $135°$ **D** $150°$

42 The diagram shows a rectangle ABCD with side
BC = 5 cm and diagonal AC = 13 cm.
The perimeter of the rectangle, in cm, is

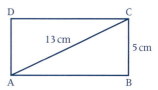

   A 34   **B** 30   **C** 26   **D** 17

43 A ladder of length 13 m is placed with its foot
on a horizontal floor and its top rests against
a vertical wall, as shown in the above diagram.
If the foot is 5 m from the wall, then the distance
of the top from the floor, in m, is

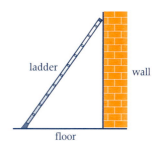

   A 18   **B** 12   **C** 9   **D** 8

44

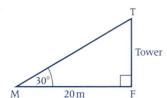

The angle of elevation of the top, T, of a vertical
tower is 30° from a point M, 20 m from the base F.
The height of the tower, in m, is

   A $20\sqrt{3}$ **B** 20   **C** $\frac{20}{\sqrt{3}}$ **D** 10

45 The bearing of P from Q is 120°. The bearing of Q from P is

   A 060°   **B** 120°   **C** 240°   **D** 300°

46 In a game, the score x occurs with the frequency shown in the table

x	1	2	4	5	20
Frequency	20	10	5	4	1

The mean score is

   A 1.5   **B** 2   **C** 2.5   **D** 4

47

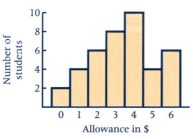

The diagram shows the daily allowance, in
dollars, of the students in a class. The
mode and median are respectively

   A $10 and $3.50   **B** $10 and $4   **C** $4 and $3.50   **D** $6 and $3.50

48 A pie chart consists of four sectors such that the second is 10° more than the
first, the third is 10° more than the second and the largest is 10° more than the
third. The size of the smallest sector is

   A 75°   **B** 80°   **C** 90°   **D** 120°

49 Two finite sets X and Y are such that $n(X) = 8$, $n(Y) = 11$, $n(X \cap Y) = 2$ and $n(U) = 20$. Hence, $n(X \cup Y)' =$

A 7   **B** 5   **C** 3   **D** 1

50

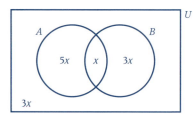

The probability that an element chosen at random from U does not belong to $A \cup B$ is

A $\frac{1}{6}$ **B** $\frac{1}{4}$ **C** $\frac{1}{3}$ **D** $\frac{1}{2}$

51 If $x \in A \cap B$ which of the following statements is true?

A $x \in A$ and $x \in B$ **B** $x \in A$ and $x \notin B$

C $x \notin A$ and $x \in B$ **D** $x \in A$ or $x \in B$

52 If $f(x) = (x - 2)(x^2 + 4)$ and $f(a) = 0$ then $a =$

A -4 **B** -2 **C** 2   **D** 4

53 A number x is doubled, increased by 3 and the result squared. This may be expressed algebraically as

A $(2x + 3)^2$ **B** $2x^2 + 3$ **C** $2(x + 3)^2$ **D** $(x + 2(3))^2$

54

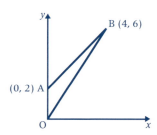

In the above diagram, the area of triangle AOB $=$

A 4 sq. units   **B** 6 sq. units   **C** 8 sq. units   **D** 12 sq. units

55 If $a * b$ denotes $a^2 - b^2$, then the positive value of a for which $a * 7 = 1 * a$ is

A 5   **B** 6   **C** 48   **D** 50

GLOSSARY

adjacent side	next to an angle
alternate segment	the segment of a circle that does not contain the angle between the tangent and chord
amount	sum of money
angle of depression	the angle that is turned through down from the horizontal to view an object below the view point
angle of elevation	the angle that is turned through up from the horizontal to view an object above the view point
anticlockwise	turning in the opposite sense of the hands of a clock
arc	a part of the circumference of a curve
area	a measure of the surface enclosed by a boundary
arithmetic average	another word for mean
associative	when an operation $*$ is such that $(a * b) * c = a * (b * c)$
bar chart	a diagram with bars representing categories, whose heights represent the quantities of the categories
bearing	a direction of one object from another object
bisect	cut in half
capacity	a measure of three-dimensional space, i.e. volume
central tendency	the mean, mode or median of a distribution of data
chord	a straight line joining two points on a curve
circumcircle	a circle passing through the vertices of a polygon
circumference	the total distance round a circle
clockwise	turning in the same sense as the hands of a clock
closed set	a set where the combination of any two elements gives an element in the set
coefficient	the multiplier of an unknown, e.g. 3 is the coefficient of $3x$
common denominator	a number that is a multiple of all the denominators of a set of fractions
common factor	a number that divides exactly into all the numbers in a set
common fraction	a fraction of the form a/b where a and b are integers
commutative	an operation on a and b, such that $a * b = b * a$
complement of a set	the members of the universal set not included in the given set
compound interest	the interest accumulated over a given time when each successive interest is added to the principle before calculating the next interest payable
converse	the converse reverses the two objects in a relationship, e.g. the converse of 'a rectangle is a quadrilateral' is 'a quadrilateral is a rectangle'

corrected number	a number rounded to a given place value
cosine	in a right-angled triangle, the cosine of an angle is equal to the length of the side adjacent to the angle over the length of the hypotenuse
cost price	price before profit is added
counter example	an example that shows that a statement is not true
cuboid	a three-dimensional solid whose faces are rectangles
cyclic quadrilateral	a quadrilateral whose vertices are on the circumference of a circle
data	a collection of facts
decimal fraction	part of a unit expressed as tenths, hundredths,..., represented by numbers to the right of the decimal point
denominator	the number on the bottom of a fraction
determinant	the determinant of a 2×2 matrix is equal to the product of entries in the leading diagonal minus the product of the entries in the other diagonal
diagonal matrix	a matrix where all the entries, except those in the leading diagonal, are zero
diameter	a chord of a circle that goes through the centre of the circle
digit	one of the figures 0, 1, 2, 3, 4, 5, 6, 7, 8, 9
direct proportion	two quantities are in direct proportion when the ratio between them is constant
directed number	a positive or negative number
distributive	an operation * is distributive over another operation # when $a * (b \# c) = a * b \# a * c$, e.g. $2 \times (3 - 5) = 2 \times 3 - 2 \times 5$
dodecahedron	a twelve-sided solid whose faces are pentagons
domain	the set of the first numbers in a relation
equation	an equality between two expressions
equilateral	equal sided
equivalent fraction	a fraction equal in value to another fraction with a different denominator
even	an integer that is divisible by 2
expression	a collection of algebraic terms
exterior angle	the angle between the extension of one side of a polygon and another side, e.g.
factor	a number that divides into another number exactly
factorise	express as a product
frequency	the number of times a value occurs, e.g if there are 5 eggs, 5 is the frequency
function	a relation where the first objects in each pair are all different
gradient of a line	the fraction given by the increase in y value over the increase in x value when moving from one point on the line to another

hexagon	a plane figure bounded by six straight lines
hypotenuse	the side opposite the right angle in a right-angled triangle
hypothesis	a statement that may or not be true
identity element	the element of a set which when combined with another element leaves the element unchanged
image	the result of a transformation or a mapping
improper fraction	a fraction whose numerator is larger than its denominator
index (indices)	the superscript of a number that tells you how many of them to multiply together
infinite	without end
intercept	the point at which a graph cuts the y-axis or the x-axis
integer	a positive or negative whole number
interior angle	the angle inside a polygon between two adjacent sides
intersection of sets	the set of elements common to both sets
inverse element	the element which when combined with another element gives the identity
inverse matrix	when the product of two matrices gives the identity matrix, one matrix is the inverse of the other
irrational number	a number that cannot be written as a/b where a and b are integers, e.g. π
isosceles	having two equal sides
leading diagonal	the entries in a matrix going from top left to bottom right
like terms	terms in an algebraic expression that have the same combination of letters
linear equation	an equation containing one unknown to the power 1
lowest common multiple	the smallest number that is a multiple of every number in a set
major arc	the larger arc between two points on the circumference of a circle
matrix	a rectangular array of numbers
mean	the sum of n values divided by n
median	the $(n+1)/2$th value of n values arranged in order of size
minor arc	the smaller arc between two points on the circumference of a circle
mixed number	the sum of a whole number and a proper fraction
mode	the value that occurs most often
multiple	a product of a given number with an integer
multiplying factor	the number that when multiplied by a quantity changes that quantity by a given amount
$\mathbb{N}$	the set of natural numbers, i.e. $\{1, 2, 3, \ldots\}$
natural number	one of the numbers 1, 2, 3, 4,...
number line	a line on which points represent numbers

numerator	the number on the top of a fraction
octagon	a plane figure bounded by eight straight lines
odd	an integer that does not divide exactly by 2
operation	a rule for combining two elements of a set, e.g. addition of numbers
opposite side	the side opposite a given angle in a right-angled triangle
parallel	a pair of lines that are always the same distance apart
parallelogram	a quadrilateral whose opposite sides are parallel
pentagon	a plane figure bounded by five straight lines
percentage	a fraction whose denominator is 100
percentage decrease	a decrease as a percentage of the original quantity
percentage increase	an increase as a percentage of the original quantity
perfect cube	a number whose cube root is exact
perfect square	a number whose square root is exact
perimeter	the distance round the edge of a plane figure
perpendicular bisector	a line at right angles to another line and cutting it in half
polyhedron	a solid whose faces are polygons
positive number	a number greater than zero
prime number	a integer whose only factors are one and itself
principal	the sum of money invested or borrowed
probability	the likelihood of an event happening expressed as a fraction
product	the multiplication of two or more quantities
profit	the difference between the cost price and the selling price
proof	an argument that convinces other people that a statement is true
proper subset	a proper subset is a set of some, but not all of the elements in another set
proportion	two quantities in the same ratio
pyramid	a solid with any polygon as the base, the other faces are triangles that meet in a common point
$\mathbb{Q}$	the set of rational numbers
quadratic	an expression such as $3x^2 - 2x + 6$
quadratic equation	an equation that can be written in the form $ax^2 + bx + c = 0$
quadratic graph	the graph of $y = $ a quadratic expression in x
quadrilateral	a plane figure bounded by four straight lines
$\mathbb{R}$	the set of real numbers
radius	the distance from the centre of a circle to its circumference
range of a relation	the set of the second objects in the ordered pairs in a relation

range of values	the numbers between which a variable can take values
ratio	a comparison of the size of two quantities
rational number	a number that can be written as a/b where a and b are integers
raw data	unorganised data
real number	any rational or irrational number
reciprocal	the reciprocal of a number a is $1/a$
rectangle	a quadrilateral whose opposite sides are equal and whose angles are right angles
recurring decimal	the numbers in the decimal places from a repeating pattern
relation	a set of ordered pairs
rhombus	a quadrilateral whose sides are all the same length
root	the numerical value of the solution of an equation
secant	a chord that is extended outside a circle
sector	part of a circle between two radii
segment	part of a circle cut off by a chord
semicircle	half a circle
set	a collection of particular objects
significant figure	the first significant figure is the first non-zero digit, the second significant figure is the next digit (which may be zero) and so on
simple interest	interest that is always calculated on the original principal
simultaneous equations	a set of equations with a common solution
sine	the sine of an angle in a right-angled triangle is the length of the side opposite the angle divided by the length of the hypotenuse
slant height	the length of a sloping edge
standard form	a number expressed as the product of a number between 1 and 10 and a power of ten, e.g. 3.6×10^6
subject of a formula	the variable that is expressed in terms of the other variables
subset	a set of some and maybe all of the elements of another set
subtend	defines an angle formed by the lines from each end of a chord of a circle to another point
supplementary angles	two angles that add up to $180°$
surd	a numerical expression containing an irrational root, e.g. $2\sqrt{3}$
tangent of an angle	the length of the side opposite the angle divided by the length of the side adjacent to the angle in a right-angled triangle
tangent to a circle	a line that touches but does not cut the circumference
tessellate	cover a flat surface with shapes without gaps

trapezium	a quadrilateral with just one pair of opposite sides parallel
triangle	a plane figure bounded by three straight lines
union of sets	the set of all the elements in two or more sets
universal set	a set large enough to include all the elements of defined sets
vector	a quantity with magnitude and direction
Venn diagram	a diagram used to represent sets
vertex	the point where two sides of a plane figure meet or where two or more edges of a solid meet
volume	the amount of three-dimensional space occupied by a solid
$\mathbb{W}$	the set of whole numbers, i.e. $\{0, 1, 2,, \ldots\}$
wedge	a solid with a constant triangular cross-section
$\mathbb{Z}$	the set of integers, i.e. $\{\ldots, -2, -1, 0, 1, 2, \ldots\}$

ANSWERS

CHAPTER 1

Exercise 1a   Page 2

1 2, 3, 5, 7
2 2, 3, 5, 7
3 11, 13, 17, 19
4 2, 3, 5, 7
5 5, 7, 11, 13
6 13, 17
7 23
8 no

Exercise 1b   Page 2

1 a yes   **c** yes   **d** yes
  b no   **d** no
2 If a number is even then 2 is a factor
3 a no   **c** yes   **e** no
  b yes   **d** yes   **f** yes
4 If a number ends in 0 or 5 then 5 is a factor
5 a 8, 16, 40, 206   **b** 40, 35, 515
6 a yes   **c** no   **e** no
  b yes   **d** yes
7 6, yes
8 a 12, yes   **b** 3, yes

Exercise 1c   Page 3

1 1, 2, 3, 6
2 1, 2, 4, 8
3 1, 2, 5, 10
4 1, 2, 4
5 1, 3, 9
6 1, 2, 3, 6, 9, 18
7 1, 3, 5, 15
8 1, 2, 4, 8, 16
9 1, 3, 7, 21
10 1, 2, 13, 26
11 1, 3, 17, 51
12 1, 19

Exercise 1d   Page 4

1 2×5
2 3×7
3 5×7
4 $2 \times 2 \times 3$
5 $2 \times 2 \times 2$
6 $2 \times 2 \times 7$
7 $2 \times 2 \times 3 \times 5$
8 $2 \times 5 \times 5$
9 $2 \times 2 \times 3 \times 3$
10 $2 \times 3 \times 11$
11 $2 \times 3 \times 3 \times 7$
12 $2 \times 2 \times 3 \times 3 \times 3$

Exercise 1e   Page 4

1 a 2, 6, 8, 10, 12, 14, 16, 18, 20
  b 3, 6, 12, 15, 18
  c 8, 12, 16, 20
  d 5, 10, 15, 20
  e 6, 12, 18
  f 8, 16
2 a 7, 14, 21, 28   **d** 10, 20, 30, 40
  b 5, 10, 15, 20   **e** 12, 24, 36, 48
  c 8, 16, 24, 32   **f** 15, 30, 45, 60
3 54, 63, 72, 81, 90, 99
4 14, 21, 28, 35, 42, 49
5 11, 22, 33, 44, 55, 66, 77, 88, 99

Exercise 1f   Page 5

1 4, 9
2 16, 25, 36, 49, 64, 81, 100
3 4, 16   **4** 4, 16, 64   **5** 9, 81
6 $6 = 2 \times 3$; $8 = 2 \times 2 \times 2$; $25 = 5 \times 5$;
  $64 = 2 \times 2 \times 2 \times 2 \times 2 \times 2$; $81 = 3 \times 3 \times 3 \times 3$;
  $125 = 5 \times 5 \times 5$. 8, 64, 125

Exercise 1g   Page 6

1 24   **4** 28   **7** 36   **10** 96
2 30   **5** 30   **8** 42   **11** 105
3 40   **6** 60   **9** 60   **12** 120

Exercise 1h   Page 6

1 4   **7** 1   **13** 5   **19** 4
2 6   **8** 12   **14** 16   **20** 9
3 6   **9** 10   **15** 6   **21** 8
4 11   **10** 10 **16** 5   **22** 7
5 13   **11** 13 **17** 8   **23** 17
6 7   **12** 1   **18** 9   **24** 13

Exercise 1i   Page 7

1 a 2, 5, 41   **d** 2, 12, 18, 36
  b 1, 5, 9, 21, 39, 41   **e** 9, 12, 18, 21, 36, 39
  c 2, 12, 18, 36   **f** 1, 2, 9, 12, 18, 36
2 a 2   **b** 5   **c** 4
3 a 6   **b** 20   **c** 12
4 a 24   **b** 14   **c** 11
5 a 6   **b** 11   **c** 5

Exercise 1j   page 7

1 a 5, 15, 21, 23, 27, 29
  b 5, 15, 20
  c 4, 5, 8, 20
  d 5, 23, 29
  e 4, 6, 8, 12, 20
  f 4, 8, 12, 20
2 a 48   **b** 8   **c** 8
3 a 2   **b** 5   **c** 72

Exercise 1k   Page 7

1 C   **2** D   **3** A   **4** A   **5** B

CHAPTER 2

Exercise 2a   Page 10

1 a yes, $5 \subset \mathbb{N}$ **b** yes, $6 \subset \mathbb{N}$
  c no, $\frac{3}{2} \not\subset \mathbb{N}$ **d** yes, $1 \subset \mathbb{N}$
  e no, $-1 \not\subset \mathbb{N}$
2 a $\frac{3}{4}$ **b** -5 **c** 2   **d** π
3 Set of negative integers
4 $\mathbb{N}$
5 Set of irrational numbers

Exercise 2b   Page 11

1 21
2 18
3 40
4 12
5 6
6 20
7 12
8 60
9 42
10 18
11 24
12 72
13 $1\frac{13}{24}$
14 $\frac{9}{10}$
15 $1\frac{29}{40}$
16 1
17 $1\frac{17}{48}$
18 $\frac{11}{12}$
19 $\frac{8}{9}$
20 $1\frac{3}{4}$
21 $2\frac{1}{5}$
22 $\frac{71}{126}$
23 $1\frac{13}{24}$
24 $1\frac{23}{42}$
25 $\frac{13}{36}$
26 $\frac{1}{36}$
27 $\frac{7}{30}$
28 $\frac{1}{20}$
29 $\frac{1}{40}$
30 $\frac{5}{18}$

31 $3\frac{29}{40}$ **34** $3\frac{11}{12}$ **37** $4\frac{2}{15}$

32 $\frac{7}{18}$ **35** $4\frac{7}{8}$ **38** $\frac{1}{8}$

33 $-\frac{9}{40}$ **36** $\frac{17}{20}$ **39** $1\frac{1}{12}$

Exercise 2c   Page 13

1 $\frac{5}{9}$ **4** $\frac{1}{10}$ **7** $4\frac{4}{7}$ **10** 3   **13** $\frac{4}{3}$

2 $1\frac{1}{3}$ **5** $\frac{10}{21}$ **8** $\frac{7}{22}$ **11** 3   **14** $\frac{8}{7}$

3 $1\frac{1}{2}$ **6** $\frac{3}{10}$ **9** 2   **12** $\frac{3}{2}$

Exercise 2d   Page 14

1 $\frac{1}{4}$ **14** $\frac{14}{81}$ **27** $-\frac{1}{2}$

2 2   **15** $\frac{2}{3}$ **28** $3\frac{7}{12}$

3 $\frac{5}{2}$ **16** $\frac{12}{49}$ **29** $3\frac{3}{140}$

4 $\frac{1}{10}$ **17** $\frac{1}{18}$ **30** $\frac{2}{5}$

5 8   **18** $4\frac{1}{2}$ **31** $\frac{22}{63}$

6 $\frac{11}{3}$ **19** $\frac{13}{30}$ **32** 14

7 $\frac{1}{100}$ **20** $\frac{69}{112}$ **33** 7

8 $\frac{9}{2}$ **21** $\frac{8}{25}$ **34** $\frac{9}{50}$

9 $\frac{4}{15}$ **22** $2\frac{1}{18}$ **35** $1\frac{2}{25}$

10 $1\frac{1}{3}$ **23** $5\frac{3}{10}$ **36** $\frac{1}{14}$

11 2   **24** $\frac{57}{110}$ **37** $\frac{21}{68}$

12 $\frac{5}{8}$ **25** $4\frac{23}{42}$ **38** $1\frac{1}{4}$

13 $6\frac{1}{4}$ **26** $\frac{7}{20}$ **39** 2

Exercise 2e   Page 16

1 $\frac{7}{20}$ **9** $\frac{11}{100}$ **17** 0.0625

2 $\frac{27}{125}$ **10** $2\frac{1}{20}$ **18** 0.54

3 $\frac{51}{250}$ **11** $1\frac{13}{125}$ **19** 1.75

4 $1\frac{9}{25}$ **12** $\frac{1}{10\,000}$ **20** 0.156 25

5 $\frac{3}{100}$ **13** 0.15   **21** 0.16

6 $\frac{3}{250}$ **14** 0.125   **22** 0.3125

7 $\frac{1}{200}$ **15** 0.6   **23** 2.375

8 $1\frac{1}{100}$ **16** 0.24   **24** 0.002

Exercise 2f   Page 17   **1** $0.\dot{3}$ **5**
 $0.\dot{1}4285\dot{7}$ **9** $0.41\dot{6}$

2 $0.\dot{2}$ **6** $0.08\dot{3}$ **10** $0.\dot{0}71\,428\dot{5}$

3 $0.8\dot{3}$ **7** $0.0\dot{9}$ **11** $0.2\dot{3}$

4 $0.0\dot{6}$ **8** $0.0\dot{5}$ **12** $0.\dot{0}76\,92\dot{3}$

Exercise 2g   Page 18

1 5.01   **17** 1.83   **33** 5.9

2 19.1   **18** 0.0068   **34** 1

3 6.17   **19** 0.96   **35** 0.02

4 8.8   **20** 0.042   **36** 0.001

5 1.82   **21** 0.008   **37** 0.6

6 26.36   **22** 0.01   **38** 7.8

7 4.832   **23** 0.25   **39** 0.5

8 1.106   **24** 0.360 72   **40** 129

9 0.002 02   **25** 3.36   **41** 11.882

10 3.2   **26** 3.355 11   **42** 3.094

11 3.3   **27** 0.000 384   **43** 1

12 0.08   **28** 7   **44** 2

13 1.21   **29** 0.3   **45** 1.69

14 0.49   **30** 2.7   **46** 0.2

15 23.02   **31** 0.008   **47** 0.4

16 0.361   **32** 0.015   **48** 8.95

Exercise 2h   Page 19

1 $<$ **6** $>$ **11** $0.79, \frac{4}{5}, 0.85$

2 $>$ **7** $>$ **12** $\frac{1}{5}, \frac{2}{7}, 0.3$

3 $<$ **8** $>$ **13** $\frac{5}{7}, 0.75, \frac{7}{9}, 0.875$

4 $<$ **9** $>$ **14** $\frac{3}{20}, 0.16, 0.2, \frac{6}{25}$

5 $>$ **10** $0.6, \frac{2}{3}, \frac{4}{5}$ **15** $1\frac{1}{8}, 1\frac{1}{5}, 1.24, 1.3$

Exercise 2i   Page 20

1 25   **12** 8010   **23** a^5

2 81   **13** 720   **24** Not possible

3 32   **14** 1102   **25** 2^2

4 125   **15** 1 100 000   **26** 7

5 64   **16** 2^7 **27** Not possible

6 144   **17** 3^7 **28** 4^3

7 1600   **18** Not possible   **29** Not possible

8 864   **19** 5^4 **30** 3^4

9 2048   **20** 2^5 **31** 3^3

10 27 783   **21** 7^7 **32** a^4

11 325   **22** 4^9 **33** Not possible

Exercise 2j   Page 21

1 a 2^6 **b** 3^4 **c** 5^4

2 a $3t^{15}$ **b** $8d^{15}$ **c** $2a^6$ **d** $125p^9$

3 a 192   **b** 6400   **c** 810   **d** 18

4 a $27x^6y^3$ **b** $5a^4b^{12}$ **c** $64u^6v^3$ **d** $2p^8q^2$

Exercise 2k   Page 22

1 $\frac{1}{2}$ **11** 4   **21** $\frac{1}{32}$ **31** 8

2 $\frac{1}{10}$ **12** $1\frac{1}{3}$ **22** $\frac{1}{10\,000}$ **32** 36

3 $\frac{1}{5}$ **13** 5   **23** $\frac{1}{100}$ **33** $1\frac{7}{9}$

4 $\frac{1}{7}$ **14** $1\frac{1}{4}$ **24** $\frac{1}{64}$ **34** $3\frac{3}{8}$

5 $\frac{1}{8}$ **15** a **25** 125   **35** $5\frac{1}{16}$

6 $\frac{1}{4}$ **16** $\frac{y}{x}$ **26** 16   **36** $12\frac{1}{4}$

7 $\frac{1}{a}$ **17** $\frac{1}{8}$ **27** 32   **37** $5\frac{1}{16}$

8 $\frac{1}{x}$ **18** $\frac{1}{25}$ **28** 81   **38** $2\frac{7}{9}$

9 3   **19** $\frac{1}{1000}$ **29** 512   **39** $123\frac{37}{81}$

10 $1\frac{1}{2}$ **20** $\frac{1}{36}$ **30** 10 000   **40** $2\frac{14}{25}$

Exercise 2l   Page 23

1 8   **6** 1   **11** $2\frac{10}{27}$ **16** $\frac{64}{125}$

2 $6\frac{1}{4}$ **7** 125   **12** $3\frac{1}{2}$ **17** $\frac{1}{12}$

3 $\frac{1}{16}$ **8** $\frac{1}{9}$ **13** 1   **18** 729

4 64   **9** 16   **14** $2\frac{314}{343}$ **19** 64

5 1   **10** 1   **15** $\frac{1}{4}$ **20** 1

Exercise 2m   Page 24

1 345   **10** 2.65×10^2 **19** 5.87×10^4

2 1200   **11** 1.8×10^{-1} **20** 2.6×10^3

3 0.0501   **12** 3.02×10^3 **21** 4.5×10^5

4 0.0047   **13** 1.9×10^{-2} **22** 7×10^{-6}

5 280   **14** 7.67×10^4 **23** 8×10^{-1}

6 0.73   **15** 3.9×10^5 **24** 5.6×10^{-4}

7 902 000   **16** 8.5×10^{-4} **25** 2.4×10^4

8 0.000 637   **17** 7×10^3 **26** 3.9×10^7

9 8 720 000   **18** 4×10^{-3} **27** 8×10^{-11}

28 a 6.25×10^{10} **c** 6.4×10^{-9}

 b 6.6049×10^8 **d** 4.9×10^{-11}

Exercise 2n Page 25

1 a 2.785   b 2.78
2 a 0.157   b 0.157
3 a 3.209   b 3.21
4 a 0.073   b 0.0733
5 a 0.151   b 0.151
6 a 0.020   b 0.0204
7 a 0.780   b 0.780
8 a 3.299   b 3.30
9 a 254.163   b 254
10 a 0.001   b 0.000 926
11 a 7.820   b 7.82
12 a 0.010   b 0.009 64
13 0.04; 0.0384
14 60 000; 47 500
15 0.05; 0.0447
16 80; 69.8
17 0.2; 0.216
18 500 000; 665 000
19 2; 2.17
20 0.2; 0.217
21 9; 8.89
22 0.08; 0.0688
23 5; 4.58
24 6; 5.38
25 60; 56.0
26 0.04; 0.0390
27 80; 69.3
28 0.03; 0.0328
29 2; 1.74
30 0.06; 0.0403
31 0.06; 0.105

Exercise 2p Page 26

1 a 30   b 42
2 a $\frac{4}{3}$ b $\frac{y}{x}$
3 a $\frac{3}{2}$ b $\frac{4}{9}$
4 $2\frac{3}{10}$
5 a 3.36   b 0.2943   c 109
6 a 16   b 1   c $\frac{1}{16}$
7 a 5^2 b 5^{12}
8 a 2.56×10^3 b 2.56×10^{-4}

Exercise 2q Page 27

1 a 24   b 30
2 a 5   b $\frac{2}{3}$
3 a $\frac{3}{4}$ b $1\frac{17}{20}$
4 $3\frac{1}{12}$
5 a 1.45   b 2.625   c 0.42
6 a $\frac{1}{4}$ b 1   c 4
7 a 5.7×10^5 b 5.7×10^{-2}

Chapter 3

Exercise 3a Page 30

1 $4 > 2$ 5 $2 > -1$ 9 $5 > -5$
2 $3 < 5$ 6 $-4 < 3$ 10 $4 > -2$
3 $-1 > -4$ 7 $-5 > -6$ 11 $-3 < 0$
4 $-5 < -2$ 8 $0 > -4$ 12 $0 < 6$

Exercise 3b Page 31

1 -5 10 7   19 4   28 2
2 -2 11 -10 20 -3 29 11
3 -6 12 6   21 -12 30 7
4 -2 13 -5 22 -6 31 0
5 1   14 6   23 -5 32 -2
6 8   15 -5 24 4   33 -1
7 -9 16 -3 25 5   34 7
8 6   17 3   26 4   35 -1
9 -6 18 -2 27 -5 36 1

Exercise 3c page 32

1 -6 7 -12 13 2   19 3
2 12   8 14   14 -2 20 5
3 56   9 4   15 -2 21 2
4 27   10 -15 16 2   22 -3
5 -8 11 9   17 -2 23 -1
6 30   12 -18 18 -5 24 -3

Exercise 3d Page 33

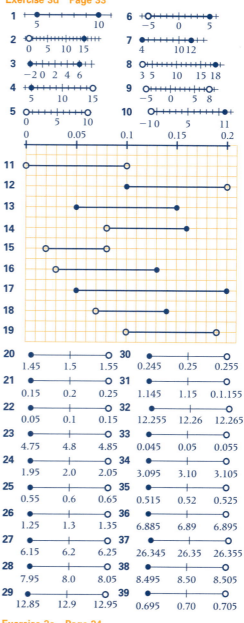

Exercise 3e Page 34

1 $5.55 \leqslant w < 5.65$ 6 $1.245 \leqslant d < 1.255$
2 $2450 \leqslant x < 2550$ 7 65 people
3 $2.75 \leqslant x \leqslant 2.85$ 8 \$254
4 $12.45 \leqslant x < 12.55$ 9 $252.5 \leqslant$ length < 257.5
5 $6500 \leqslant x < 7400$ 10 97.5 m

Exercise 3f Page 36

1 3	**7** 5	**13** -3	**19** -2
2 -12	**8** 36	**14** -28	**20** 0
3 25	**9** 13	**15** 81	**21** 1
4 2	**10** -30	**16** -4	**22** -5
5 3	**11** -4	**17** 2	**23** -14
6 -12	**12** 3	**18** 20	**24** 7

25 1.45 mm, 1.54 mm **27** 445, 454
26 65, 74 **28** 0.745 m, 0.754 m

Chapter 4

Exercise 4a Page 39

1 Not possible	**6** $p+q$
2 $2a$	**7** $4x-2y$
3 Not possible	**8** $5u$
4 $7v$	**9** $3b-a$
5 $2x$	**10** $4c+2d$

Exercise 4b Page 39

1 xy	**9** Not possible	**17** $6st$
2 a^2	**10** Not possible	**18** $2p^2$
3 $6s^2$	**11** $-mn$	**19** $-4q$
4 $12x^2$	**12** Not possible	**20** $r+4s$
5 $\dfrac{u}{v}$	**13** $-2a$	**21** $\dfrac{4p}{q}$
6 $\dfrac{-a}{b}$	**14** $4p^3$	**22** $6st$
7 1	**15** $\dfrac{2u}{w}$	**23** $-2b^2$
8 $\dfrac{3b}{c}$	**16** Not possible	**24** $\dfrac{x}{y}$

25 $3b-2a$	**33** $5q-p$
26 a^2-a	**34** $a^2+ab-2a+2b$
27 $3a-3b$	**35** x^2+y^2-2xy
28 $6a-4c-2b$	**36** $2b-6c$
29 $2z-y$	**37** $2p-2q$
30 $6x+4y+2z$	**38** w^2+x^2
31 $p+3q-2r$	**39** $8n-2m$
32 $x-y$	

Exercise 4c Page 41

1 $p=-\frac{2}{3}$	**11** $x=-\frac{1}{2}$	**21** $x=6\frac{1}{2}$
2 $s=\frac{1}{2}$	**12** $x=-5$	**22** $x=-\frac{1}{6}$
3 $x=3\frac{1}{2}$	**13** $x=-\frac{1}{7}$	**23** $x=2\frac{1}{2}$
4 $a=\frac{1}{5}$	**14** $y=3\frac{1}{3}$	**24** $x=1\frac{3}{4}$
5 $x=1$	**15** $x=2\frac{4}{7}$	**25** $x=\frac{3}{14}$
6 $y=1$	**16** $x=3\frac{2}{3}$	**26** $b=19$
7 $x=2$	**17** $a=-11$	**27** $x=\frac{5}{6}$
8 $a=4$	**18** $p=0$	**28** $x=-1$
9 $x=\frac{1}{2}$	**19** $w=2$	**29** $x=2$
10 $x=2$	**20** $x=5$	**30** $x=\frac{1}{2}$

Exercise 4d Page 43

1 $a=b+c$	**7** $x=\dfrac{y}{2}$
2 $m=2(n+p)$	**8** $a=\dfrac{b}{2c}$
3 $z=xy$	**9** $k=2u+3v$
4 $a=2bc$	**10** $x=2y-z$
5 $v=n^2$	**11** $n=p+p^2$
6 $d=e-f$	**12** $v=u+at$

13 $R=Np$		**17** $P=\dfrac{x+y}{50}$
14 $y=nx$		**18** $b=\dfrac{ac}{1000}$
15 $X=xy$		**19** $n=1+5m$
16 $N=y+z$		**20** $R=\dfrac{x}{10}+\dfrac{y}{20}$

Exercise 4e Page 44

1 $p=8$	**7** $x=24$	**13** $r=2\frac{2}{3}$
2 $v=2$	**8** $p=6$	**14** $n=\frac{1}{2}$
3 $z=\frac{3}{4}$	**9** $s=10$	**15** $a=2$
4 $a=2$	**10** $v=-5$	**16** $V=32$
5 $x=25$	**11** $p=4$	**17** $P=21$
6 $C=30$	**12** $a=9$	**18** $a=6$

Exercise 4f Page 45

1 $s=p-r$		**19** $w=100z$
2 $y=x-3$		**20** $p=qn$
3 $b=a+c$		**21** $s=\dfrac{P-r}{2}$
4 $Y=X+Z$		**22** $t=\dfrac{u-v}{3}$
5 $s=r-2t$		**23** $c=\dfrac{b-a}{4}$
6 $m=k-l$		**24** $v=\dfrac{V-3u}{2}$
7 $v=u+5$		**25** $w=\dfrac{x+y}{2}$
8 $y=z-x$		**26** $t=\dfrac{l-k}{4}$
9 $P=N+Q$		**27** $y=\dfrac{x-w}{6}$
10 $u=v-10t$		**28** $s=\dfrac{lt-N}{2}$
11 $y=\dfrac{x}{2}$		**29** $y=\dfrac{4x}{3}$
12 $t=2v$		**30** $t=\dfrac{u-v}{5}$
13 $b=\dfrac{a}{c}$		**31** $I=10(A-P)$
14 $u=3t$		**32** $y=3(x-z)$
15 $m=kl$		**33** $R=\dfrac{IV}{2}$
16 $b=\dfrac{a}{3}$		**34** $r=\dfrac{p+w}{2}$
17 $N=10X$		**35** $c=2(a-b)$
18 $u=\dfrac{v}{t}$		**36** $r=5(q-p)$

37 $u=v-at;\ u=140$
38 $B=A-\dfrac{C}{100};\ B=17.5$
39 $C=NP;\ C=40$
40 $x=2(z+3t);\ x=-10$
41 a $a=b+2c$ **b** $a=4$ **c** $b=a-2c$
42 a $x=2yz$ **b** $x=12$ **c** $y=\dfrac{x}{2z}$

43 a $d = e^2 + 2f$ **b** $f = \dfrac{d - e^2}{2}$

  c $f = \frac{1}{2}$

44 a $R = \dfrac{3xn}{25}$ **b** $R = 4.8$

Exercise 4g   Page 47

1 a -4 **b** -1 **c** -2
2 a $4x$ **b** $6b$ **c** $-3x^3$
3 a $a + b$ **b** $a + 5b$
4 a $x = 1\frac{1}{4}$ **b** $x = 4\frac{2}{3}$
5 a $r = \dfrac{v - u}{4}$ **b** $r = \dfrac{ps}{5}$
6 a $37\frac{1}{2}$ **b** 40

Exercise 4h   Page 47

1 a 13   **b** 2   **c** 4
2 a $10a - 3b$ **b** $4x + x^2$ **c** $12ab$
3 a $3y - 2x$ **b** $2y - 6x$
4 a $a = -1$ **b** $x = \frac{7}{8}$
5 a $d = \dfrac{C}{\pi}$ **b** $d = \dfrac{a + s}{7}$
6 a 56   **b** -86

Chapter 5
Exercise 5a   Page 50

	Fraction	Percentage	Decimal
1	$\frac{3}{5}$	60%	0.6
2	$\frac{2}{5}$	40%	0.4
3	$\frac{11}{20}$	55%	0.55
4	$\frac{17}{20}$	85%	0.85
5	$\frac{27}{50}$	54%	0.54
6	$\frac{6}{25}$	24%	0.24
7	$\frac{23}{25}$	92%	0.92
8	$\frac{21}{25}$	84%	0.84
9	$\frac{37}{40}$	$92\frac{1}{2}\%$	0.925
10	$\frac{2}{3}$	$66\frac{2}{3}\%$	0.6

11 24%   **21** 40%   **31** 1.5%   **41** 49.28
12 64%   **22** 225%   **32** 2.5%   **42** 348
13 20%   **23** 75   **33** 2.4%   **43** 31.59
14 40%   **24** 92 c   **34** 60%   **44** 172
15 25%   **25** 0.61 cm **35** 30%   **45** 64.68
16 34%   **26** 0.54 km **36** 89.6% **46** 294
17 5%   **27** 189 g **37** 949   **47** 5.74
18 2.5%   **28** 42 m^2 **38** 1007 **48** 59.4 kg
19 2%   **29** 2.5%   **39** 627   **49** $19 350
20 27%   **30** 2%   **40** 381   **50** 51

Exercise 5b   Page 54

1 25%   **11** $60.90   **21 b** by 70 c
2 30%   **12** $18   **22** the same
3 25%   **13** $27   **23** $138
4 10%   **14** $80   **24** $368
5 20%   **15** $8   **25** $96.60
6 20%   **16** $1440   **26** $174.80
7 15%   **17** $12   **27** $2760
8 24%   **18** $1620   **28** $9.20
9 $56   **19** $14   **29** $2392
10 $72   **20 b** by $8 **30** $1725

Exercise 5c   Page 55

1 $1500   **9** $4000   **17** $368
2 $2400   **10** $7680   **18** $51.20
3 $1950   **11** $5400   **19** $232
4 $3750   **12** $3300   **20** $13540
5 $1935   **13** $10 620 **21** $3825
6 $2478   **14** $6336   **22** $1615
7 $3300   **15** $12 810 **23** $1430
8 $2240   **16** $84   **24** $3225
25 a $1220   **b** $1480

Exercise 5d   Page 57

1 $70   **10** $40   **19** $240
2 $40   **11** $80   **20** $250
3 $16   **12** $200   **21** $520
4 $6   **13** $18   **22** $184
5 $2.70   **14** $1360   **23** $72
6 $32   **15** $50   **24** $1950
7 $800   **16** $160   **25** $320
8 $900   **17** $17   **26** 25 cm
9 $800   **18** $160

Exercise 5e   Page 59

1 $127 680   **6** $11250   **11 a** $1440
2 $41.07   **7** 2125 cm^3 **b** $1320
3 $33\frac{1}{3}\%$ **8** $3440   **12** $840
4 40%   **9** 15 km/h
5 $1200   **10** 44 275

Exercise 5f   Page 61

1 $32   **6** $126   **11** $440
2 $112   **7** $201.60   **12** 7 years
3 $284.70   **8** $375.30   **13** 6.5%
4 $11872   **9** $500   **14** $300
5 $162.50   **10** $168.80 **15** 4%
16 a $170.40   **b** $392   **c** $81620
17 a 3 years   **b** 5 years   **c** 4 years

Exercise 5g   Page 62

1 $42   **4** $191.77   **7** $252.68   **10** $4374
2 $76.32   **5** $143.99   **8** $96 800   **11** $12 800
3 $103.88 **6** $206.72   **9** $76

Chapter 6
Exercise 6a   Page 67

1 3, 2   **4** 1, 7   **7** $-2, 1$ **10** 6, 0
2 2, 4   **5** 4, -3 **8** 5, 1   **11** $-1, -2$
3 3, 5   **6** 2, 5   **9** $3, 1\frac{1}{2}$ **12** 0, 9

Exercise 6b   Page 68

1 3, 1   **11** 4, -1 **21** 3, 0
2 4, 2   **12** 6, 2   **22** $1\frac{1}{2}, 2$
3 3, 4   **13** $5, 1\frac{1}{2}$ **23** $-3, 2$
4 3, -1 **14** 4, 3   **24** 4, -2
5 $3, \frac{1}{2}$ **15** $\frac{1}{2}, 4$ **25** 6, 2
6 9, 1   **16** 4, -2 **26** 4, 3
7 4, -2 **17** $-3, 1$ **27** $-1, 4$
8 1, 0   **18** $2, \frac{1}{3}$ **28** $-1, -2$
9 0, 6   **19** 3, 2   **29** 5, 4
10 2, 3   **20** 4, 5   **30** 2, -4

Exercise 6c   Page 69

1 3, 1
2 1, 2
3 $\frac{1}{3}$, 1
4 −12, 27
5 0, 1
6 4, 3
7 1, 2
8 2, 1
9 3, −1
10 0, 3
11 1, −1
12 3, $\frac{1}{2}$

Exercise 6d   Page 70

1 3, 2
2 1, 5
3 3, 1
4 $1\frac{1}{2}$, 0
5 0, 6
6 3, −1
7 1, 4
8 1, 1
9 2, 2
10 3, −1
11 4, 2
12 −3, 0
13 2, $\frac{2}{3}$
14 −1, 2
15 3, −2
16 2, −2
17 0, 4
18 −1, −2
19 1, 1
20 3, 1
21 2, −1
22 8, 4
23 −3, 4
24 3, $-3\frac{1}{2}$
25 3, 4
26 2, 5
27 3, 2
28 −1, −3

Exercise 6e   Page 72

1 1, 4
2 −1, 5
3 3, −2
4 6, 28
5 2, 3
6 −1, −1
7 $3\frac{1}{2}$, $2\frac{1}{2}$
8 1, −2
9 5, 0
10 0, 4
11 −1, −1
12 −4, −5

Exercise 6f   Page 72

1 2, 4
2 5, 3
3 1, 1
4 −2, 7
5 4, 6
6 1, 1
7 1, 10
8 $2\frac{1}{3}$, $-\frac{2}{3}$
9 1, 5
10 −12, −4
11 2, 6
12 $4\frac{1}{2}$, $7\frac{1}{2}$

Exercise 6h   Page 74

1 1, 2
2 3, 1
3 2, −2
4 $\frac{1}{2}$, 1
5 3, 4
6 −2, −1
7 1, 10
8 $2\frac{1}{3}$, $-\frac{2}{3}$
9 1, 5
10 −12, −4
11 2, 6
12 $4\frac{1}{2}$, $7\frac{1}{2}$

Exercise 6i   Page 75

1 12, 8
2 11, 5
3 8, 2
4 10, 3
5 10, 6
6 11, 5
7 3 choc ices, 7 orange ices
8 54, 36
9 cup $1.35, saucer $1.80
10 Petties $1.40 each, Roti $2.10 each
11 Harry 32, Adam 10, Sam 20
12 3, 5
13 AB = $9\frac{1}{2}$ cm, BC = 6 cm
14 $m = 2$, $c = 4$, $y = 2x + 4$

Exercise 6j   Page 77

1 $1\frac{1}{2}$, $4\frac{1}{2}$
2 $1\frac{1}{3}$, $3\frac{2}{3}$
3 $1\frac{1}{2}$, $5\frac{1}{2}$
4 $-\frac{1}{2}$, $1\frac{1}{2}$
5 $\frac{1}{2}$, 2
6 $1\frac{1}{2}$, $3\frac{1}{2}$
7 $2\frac{2}{5}$, $1\frac{4}{5}$
8 $-\frac{2}{5}$, $1\frac{3}{5}$
9 $2\frac{2}{5}$, $1\frac{1}{5}$
10 $\frac{1}{3}$, $1\frac{2}{3}$

Chapter 7

Exercise 7a   Page 80

1 2 × 2
2 2 × 3
3 2 × 1
4 1 × 3
5 1 × 1
6 3 × 2
7 a 6   b 8   c 2   d 7
8 3 1 7; 7   a 7   b 6   c 4
    2

9 $\begin{pmatrix} 0 & 0 & 0 \\ 1 & 1 & 1 \\ 2 & 2 & 2 \end{pmatrix}$

10 $\begin{pmatrix} 3 & 1 \\ 3 & 1 \\ 3 & 1 \end{pmatrix}$

Exercise 7b   Page 82

1 $\begin{pmatrix} 12 \\ 15 \end{pmatrix}$
2 $\begin{pmatrix} 15 & 4 \\ 7 & 1 \end{pmatrix}$
3 Not possible
4 (9, 5)
5 $\begin{pmatrix} 11 & 2 & 2 \\ 6 & 7 & 7 \end{pmatrix}$
6 $\begin{pmatrix} 11 & 11 \\ 11 & 5 \end{pmatrix}$
7 (5 3 5)
8 Not possible
9 $\begin{pmatrix} 6 & 8 \\ 7 & 7 \\ 7 & 7 \end{pmatrix}$
10 (10   8)
11 $\begin{pmatrix} 1 & 8 \\ -4 & 7 \end{pmatrix}$
12 $\begin{pmatrix} -2 \\ 2 \end{pmatrix}$

13 $\begin{pmatrix} -2 & 7 \\ -5 & 3 \end{pmatrix}$
14 (4   6)
15 $\begin{pmatrix} -3 \\ -3 \\ 6 \end{pmatrix}$
16 Not possible
17 $\begin{pmatrix} 2 & 10 \\ 5 & -3 \end{pmatrix}$
18 $\begin{pmatrix} 5 & -5 \\ 3 & 0 \end{pmatrix}$
19 $\begin{pmatrix} 0 & 8 \\ 8 & -2 \end{pmatrix}$
20 Not possible
21 Not possible
22 (1   6   −3)
23 $\begin{pmatrix} 2 & 3 & 4 \\ 5 & 0 & -12 \end{pmatrix}$

Exercise 7c   Page 83

1 $\begin{pmatrix} 3 \\ 6 \\ 12 \end{pmatrix}$
2 $\begin{pmatrix} 2 & 8 & 0 \\ 4 & -2 & 6 \end{pmatrix}$
3 $\begin{pmatrix} 1 & 2 \\ \frac{1}{2} & 3 \\ 1\frac{1}{2} & 4 \end{pmatrix}$
4 $\begin{pmatrix} 6 & 24 \\ 18 & -12 \end{pmatrix}$
5 $\begin{pmatrix} -6 & -30 \\ 6 & 12 \end{pmatrix}$
6 $\begin{pmatrix} 4 & 0 \\ \frac{2}{3} & 1\frac{1}{3} \\ 2 & 3\frac{1}{3} \end{pmatrix}$

7 $\begin{pmatrix} 2 & -2 \\ 1 & 3 \end{pmatrix}$
8 Not possible
9 $\begin{pmatrix} -3 \\ 3 \\ 0 \end{pmatrix}$
10 $\begin{pmatrix} 2 & 4 & 2 \\ -3 & -3 & -1 \end{pmatrix}$
11 Not possible
12 $\begin{pmatrix} -3 & -1 & 2 \\ 9 & 5 & 4 \\ 1 & 11 & 5 \end{pmatrix}$

Exercise 7d   Page 84

1 $\begin{pmatrix} -1 & 8 \\ 6 & 1 \end{pmatrix}$
2 $\begin{pmatrix} 3 & 0 \\ 0 & 3 \end{pmatrix}$
3 Not possible
4 $\begin{pmatrix} 7 & -1 \\ 5 & -1 \end{pmatrix}$
5 $\begin{pmatrix} 1\frac{1}{3} \\ 1\frac{2}{3} \\ -\frac{1}{3} \end{pmatrix}$
6 $\begin{pmatrix} -3 \\ -3 \\ -3 \end{pmatrix}$

7 $\begin{pmatrix} 2 & 8 \\ 6 & 4 \end{pmatrix}$
8 $\begin{pmatrix} -1 & 2 \\ 1\frac{1}{2} & -\frac{1}{2} \end{pmatrix}$
9 $\begin{pmatrix} 8 \\ 9 \\ 3 \end{pmatrix}$
10 $\begin{pmatrix} 24 & 8 & -4 \\ 16 & 12 & 16 \end{pmatrix}$
11 Not possible
12 $\begin{pmatrix} 8 & -2 & -2 \\ 1 & 4 & 4 \end{pmatrix}$

Exercise 7e   Page 85

1 B 2×1, C 2×2, D 2×2, E 1×3, F 1×2,
   G 2×3

2 $\begin{pmatrix} 9 & 4 & 4 \\ 7 & 1 & 7 \end{pmatrix}$ **8** Not possible

3 Not possible

4 $\begin{pmatrix} 4 & -1 \\ 0 & 6 \end{pmatrix}$

9 $\begin{pmatrix} 1 & -2 & 2 \\ 5 & -3 & 1 \end{pmatrix}$

10 $\begin{pmatrix} 24 \\ 6 \end{pmatrix}$

5 $\begin{pmatrix} 12 & 9 & 3 \\ 3 & 6 & 9 \end{pmatrix}$

11 $\begin{pmatrix} 4\frac{1}{2} & 1\frac{1}{2} \\ \frac{3}{4} & 3 \end{pmatrix}$

6 Not possible

7 $(1\frac{1}{2} \quad 1)$

12 Not possible
13 Not possible

Exercise 7f   Page 86

1 $\begin{pmatrix} 29 \\ 27 \end{pmatrix}$ **4** $\begin{pmatrix} 9 \\ 2 \end{pmatrix}$ **7** $\begin{pmatrix} 26 \\ 10 \end{pmatrix}$

2 $\begin{pmatrix} 14 \\ 11 \end{pmatrix}$ **5** $\begin{pmatrix} 9 \\ 5 \end{pmatrix}$ **8** $\begin{pmatrix} 58 \\ 19 \end{pmatrix}$

3 $\begin{pmatrix} 5 \\ 7 \end{pmatrix}$ **6** $\begin{pmatrix} 18 \\ 14 \end{pmatrix}$ **9** $\begin{pmatrix} 56 \\ 49 \end{pmatrix}$

Exercise 7g   page 87

1 $\begin{pmatrix} 7 \\ 10 \end{pmatrix}$ **11** $\begin{pmatrix} 16 & 14 \\ 11 & 14 \end{pmatrix}$

2 $\begin{pmatrix} 14 \\ & 22 \end{pmatrix}$ **12** $\begin{pmatrix} 8 & 18 \\ 18 & 36 \end{pmatrix}$

3 $\begin{pmatrix} 37 \\ & 2 \end{pmatrix}$ **13** $\begin{pmatrix} 44 & 40 \\ 18 & 31 \end{pmatrix}$

4 $\begin{pmatrix} 23 & 16 \\ & \end{pmatrix}$ **14** $\begin{pmatrix} 21 & 7 \\ 17 & 9 \end{pmatrix}$

5 $\begin{pmatrix} 19 \\ 22 \end{pmatrix}$ **15** $\begin{pmatrix} 0 & 14 \\ 10 & 8 \end{pmatrix}$

6 $\begin{pmatrix} & 16 \\ 12 & \end{pmatrix}$ **16** $\begin{pmatrix} 15 & 20 \\ 5 & 0 \end{pmatrix}$

7 $\begin{pmatrix} 12 \\ 3 \end{pmatrix}$ **17** $\begin{pmatrix} 3 & -4 \\ 13 & 6 \end{pmatrix}$

8 $\begin{pmatrix} 17 & 19 \\ 5 & \end{pmatrix}$ **18** $\begin{pmatrix} -24 & -17 \\ -10 & -9 \end{pmatrix}$

9 $\begin{pmatrix} 22 & 52 \\ 10 & 22 \end{pmatrix}$ **19** $\begin{pmatrix} 21 & 11 \\ 9 & 2 \end{pmatrix}$

10 $\begin{pmatrix} 44 & 32 \\ 8 & 7 \end{pmatrix}$ **20** $\begin{pmatrix} -16 & 1 \\ -6 & -1 \end{pmatrix}$

Exercise 7h   Page 88

1 $\begin{pmatrix} 20 & 13 \\ 8 & 5 \end{pmatrix}$ **7** $\begin{pmatrix} 15 & 17 \\ 31 & 35 \end{pmatrix}$

2 $\begin{pmatrix} 10 & 7 \\ 22 & 15 \end{pmatrix}$ **8** $\begin{pmatrix} 46 & 31 \\ 6 & 4 \end{pmatrix}$

3 $\begin{pmatrix} 31 & 35 \\ 15 & 17 \end{pmatrix}$ **9** $\begin{pmatrix} 4 & 2 \\ 8 & 6 \end{pmatrix}$

4 $\begin{pmatrix} 44 & 29 \\ 6 & 4 \end{pmatrix}$ **10** $\begin{pmatrix} 4 & 2 \\ 8 & 6 \end{pmatrix}$

5 $\begin{pmatrix} 8 & 6 \\ 4 & 2 \end{pmatrix}$ **11** $\begin{pmatrix} 14 & 16 \\ 2 & 2 \end{pmatrix}$

6 $\begin{pmatrix} 8 & 6 \\ 4 & 2 \end{pmatrix}$ **12** $\begin{pmatrix} 14 & 16 \\ 2 & 2 \end{pmatrix}$

13 One of the two matrices was **D**

Exercise 7i   Page 90

1 $\begin{pmatrix} 7 \\ 10 \end{pmatrix}$ **6** $\begin{pmatrix} 24 \\ 33 \\ 42 \end{pmatrix}$

2 $\begin{pmatrix} 13 \\ 32 \end{pmatrix}$ **7** $(13 \quad 31 \quad 27)$

3 (10) **8** $\begin{pmatrix} 10 & 11 \\ 36 & 30 \\ 31 & 28 \end{pmatrix}$

4 $\begin{pmatrix} 20 & 10 \\ 70 & 23 \end{pmatrix}$ **9** $\begin{pmatrix} 21 & 39 & 8 \\ 17 & 26 & 7 \end{pmatrix}$

5 $\begin{pmatrix} 32 & 26 & 16 \\ 20 & 19 & 11 \end{pmatrix}$ **10** (15)

Exercise 7j   Page 91

1 $2 \times \boxed{2 \quad 2} \times 1 = 2 \times 1;$ $\begin{pmatrix} 7 \\ 6 \end{pmatrix}$

2 $2 \times \boxed{3 \quad 3} \times 1 = 2 \times 1;$ $\begin{pmatrix} 22 \\ 12 \end{pmatrix}$

3 $1 \times \boxed{2 \quad 2} \times 1 = 1 \times 1;$ (10)

4 $2 \times \boxed{3 \quad 3} \times 2 = 2 \times 2;$ $\begin{pmatrix} 20 & 10 \\ 70 & 23 \end{pmatrix}$

5 $2 \times \boxed{2 \quad 2} \times 2 = 2 \times 2;$ $\begin{pmatrix} 11 & 20 \\ 24 & 43 \end{pmatrix}$

6 $2 \times \boxed{1 \quad 1} \times 2 = 2 \times 2;$ $\begin{pmatrix} 3 & 4 \\ 6 & 8 \end{pmatrix}$

7 $1 \times \boxed{2 \quad 2} \times 2 = 1 \times 2;$ $(21 \quad 36)$

8 $3 \times \boxed{1 \quad 1} \times 3 = 3 \times 3;$ $\begin{pmatrix} 4 & 5 & 6 \\ 8 & 10 & 12 \\ 12 & 15 & 18 \end{pmatrix}$

9 $\begin{pmatrix} 16 \\ 6 \end{pmatrix}$ **14** Not possible

10 Not possible   **15** (30)

11 $\begin{pmatrix} 11 & 20 \\ 24 & 43 \end{pmatrix}$ **16** Not possible

12 Not possible   **17** $(3 \quad 24)$

13 $\begin{pmatrix} 15 & 4 & 3 \\ 48 & 13 & 12 \end{pmatrix}$ **18** $\begin{pmatrix} 6 & 12 & 15 \\ 8 & 16 & 20 \\ 2 & 4 & 5 \end{pmatrix}$

Exercise 7k   page 92

1 $\begin{pmatrix} 6 \\ 5 \end{pmatrix}$ **6** $(-38 \quad 12)$

2 $\begin{pmatrix} 10 \\ -19 \end{pmatrix}$ **7** (-26)

3 $(-2 \quad -6)$ **8** $\begin{pmatrix} -24 & -4 & 12 \\ 6 & 1 & -3 \\ 6 & 1 & -3 \end{pmatrix}$

4 $\begin{pmatrix} 1 \\ 1 \\ -22 \end{pmatrix}$ **9** $\begin{pmatrix} 7 & 18 & -1 \\ -7 & -18 & 1 \end{pmatrix}$

5 $\begin{pmatrix} 8 & -26 \\ -16 & -17 \end{pmatrix}$ **10** $\begin{pmatrix} 12 & 18 \\ 8 & 12 \\ -6 & -9 \end{pmatrix}$

Exercise 7l   page 93

1 $\begin{pmatrix} 13 & -8 \\ 7 & -2 \end{pmatrix}$ **9** Not possible

2 $\begin{pmatrix} 2 & -1 \\ 12 & 9 \end{pmatrix}$ **10** (11)

3 $\begin{pmatrix} 10 \\ 5 \end{pmatrix}$ **11** $(14 \quad 6)$

4 Not possible   **12** Not possible

5 $\begin{pmatrix} 3 & 4 \\ 6 & 8 \end{pmatrix}$ **13** Not possible

6 Not possible   **14** $(9 \quad 12)$

7 Not possible   **15** Not possible

8 Not possible

16 $\mathbf{AA} = \begin{pmatrix} 19 & 18 \\ 6 & 7 \end{pmatrix}$ $\mathbf{DH} = \begin{pmatrix} 3 \\ 6 \end{pmatrix}$

$\mathbf{AC} = \begin{pmatrix} 21 & 8 \\ 4 & 2 \end{pmatrix}$ $\mathbf{EA} = (16 \quad 17)$

$\mathbf{BB} = \begin{pmatrix} -5 & -2 \\ 3 & -6 \end{pmatrix}$ $\mathbf{EB} = (15 \quad -6)$

$\mathbf{BC} = \begin{pmatrix} 8 & 2 \\ 18 & 6 \end{pmatrix}$ $\mathbf{FE} = \begin{pmatrix} 18 & 24 \\ 3 & 4 \\ -12 & -16 \end{pmatrix}$

$\mathbf{BD} = \begin{pmatrix} -3 \\ 3 \end{pmatrix}$ $\mathbf{FH} = \begin{pmatrix} 18 \\ 3 \\ -12 \end{pmatrix}$

$\mathbf{CA} = \begin{pmatrix} 26 & 22 \\ -4 & -3 \end{pmatrix}$ $\mathbf{GF} = \begin{pmatrix} -4 \\ 15 \\ 5 \end{pmatrix}$

$\mathbf{CB} = \begin{pmatrix} 12 & -12 \\ -1 & 2 \end{pmatrix}$ $\mathbf{GG} = \begin{pmatrix} 18 & -2 & 17 \\ -13 & 35 & 11 \\ 23 & 4 & 27 \end{pmatrix}$

$\mathbf{CC} = \begin{pmatrix} 34 & 12 \\ -6 & -2 \end{pmatrix}$ $\mathbf{HH} = (9)$

$\mathbf{CD} = \begin{pmatrix} 10 \\ -1 \end{pmatrix}$

Exercise 7m   Page 93

1 $\begin{pmatrix} 8 & 2 \\ -21 & -8 \end{pmatrix}$ **7** $(-2 \quad -3)$

2 $\begin{pmatrix} 7 & 4 \\ -3 & 3 \end{pmatrix}$ **8** $\begin{pmatrix} -2 & 1 \\ 8 & 12 \end{pmatrix}$

3 $\begin{pmatrix} -5 & 0 \\ -5 & 3 \end{pmatrix}$ **9** $\begin{pmatrix} 5 & 3 & 3 \\ 11 & 5 & 9 \\ 2 & 2 & 0 \end{pmatrix}$

4 $\begin{pmatrix} 5 & 0 \\ 5 & -3 \end{pmatrix}$ **10** Not possible

5 Not possible   **11** Not possible
6 Not possible   **12** $(4 \quad 2)$

Exercise 7n   Page 94

1 $\begin{pmatrix} 12 & 10 \\ -2 & 13 \end{pmatrix}$ **3** $\begin{pmatrix} 13 & -6 \\ 18 & 1 \end{pmatrix}$

2 $\begin{pmatrix} 78 & -10 \\ 31 & -13 \end{pmatrix}$ **4** $\begin{pmatrix} 68 & 16 \\ 61 & 4 \end{pmatrix}$

---

5 $\begin{pmatrix} 48 & 40 \\ 38 & 17 \end{pmatrix}$ **9** $\begin{pmatrix} 55 & -11 \\ 66 & -22 \end{pmatrix}$

6 $\begin{pmatrix} 34 & -25 \\ 75 & -16 \end{pmatrix}$ **10** $\begin{pmatrix} -16 & 56 \\ -56 & 20 \end{pmatrix}$

7 $\begin{pmatrix} 78 & 8 \\ 31 & 28 \end{pmatrix}$ **11** $\begin{pmatrix} -64 & 0 \\ 0 & -64 \end{pmatrix}$

8 $\begin{pmatrix} 50 & 27 \\ 32 & 56 \end{pmatrix}$ **12** $\begin{pmatrix} 68 & -12 \\ 61 & -7 \end{pmatrix}$

Exercise 7p   Page 95

1 2×2 and 2×1
2 Yes
3 **A**, **C** are compatible but not **C**, **A**

4 $\begin{pmatrix} 23 & -11 \\ 19 & -13 \end{pmatrix}$

5 $\mathbf{A}^2 = \begin{pmatrix} 27 & 18 \\ 9 & 18 \end{pmatrix}$. It is not possible to find $\mathbf{C}^2$

6 Not possible

7 $\begin{pmatrix} 9 & -3 \\ 12 & -9 \end{pmatrix}$ **9** 4

8 $\begin{pmatrix} 13 & 3 \\ 6 & 5 \end{pmatrix}$ **10** BC

Exercise 7q   Page 95

1 $\begin{pmatrix} 4 & 2 & -2 \\ 8 & 6 & 2 \end{pmatrix}$ **6** 3
                                7 1
2 Not possible   **8** QP
3 Not possible
4 2×3 and 2×2 **9** $\begin{pmatrix} 17 \\ 13 \end{pmatrix}$
5 No
10 $\mathbf{P}^2$ is not possible to find. $\mathbf{Q}^2 = \begin{pmatrix} 7 & 14 \\ -7 & 14 \end{pmatrix}$

Chapter 8
Exercise 8a   Page 98

1 Yes, 3×3 **3** Yes, 2×2 **5** Yes, 2×2
2 No   **4** No   **6** Yes, 3×3

Exercise 8b   Page 98

1 $\begin{pmatrix} 4 & 7 \\ 7 & 11 \end{pmatrix}$ **7** $\begin{pmatrix} 11 & 7 \\ 7 & 4 \end{pmatrix}$

2 $(7 \quad -6)$ **8** Not possible

3 $\begin{pmatrix} 7 & 10 & 1 \\ 15 & 26 & 1 \end{pmatrix}$ **9** $\begin{pmatrix} 4 & 28 & 4 \\ 3 & 18 & 3 \\ 2 & 12 & 2 \end{pmatrix}$

4 Not possible   **10** (24)

5 Not possible   **11** $\begin{pmatrix} 1 \\ -1 \end{pmatrix}$

6 $\begin{pmatrix} 26 & 13 \\ -4 & -2 \end{pmatrix}$ **12** $(34 \quad 6)$

Exercise 8c   Page 99

1 $\begin{pmatrix} 4 & 2 \\ 3 & 4 \end{pmatrix}$ **3** $(0 \quad 0)$ **5** $(3 \quad 2)$

2 $\begin{pmatrix} 4 \\ 5 \end{pmatrix}$ **4** $\begin{pmatrix} 0 & 0 \\ 0 & 0 \end{pmatrix}$ **6** $\begin{pmatrix} 3 & 2 & -1 \\ 4 & 3 & -1 \end{pmatrix}$

Exercise 8d Page 100

1 $\begin{pmatrix} 1 & 0 \\ 0 & 1 \end{pmatrix}$

2 $\begin{pmatrix} 5 & 0 \\ 0 & 5 \end{pmatrix}$

3 $\begin{pmatrix} 5 & 0 \\ 0 & 5 \end{pmatrix}$

4 $\begin{pmatrix} 2 & 0 \\ 0 & 2 \end{pmatrix}$

5 $\begin{pmatrix} 3 & 0 \\ 0 & 3 \end{pmatrix}$

6 $\begin{pmatrix} 3 & 0 \\ 0 & 3 \end{pmatrix}$

7 $\begin{pmatrix} -1 & 0 \\ 0 & -1 \end{pmatrix}$

8 $\begin{pmatrix} 1 & 0 \\ 0 & 1 \end{pmatrix}$

10 $\begin{pmatrix} 6 & -2 \\ -8 & 3 \end{pmatrix}$

11 $\begin{pmatrix} 3 & -1 \\ -2 & 2 \end{pmatrix}$

12 $\begin{pmatrix} 3 & 1 \\ 20 & 6 \end{pmatrix}$

13 $\begin{pmatrix} 5 & 3 \\ -1 & 1 \end{pmatrix}$

14 $\begin{pmatrix} 4 & -7 \\ -3 & 5 \end{pmatrix}$

15 $\begin{pmatrix} 6 & -3 \\ -9 & 5 \end{pmatrix}$

16 $\begin{pmatrix} 8 & -4 \\ -9 & 5 \end{pmatrix}$

17 $\begin{pmatrix} 4 & -2 \\ -14 & 6 \end{pmatrix}$

18 $\begin{pmatrix} 1 & -2 \\ -2 & 1 \end{pmatrix}$

19 $\begin{pmatrix} -2 & 3 \\ -3 & 4 \end{pmatrix}$

Exercise 8e Page 101

1 $\begin{pmatrix} 2 & -1 \\ -7 & 4 \end{pmatrix}$

2 $\begin{pmatrix} 2 & -3 \\ -7 & 11 \end{pmatrix}$

3 $\begin{pmatrix} -2 & 3 \\ -7 & 10 \end{pmatrix}$

4 $\begin{pmatrix} 7 & -5 \\ -4 & 3 \end{pmatrix}$

5 $\begin{pmatrix} 7 & -4 \\ -12 & 7 \end{pmatrix}$

6 $\begin{pmatrix} 2 & -1 \\ -1 & 1 \end{pmatrix}$

Exercise 8f Page 102

1 $\begin{pmatrix} 1\frac{1}{2} & -1 \\ -4 & 3 \end{pmatrix}$

2 $\begin{pmatrix} \frac{1}{3} & -\frac{2}{3} \\ -1 & 3 \end{pmatrix}$

3 $\begin{pmatrix} 1\frac{1}{2} & -\frac{1}{2} \\ -2\frac{1}{2} & 1 \end{pmatrix}$

4 $\begin{pmatrix} \frac{1}{3} & \frac{2}{3} \\ -\frac{1}{3} & \frac{1}{3} \end{pmatrix}$

5 $\begin{pmatrix} \frac{1}{2} & 0 \\ 0 & \frac{1}{3} \end{pmatrix}$

6 $\begin{pmatrix} 4 & -1 \\ -5\frac{1}{2} & 1\frac{1}{2} \end{pmatrix}$

7 $\begin{pmatrix} 4 & 3 \\ 5 & 4 \end{pmatrix}$

8 $\begin{pmatrix} 4 & -3 \\ -5 & 4 \end{pmatrix}$

9 $\begin{pmatrix} 1 & 1 \\ \frac{1}{2} & \frac{3}{4} \end{pmatrix}$

10 $\begin{pmatrix} -1 & 1 \\ 3\frac{1}{2} & -3 \end{pmatrix}$

11 $\begin{pmatrix} -1 & 2 \\ 2\frac{1}{2} & -4\frac{1}{2} \end{pmatrix}$

12 $\begin{pmatrix} -\frac{1}{5} & \frac{1}{5} \\ \frac{1}{5} & \frac{4}{5} \end{pmatrix}$

13 $\begin{pmatrix} -1 & 1 \\ 2 & -1\frac{1}{2} \end{pmatrix}$

14 $\begin{pmatrix} -1 & -1\frac{1}{3} \\ -1 & -1 \end{pmatrix}$

15 $\begin{pmatrix} -3 & 2 \\ 4 & -2\frac{1}{2} \end{pmatrix}$

16 a Yes **b** No **c** Yes

17 a Yes **b** Yes **c** Yes

18 $\begin{pmatrix} 1 & -1 \\ -1 & 1\frac{1}{5} \end{pmatrix}$

19 No inverse

20 $\begin{pmatrix} \frac{1}{5} & 0 \\ 0 & \frac{1}{5} \end{pmatrix}$

21 $\begin{pmatrix} -4 & 7 \\ 3 & -5 \end{pmatrix}$

22 $\begin{pmatrix} 2 & -1 \\ -3 & 1\frac{2}{3} \end{pmatrix}$

23 No inverse

Exercise 8g Page 103

1 1, $\begin{pmatrix} 2 & -3 \\ -3 & 5 \end{pmatrix}$

2 2, $\begin{pmatrix} 1\frac{1}{2} & -1 \\ -1 & 1 \end{pmatrix}$

3 I

4 $\begin{pmatrix} 16 & 19 \\ 10 & 12 \end{pmatrix}$

5 $\begin{pmatrix} 6 & -9\frac{1}{2} \\ -5 & 8 \end{pmatrix}$

6 $\begin{pmatrix} 6 & -5 \\ -9\frac{1}{2} & 8 \end{pmatrix}$

7 $\begin{pmatrix} 6 & -9\frac{1}{2} \\ -5 & 8 \end{pmatrix}$

8 $\begin{pmatrix} 34 & 21 \\ 21 & 13 \end{pmatrix}$

9 $\begin{pmatrix} 13 & -21 \\ -21 & 34 \end{pmatrix}$

10 $\begin{pmatrix} 13 & -21 \\ -21 & 34 \end{pmatrix}$

11 $\begin{pmatrix} 2 & 1 \\ -3\frac{1}{2} & -1\frac{1}{2} \end{pmatrix}$

12 $\begin{pmatrix} -2\frac{1}{2} & 4 \\ -2 & 3 \end{pmatrix}$

Exercise 8h Page 104

1 9

2 17

3 0

4 19

5 −14

6 10

7 −1

8 −8

9 9

10 5

11 5

12 −9

Exercise 8i Page 105

1 $x + 2y = 3$
$3x + 2y = 5$

2 $4x + 2y = 12$
$5x + 3y = 15$

3 $9x + 2y = 24$
$4x + y = 11$

4 $6p - q = -8$
$2p + q = 0$

5 $\begin{pmatrix} 3 & 2 \\ 1 & 1 \end{pmatrix}\begin{pmatrix} x \\ y \end{pmatrix} = \begin{pmatrix} 8 \\ 3 \end{pmatrix}$

6 $\begin{pmatrix} 4 & -3 \\ 2 & 1 \end{pmatrix}\begin{pmatrix} x \\ y \end{pmatrix} = \begin{pmatrix} 1 \\ 3 \end{pmatrix}$

7 $\begin{pmatrix} 4 & 3 \\ 5 & 4 \end{pmatrix}\begin{pmatrix} x \\ y \end{pmatrix} = \begin{pmatrix} 5 \\ 6 \end{pmatrix}$

8 $\begin{pmatrix} 3 & -2 \\ 1 & -1 \end{pmatrix}\begin{pmatrix} x \\ y \end{pmatrix} = \begin{pmatrix} 1 \\ 0 \end{pmatrix}$

9 $\begin{pmatrix} 7 & -2 \\ 3 & 4 \end{pmatrix}\begin{pmatrix} x \\ y \end{pmatrix} = \begin{pmatrix} 3 \\ 11 \end{pmatrix}$

10 $\begin{pmatrix} 5 & 1 \\ 4 & -3 \end{pmatrix}\begin{pmatrix} x \\ y \end{pmatrix} = \begin{pmatrix} -8 \\ -14 \end{pmatrix}$

Exercise 8j Page 106

1 $x = 1$, $y = 2$

2 $x = 2$, $y = 3$

3 $x = 1$, $y = -1$

4 $x = 2$, $y = -1$

5 $x = 3$, $y = 0$

6 $x = 1$, $y = 2$

7 $x = 4$, $y = 2$

8 $x = 1$, $y = -2$

9 $x = 4$, $y = 2$

10 $p = 1$, $q = 1$

11 $s = -2$, $t = 3$

12 $\begin{pmatrix} 1 & 1 \\ 1 & 2 \end{pmatrix}\begin{pmatrix} x \\ y \end{pmatrix} = \begin{pmatrix} 2 \\ 3 \end{pmatrix}$; $x = 1$, $y = 1$

13 $\begin{pmatrix} 4 & -1 \\ 1 & 1 \end{pmatrix}\begin{pmatrix} x \\ y \end{pmatrix} = \begin{pmatrix} 5 \\ 5 \end{pmatrix}$; $x = 2$, $y = 3$

14 $\begin{pmatrix} 5 & 4 \\ 1 & 1 \end{pmatrix}\begin{pmatrix} x \\ y \end{pmatrix} = \begin{pmatrix} 1 \\ 0 \end{pmatrix}$; $x = 1$, $y = -1$

15 $\begin{pmatrix} 2 & 3 \\ 3 & 5 \end{pmatrix}\begin{pmatrix} x \\ y \end{pmatrix} = \begin{pmatrix} 15 \\ 23 \end{pmatrix}$; $x = 6$, $y = 1$

16 $\begin{pmatrix} 9 & 2 \\ 3 & 1 \end{pmatrix}\begin{pmatrix} x \\ y \end{pmatrix} = \begin{pmatrix} 11 \\ 5 \end{pmatrix}$; $x = \frac{1}{3}$, $y = 4$

17 $\begin{pmatrix} 2 & 3 \\ 3 & 2 \end{pmatrix}\begin{pmatrix} x \\ y \end{pmatrix} = \begin{pmatrix} 7 \\ 8 \end{pmatrix}$; $x = 2$, $y = 1$

18 $\begin{pmatrix} 5 & 2 \\ 3 & -1 \end{pmatrix}\begin{pmatrix} x \\ y \end{pmatrix} = \begin{pmatrix} 16 \\ 3 \end{pmatrix}$; $x = 2$, $y = 3$

19 $\begin{pmatrix} 1 & 4 \\ 2 & 3 \end{pmatrix}\begin{pmatrix} x \\ y \end{pmatrix} = \begin{pmatrix} 11 \\ 7 \end{pmatrix}$; $x = -1$, $y = 3$

Exercise 8k   Page 107

1 $\begin{pmatrix} 5 & 6 \\ -3 & 0 \end{pmatrix}$ **4** $\begin{pmatrix} -5 & -6 \\ 3 & 0 \end{pmatrix}$ **7** $\begin{pmatrix} 3 & -1 \\ -5 & 1 \end{pmatrix}$

2 $\begin{pmatrix} 7 & 2 \\ -3 & 2 \end{pmatrix}$ **5** $\begin{pmatrix} 2 & 1\frac{1}{2} \\ 1 & \frac{1}{2} \end{pmatrix}$ **8** $\begin{pmatrix} -\frac{1}{2} & 1\frac{1}{2} \\ 1 & -2 \end{pmatrix}$

3 $\begin{pmatrix} 15 & 19 \\ 9 & 9 \end{pmatrix}$ **6** $\begin{pmatrix} 3 & -6 \\ 0 & 3 \end{pmatrix}$

Exercise 8l   Page 108

1 $\begin{pmatrix} 5 & 4 & 3 \\ 10 & -8 & 4 \end{pmatrix}$ **4** $\begin{pmatrix} \frac{4}{7} & -\frac{3}{7} \\ -\frac{3}{7} & \frac{4}{7} \end{pmatrix}$

2 $\begin{pmatrix} 1 & 3\frac{1}{2} \\ 1\frac{1}{2} & -\frac{1}{2} \end{pmatrix}$ **5** (-9)

3 24   **6** $\begin{pmatrix} 13 & 33 \\ 6 & 22 \end{pmatrix}$

Exercise 8m   Page 108

1 $\begin{pmatrix} 5 & 3 \\ -1 & 4 \end{pmatrix}$ **3** $\begin{pmatrix} 1 & 1 \\ 2 & 3 \end{pmatrix}$ **5** $(6 \quad 10)$

2 2   **4** $\begin{pmatrix} 3 \\ -1 \end{pmatrix}$ **6** $(3 \quad -1\frac{1}{2})$

Chapter 9

Exercise 9a   Page 112

1 $x = 4$ **3** $y = -3$

2 $y = 5$ **4** $x = -2$

5   **7**

6   **8**

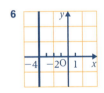

Exercise 9b   Page 114

1

x	-2	0	4
y	2	4	8

3

x	-3	0	3
y	7	4	1

2

x	-2	0	3
y	-3	1	7

4

x	-2	0	4
y	8	2	-10

13 a $1\frac{1}{2}$ **b** 0.4   **c** -1.6
14 a 0   **b** -0.8 **c** -3.4
15 a -2.6 **b** -1.8 **c** 1.2
16 a 3.6   **b** 0.6   **c** 1.2
17 a $-2\frac{1}{2}$ **b** 4.4   **c** 2.4
18 a 4.8   **b** 1.2   **c** -112
19 a -1.4 **b** 1.4   **c** 3.5
20 a 8.6   **b** 2.8   **c** 3

Exercise 9c   Page 116

1 Yes, No   **3** No, No   **5** Yes, Yes
2 Yes, No   **4** No, Yes   **6** No, Yes

Exercise 9d   Page 117

1 Lines are parallel; coefficient of x is 2 in each equation
2 Lines are parallel; coefficient of x is -3 in each equation
3 Lines are parallel; coefficient of x is $\frac{1}{2}$ in each equation
4 Lines are parallel; coefficient of x is 1 in each equation
5 Lines **a** and **c** are parallel
6 Lines are parallel; coefficient of x is -1 in each equation

Exercise 9e   Page 118

1 4   **3** 1   **5** $-\frac{9}{5}$
2 -2 **4** $-\frac{3}{4}$ **6** $\frac{2}{3}$
7 0
8 y-axis. You find yourself dividing by zero
9 a parallel to the y-axis
   b zero gradient
   c zero gradient
   d parallel to the y-axis

Exercise 9f   Page 120

1 2   **3** 2   **5** 4
2 1   **4** -2
6 a 4   **b** -3 **c** 1   **d** $\frac{1}{2}$

Exercise 9g   Page 122

1 2, 4   **5** -2, 3   **9** $-\frac{1}{3}$, 4   **13** -0.4, 9
2 5, 3   **6** -4, 2   **10** 3, -7 **14** 5, 4
3 3, -4 **7** 5, 2   **11** -3, 7   **15** 2, $2\frac{1}{2}$
4 1, -6 **8** $\frac{1}{2}$, -1 **12** $\frac{1}{3}$, 7   **16** $\frac{1}{3}$, -2
17 $\frac{2}{5}$, 1   **22** $y = 2x - 5$
18 $-\frac{3}{4}$, 2   **23** $y = \frac{1}{2}x + 6$
19 $y = 2x + 7$ **24** $y = -2x + 1$
20 $y = 3x + 1$ **25** $y = x - 2$
21 $y = x + 3$ **26** $y = -\frac{1}{2}x + 4$

Exercise 9h   Page 123

1 $y = 3x + 1$, $y = 5 + 3x$, $y = 3x - 4$
2 $y = 2 - x$, $y = 4 - x$, $2y = 3 - 2x$, $y = -x + 1$, $y = -x$
3 $3y = x$, $y = \frac{1}{3}x + 2$, $y = \frac{1}{3} + \frac{1}{3}x$, $y = \frac{1}{3}x - 4$
4 $y = \frac{1}{2}x + 2$ and $y = \frac{1}{2}x - 1$; $y = 2 - \frac{1}{2}x$ and $2y = 3 - x$
5 2; $y = 2x + 3$
6 -3; $y = -3x + 1$
7 $y = 4x$
8 e.g. $y = 6 - x$, $y = -x$, $y = -2 - x$
9 a $y = 4x + 4$ **c** $y = \frac{1}{2}x + 4$
   b $y = -3x + 4$
10 a $y = \frac{1}{3}x + 6$ **c** $y = \frac{1}{3}x - 3$
   b $y = \frac{1}{3}x$
11 a $y = 2x + 2$ **c** $y = 2x - 4$
   b $y = 2x + 10$

12 $y = 3 + 2x$ and $y = 2x - 3$
13 $-3, 4$; $4, -3$; $y = -3x - 3$
14 a $y = -4x$ **b** $y = -4x - 7$

Exercise 9i Page 125

1 $-\frac{3}{5}$ **4** -1 **7** $\frac{1}{3}$
2 $-\frac{1}{3}$ **5** -2 **8** 2
3 $\frac{1}{4}$ **6** $-\frac{1}{3}$ **9** -1 in each case

Exercise 9j Page 125

1 $-\frac{3}{4}$ **3** $\frac{1}{2}$ **5** 2
2 $-\frac{3}{5}$ **4** -2 **6** $\frac{3}{4}$
7 a $(2, 0), (0, 4)$ **b** $(12, 0), (0, -9)$
8 a $\frac{x}{6} + \frac{y}{5} = 1$ **b** $\frac{x}{4} - \frac{y}{3} = 1$
9 $-\frac{1}{3}$

Exercise 9k Page 126

1 $-\frac{3}{5}, 3$ **8** $-\frac{3}{5}, 3$ **15** $-2, 4$
2 $-\frac{1}{3}, 2$ **9** $\frac{1}{2}, -2$ **16** $-\frac{2}{5}, 3$
3 $\frac{1}{4}, -2$ **10** $-3, 6$ **17** $-\frac{1}{2}, 5$
4 $\frac{1}{3}, -2$ **11** $-\frac{4}{3}, 4$ **18** $2, \frac{5}{2}$
5 $3, 6$ **12** $\frac{4}{3}, -4$ **19** $2, -4$
6 $-\frac{1}{3}, 2$ **13** $4, 2$ **20** $-1, -3$
7 $-\frac{3}{4}, 3$ **14** $-1, 4$ **21** $-\frac{3}{4}, 3$

Exercise 9l Page 128

1 $-\frac{4}{3}, 4$; $y = -\frac{4}{3}x + 4$ **7** $\frac{1}{6}, 1$; $y = \frac{1}{6}x + 1$
2 $-2, 7$; $y = -2x + 7$ **8** $\frac{4}{5}, -3$; $y = \frac{4}{5}x - 3$
3 $\frac{3}{5}, 1$; $y = \frac{3}{5}x + 1$ **9** $\frac{5}{3}, -4$; $y = \frac{5}{3}x - 4$
4 $-\frac{4}{3}, 2$; $y = -\frac{4}{3}x + 2$ **10** $-1, -5$; $y = -x - 5$
5 $\frac{7}{2}, -4$; $y = \frac{7}{2}x - 4$ **11** $2, 12$; $y = 2x + 12$
6 $\frac{1}{3}, -1$; $y = \frac{1}{3}x - 1$ **12** $\frac{5}{6}, 6$; $y = \frac{5}{6}x + 6$
13 AB, $5y = 2x + 20$; AC, $5x + 3y = 12$
14 $3, y = 3x - 11$ **19** $-1, y = -x + 3$
15 $-3, y = -3x + 7$ **20** $-1, y = -x + 1$
16 $\frac{5}{2}, y = \frac{5}{2}x - \frac{1}{2}$ **21** $2, y = 2x - 11$
17 $2, y = 2x + 7$ **22** $\frac{1}{5}, y = \frac{1}{5}x - \frac{6}{5}$
18 $5, y = 5x - 21$
23 $-\frac{5}{4}, \frac{x}{4} + \frac{y}{5} = 1$ or $y = -\frac{5}{4}x + 5$
24 $-\frac{2}{3}, \frac{x}{3} + \frac{y}{2} = 1$ or $y = -\frac{2}{3}x + 2$
25 $\frac{2}{3}, \frac{x}{3} - \frac{y}{2} = 1$ or $y = \frac{2}{3}x - 2$
26 $-3, \frac{x}{2} + \frac{y}{6} = 1$ or $y = -3x + 6$
27 $3, y = 3x - 10$ **29** $\frac{7}{2}, y = \frac{7}{2}x - 6$
28 $-1, y = -x + 4$ **30** $-1, y = -x + 3$
31 $\frac{5}{2}, -\frac{x}{2} + \frac{y}{5} = 1$ or $y = \frac{5}{2}x + 5$
32 $\frac{2}{11}, y = \frac{2}{11}x + \frac{21}{11}$ **34** $-\frac{1}{4}, y = -\frac{1}{4}x + \frac{11}{4}$
33 $1, y = x - 1$

Exercise 9m Page 129

1 $y = 3x - 16$ **5** Midpoint is $(5, 3)$; $y = -2x + 13$
2 Square   **6** $y = -x + 4$
3 Rhombus   **7** Square
4 $(\frac{1}{2}, 3)$

Exercise 9n Page 130

1 2 **3** $(4, 0)$ **5** $y = 5x$ **7** Yes
2 $(0, 4)$ **4** 12 **6** $(12, 0)$ **8** $\frac{3}{5}$

Exercise 9p Page 130

1 -3 **5** $(0, 6), (6, 0)$
2 No   **6** $-\frac{3}{2}$
3 $y = -4x$ **7** $y = \frac{1}{2}x$
4 $(0, 4)$ **8** $(2, 0), (0, 3)$

Chapter 10

Exercise 10a Page 134

1 a 200 tonnes   **b** 3.16 cm
2 a $168.9   **b** 7.27 years
3 a 3.6   **b** 2.3
4 a 3 cm   **b** 67 cm^2
5 a 165 cm   **b** 2.21 litres

Exercise 10b Page 136

1 a 59.5 m   **b** 4.47 s
2 a 4.8   **b** 7.5
3 a 1.67   **b** 1.09
4 2.65, 5.29
5 a 3.33   **b** -1.43

Exercise 10c Page 138

1 The graph passes through the origin O, which is also the lowest value for y.
2 a 1.73 or -1.73 **b** No
3 a 2 and -2 **b** 1 and -1, Yes
4 They all have the same shape.
They all have the same shape but cross the y-axis at different points.
5 a When $x = 0$ and $x = 3$
b -0.79 and 3.79
6 a When $x = 0$ and 1.5
b $-1\frac{1}{8}$ when $x = \frac{3}{4}$
7 a -2 when $x = -1$ **c** 0 and -2
b 10.5
8 a -4 when $x = 1$
b i -1.24 and 3.24 **ii** -2.46 and 4.46
9 a 6.25 when $x = 0.5$
b i -2.37 and 3.37 **ii** -1 and 2

Review test 1 Page 142

1 A   **3** C   **5** C   **7** C   **9** A
2 D   **4** B   **6** A   **8** B   **10** A
11 a $-12/1$ **b** 2
12 a $3\frac{1}{2}$%   **b** $40 500
13 a 5.9
b $v = 30°$, triangles PST and TRQ
14 a 11   **b** $\{-3, -2, -1, 0, 1\}$
15 a $\begin{pmatrix} 2 & 3 \\ 3 & -1 \end{pmatrix} \begin{pmatrix} x \\ y \end{pmatrix} = \begin{pmatrix} 7 \\ 5 \end{pmatrix}$
b $y = 4x - 22$

Chapter 11

Exercise 11a Page 145

1 40 cm^2 **6** 30 sq. units
2 10 cm^2 **7** 12 sq. units
3 17 cm^2 **8** 16 sq. units
4 1935 cm^2 **9** $24\frac{1}{2}$ sq. units
5 12 sq. units   **10** 4 cm

11 450 mm^2

12 5 cm

13 $5 \text{ m}, 25 \text{ m}^2$

14 4 cm

15 a 175 cm^2 **b** 5.83 cm

16 a 12 cm^2 **b** 3.43 cm

17 a 40 cm^2 **b** 6.67 cm

18 a 7 cm^2 **b** 2 cm

Exercise 11b   Page 147

1 60 cm^2 **3** 30 cm^2 **5** 135 cm^2

2 40 cm^2 **4** 45 cm^2 **6** 2775 cm^2

Exercise 11c   Page 149

1 42 cm^2 **6** 33 sq. units

2 94.5 cm^2 **7** 56 sq. units

3 21 cm^2 **8** 16 sq. units

4 8.75 cm^2 **9** 84 sq. units

5 30 sq. units   **10** 47 sq. units

Exercise 11d   Page 150

1 Area of each parallelogram is 35 cm^2

2 Area of each triangle is 28 cm^2

3 Each parallelogram has a base of length 4 units and height of 3 units. The areas are each equal to 12 sq. units.

4 Each base is 6 units long. Each height is 3 units. The areas are each equal to 9 sq. units.

8 Ratio of heights is $4 : 5 : 7 : 9$. Ratio of areas is $4 : 5 : 7 : 9$. The ratio of areas is equal to the ratio of heights.

9 The y coordinate of D is 9 (or -7 if drawn below the x-axis)

10 The y coordinate of E is 3 (or -1 if drawn below the x-axis)

11 a $4 : 5 : 3 : 9$ **b** $4 : 5 : 3 : 9$
The triangles have equal heights and the ratio of their areas is equal to the ratio of their bases

Exercise 11e   Page 151

7 12 cm

8 $2 : 1$

9 14 cm

10 8 cm

11 $30°$

12 $\triangle \text{BEC} = 27 \text{ cm}^2$
$\triangle \text{DEC} = 12 \text{ cm}^2$

13 $132°$

14 6 cm

15 36 cm^2

Chapter 12

Exercise 12a   Page 156

1 46

2 40

3 20

4 41.4

5 a 255   **b** 51

6 a 366   **b** 61   **11** \$5.75

7 765 cm, 153 cm   **12** 80 cl

8 558   **13** 8800 t

9 $29°$

10 52 c

14 a \$602.50   **b** \$794.40   **c** \$191.90

15 a 130   **b** 192   **c** 62

16 a 225   **b** 25

17 17 y 5 m, 43 y 9 m

18 42.5 kg   **19** 66.4 kg   **20** 2

Exercise 12b   Page 159

1 60 km/h   **7** 50 m.p.h.   **13** 155 km

2 30 km/h   **8** 8 m.p.h.   **14** 81 miles

3 36 km/h   **9** 45 m.p.h.   **15** 3 hours

4 90 km/h   **10** 54 m.p.h.   **16** 5 hours

5 18 km/h   **11** 260 miles   **17** 20 min

6 $262\frac{1}{2}$ km/h   **12** 80 miles   **18** 45 min

Exercise 12c   Page 161

1 225 km//h   **5** 90 km/h

2 100 km/h   **6** 55 km/h

3 140 km/h   **7** 80 km/h

4 9 km/h   **8** c

Chapter 13

Exercise 13a   Page 165

1 $2x + 2$ **7** $5 - 5b$ **13** $15xy + 5xz$

2 $3x - 3$ **8** $6a - 2$ **14** $16xy + 12yz$

3 $4x + 12$ **9** $8 + 12b$ **15** $6np - 10nq$

4 $5a + 20$ **10** $5ab - 5ac$ **16** $16rt - 8rs$

5 $3b + 21$ **11** $4ab - 8ac$ **17** $3ab - 15ac$

6 $3 - 3a$ **12** $6a^2 + 3ab$ **18** $12xy + 8xz$

Exercise 13b   Page 165

1 $ac + ad + bc + bd$

2 $ps + pt + qs + qt$

3 $2ac + 4ad + bc + 2bd$

4 $5xz + 15x + 2yz + 6y$

5 $xz - 4x + yz - 4y$

6 $ac + ad - bc - bd$

7 $xy + xz + y^2 + yz$

8 $6ac + 2ad + 3bc + bd$

9 $5xz + 10x + 4yz + 8y$

10 $15x - 3xz - 10y + 2yz$

11 $2ps - 3pt + 2qs - 3qt$

12 $ac - ad - 2bc + 2bd$

13 $6uw - 30ur - 5vw + 25vr$

14 $6ac - 9ad + 8bc - 12bd$

15 $9xz + 6x + 6yz + 4y$

16 $12pr - 9ps - 4qr + 3qs$

17 $9ac + 12ad - 12bc - 16bd$

18 $21x - 14xz - 6y + 4yz$

Exercise 13c   Page 166

1 $x^2 + 7x + 12$ **15** $x^2 - 7x + 12$

2 $x^2 + 6x + 8$ **16** $x^2 - 12x + 32$

3 $x^2 + 7x + 6$ **17** $b^2 - 6b + 8$

4 $x^2 + 7x + 10$ **18** $a^2 - 8a + 16$

5 $x^2 + 11x + 24$ **19** $x^2 + x - 6$

6 $a^2 + 9a + 20$ **20** $x^2 + x - 20$

7 $b^2 + 9b + 14$ **21** $x^2 - 3x - 28$

8 $c^2 + 10c + 24$ **22** $a^2 - 7a - 30$

9 $p^2 + 15p + 36$ **23** $p^2 - 25$

10 $x^2 - 5x + 6$ **24** $x^2 + 5x - 14$

11 $x^2 - 12x + 35$ **25** $x^2 + x - 30$

12 $a^2 - 10a + 16$ **26** $x^2 + 9x - 10$

13 $x^2 - 13x + 30$ **27** $b^2 - 15b + 56$

14 $b^2 - 10b + 25$

Exercise 13d   Page 168

1 $x^2 + 9x + 20$ **13** $a^2 - 3a - 70$

2 $a^2 + 7a + 10$ **14** $y^2 + 8y - 20$

3 $x^2 - 9x + 20$ **15** $z^2 - 11z - 12$

4 $a^2 - 7a + 10$ **16** $p^2 - 11p - 26$

5 $x^2 + 14x + 48$ **17** $x^2 - 6x + 5$

6 $a^2 + 17a + 70$ **18** $b^2 + 16b + 63$

7 $x^2 - 14x + 48$ **19** $a^2 - 16$

8 $a^2 - 17a + 70$ **20** $r^2 - 12r - 28$

9 $a^2 - 3a - 10$ **21** $p^2 + 14p + 24$

10 $y^2 - 3y - 18$ **22** $t^2 - 7t - 60$

11 $z^2 - 6z - 40$ **23** $c^2 + 3c - 40$

12 $p^2 - 3p - 40$ **24** $x^2 - 25$

Exercise 13e   Page 168

1 $2x^2 + 3x + 1$
2 $5x^2 + 12x + 4$
3 $5x^2 + 17x + 6$
4 $3x^2 + 19x + 20$
5 $3x^2 + 5x + 2$
6 $3x^2 + 11x + 6$
7 $4x^2 + 7x + 3$
8 $7x^2 + 23x + 6$
9 $6x^2 + 13x + 6$
10 $12x^2 - 25x + 12$
11 $10x^2 - 3x - 18$
12 $21a^2 - 58a + 21$
13 $10x^2 + 31x + 15$
14 $21x^2 - 20x + 4$
15 $12x^2 - 5x - 2$
16 $6b^2 - 5b - 25$
17 $4a^2 - 9$
18 $9b^2 - 49$
19 $49y^2 - 25$
20 $20a^2 + a - 12$
21 $16x^2 - 9$
22 $25y^2 - 4$
23 $9x^2 - 1$
24 $16x^2 - 8x - 35$
25 $6x^2 + 5x + 1$
26 $-5x^2 + 8x + 4$
27 $-6x^2 + 19x - 3$
28 $-35a^2 + 29a - 6$
29 $8 + 10x - 3x^2$
30 $4x^2 + 7x - 15$
31 $15x^2 + 26x + 8$
32 $-14x^2 + 13x + 12$
33 $-20x^2 + 27x - 9$
34 $12 - p - p^2$
35. $x^2 - 3x - 10$
36 $4x^2 + 9x - 9$

Exercise 13f   Page 170

1 $x^2 + 2x + 1$
2 $x^2 + 4x + 4$
3 $a^2 + 6a + 9$
4 $b^2 + 8b + 16$
5 $t^2 + 20t + 100$
6 $x^2 + 24x + 144$
7 $x^2 + 16x + 64$
8 $p^2 + 14p + 49$
9 $x^2 + 2xy + y^2$
10 $y^2 + 2yz + z^2$
11 $c^2 + 2cd + d^2$
12 $m^2 + 2mn + n^2$
13 $p^2 + 2pq + q^2$
14 $a^2 + 2ab + b^2$
15 $e^2 + 2ef + f^2$
16 $u^2 + 2uv + v^2$
17 $4x^2 + 4x + 1$
18 $16b^2 + 8b + 1$
19 $25x^2 + 20x + 4$
20 $36c^2 + 12c + 1$
21 $9a^2 + 6a + 1$
22 $4x^2 + 20x + 25$
23 $9a^2 + 24a + 16$
24 $16y^2 + 24y + 9$
25 $x^2 + 4xy + 4y^2$
26 $9x^2 + 6xy + y^2$
27 $4x^2 + 20xy + 25y^2$
28 $9a^2 + 12ab + 4b^2$
29 $9a^2 + 6ab + b^2$
30 $p^2 + 8pq + 16q^2$
31 $49x^2 + 28xy + 4y^2$
32 $9s^2 + 24st + 16t^2$
33 $x^2 - 4x + 4$
34 $x^2 - 12x + 36$
35 $a^2 - 20a + 100$
36 $x^2 - 2xy + y^2$
37 $x^2 - 6x + 9$
38 $x^2 - 14x + 49$
39 $a^2 - 2ab + b^2$
40 $u^2 - 2uv + v^2$
41 $9x^2 - 6x + 1$
42 $25z^2 - 10z + 1$
43 $100a^2 - 180a + 81$
44 $16x^2 - 24x + 9$
45 $4a^2 - 4a + 1$
46 $16y^2 - 8y + 1$
47 $49b^2 - 28b + 4$
48 $25x^2 - 30x + 9$
49 $4y^2 - 4yx + x^2$
50 $25x^2 - 10xy + y^2$
51 $9m^2 - 12mn + 4n2$
52 $49x^2 - 42xy + 9y^2$
53 $a^2 - 6ab + 9b^2$
54 $m^2 - 16mn + 64n^2$
55 $25a^2 - 20ab + 4b^2$
56 $9p^2 - 30pq + 25q^2$

Exercise 13g   Page 171

1 $x^2 - 16$
2 $b^2 - 36$
3 $c^2 - 9$
4 $x^2 - 144$
5 $x^2 - 25$
6 $a^2 - 49$
7 $q^2 - 100$
8 $x^2 - 64$
9 $4x^2 - 1$
10 $9x^2 - 1$
11 $49a^2 - 4$
12 $25a^2 - 16$
13 $25x^2 - 1$
14 $4a^2 - 9$
15 $100 m^2 - 1$
16 $9x^2 - 16y^2$
17 $4a^2 - 25b^2$
18 $1 - 4a^2$
19 $49y^2 - 9z^2$
20 $100a^2 - 81b^2$
21 $25a^2 - 16b^2$
22 $1 - 9x^2$
23 $9 - 25x^2$
24 $25m^2 - 64n^2$

Exercise 13h   Page 172

1 $2x^2 + 9x + 12$
2 $2x^2 + 9x + 2$
3 $x^2 + 15x + 32$
4 $a^2 - 9a + 36$
5 $2a^2 - 10a - 3$
6 $x^2 + 13x + 25$
7 $x^2 - 2x - 21$
8 $x^2 - 2x - 23$

9 $16x^2 + 6x - 10$
10 $12x^2 + 8x - 20$
11 $x^2y^2 - 6xy + 9$
12 $25 - 10yz + y^2z^2$
13 $x^2y^2 + 8xy + 16$
14 $9p^2q^2 + 48pq + 64$
15 $a^2 - 2abc + b^2c^2$
16 $a^2b^2 - 4ab + 4$
17 $36 - 12pq + p^2q^2$
18 $m^2n^2 + 6mn + 9$
19 $u^2v^2 - 4uvw + 4w^2$

Exercise 13i   page 173

1 $5x + 10$
2 $24pq - 16pr$
3 $6a^2 - 13ab - 5b^2$
4 $12x^2 - 17x - 5$
5 $x^2 + 16x + 60$
6 $x^2 - 20x + 96$
7 $16y^2 - 16y - 21$
8 $16y^2 - 81$
9 $25x^2 - 70a + 49$
10 $4a^2 - 28ab + 49b^2$

Exercise 13j   Page 173

1 $8 - 20x$
2 $16a - 24a^2$
3 $12a^2 - 35a - 33$
4 $x^2 + 2x - 99$
5 $-20x^2 - 48x + 5$
6 $y^2 + 4yz + 4z^2$
7 $36y^2 + 24yz - 5z^2$
8 $16a^2 + 8a + 1$
9 $25a^2 - 70a + 49$
10 $36z^2 - 156zy + 169y^2$

Exercise 13k   Page 174

1 $6 - 3a$
2 $8ab + 4ac$
3 $10ac + 25ad + 4bc + 10bd$
4 $x^2 - 19x + 84$
5 $a^2 + 16a + 63$
6 $a^2 - a - 20$
7 $6x^2 + 11x + 3$
8 $25x^2 - 4$
9 $9x^2 - 42x + 49$
10 $25x^2 - 4y^2$

Chapter 14

Exercise 14a   Page 177

1 $4(x+1)$
2 $3(4x-1)$
3 $2(3a+1)$
4 $5(a-2b)$
5 $3(t-3)$
6 $5(2a-1)$
7 $4(3a+1)$
8 $2(a+2b)$
9 $7(2x-1)$
10 $x(x+2)$
11 $x(x-7)$
12 $a(a+6)$
13 $x(2x+1)$
14 $2t(2-t)$
15 $x(x+5)$
16 $x(x-4)$
17 $b(b+4)$
18 $a(4a-1)$
19 $2x(x-3)$
20 $2z(z^2+2)$
21 $5a(5a-1)$
22 $4x(3x+4)$
23 $5b(a-2c)$
24 $3y(y+9)$
25 $2a(a-6)$
26 $2p(3p+1)$
27 $3y(3y-2)$
28 $2(x^2+2x+3)$
29 $5(2a^2-a+4)$
30 $b(a+4c-3d)$
31 $4(2x-y+3z)$
32 $3a(3b-2c-d)$
33 $3(x^2-2x+3)$
34 $4(a^2+2a-1)$
35 $x(5y+4z+3)$
36 $5b(a+2c+d)$
37 $2y(x-2z+4w)$
38 $x^2(x+1)$
39 $x^2(1-x)$
40 $5a^2(4-a)$
41 $4x^2(3x-4)$
42 $4x^2(x^2+3)$
43 $a^2(1+a)$
44 $b^2(b-1)$
45 $2x^2(2x-1)$
46 $9a^2(3-2a)$
47 $5x^2(2-3x^2)$
48 $4(3x+2)$
49 $4x(2x+3)$
50 $3(3x^2-2x+4)$
51 $5x(x^2-2)$
52 $4q(2p+r)$
53 $x(x-8)$
54 $3(4+3y^2)$
55 $4x(3y+4z+2)$
56 $2x(2x^2+3)$
57 $\frac{1}{2}h(a+b)$
58 $m(g-a)$
59 $\frac{1}{2}m(v^2+u^2)$
60 $P\left(1+\dfrac{RT}{100}\right)$
61 $\pi r(2r+h)$
62 $\pi(R^2+r^2)$
63 $2g(h_1-h_2)$
64 $m(\frac{1}{2}v^2-gh)$

65 $\dfrac{\pi r^2}{3}(4r-h)$ **67** $\dfrac{1}{2}mu(u+1)$

66 $\pi r(3r+2h)$ **68** $\dfrac{1}{4}c(2b-a)$

Exercise 14b   Page 179

1 $(x+3)(y+3)$
2 $(a+2b)(1+b)$
3 $(a+b)(a+c)$
4 $(x-3)(y+2)$
5 $(x+1)(y+z)$
6 $(x+2)(y+4)$
7 $(a+b)(c+4)$
8 $(x+4)(y-2)$
9 $(p+q)(r+s)$
10 $(y+4)(x-3)$
11 $(x+2)(y-5)$
12 $(p+q)(r-s)$
13 $(a+2)(b-3)$
14 $(p-q)(r+s)$
15 $(p+4)(q+2)$
16 $(2+a)(3+b)$
17 $(p-q)(r-s)$
18 $(3a-b)(3-b)$
19 $(2a-b)(1-b)$
20 $(a-2)(a+2b)$
21 $(2-x)(3-y)$
22 $(a-2)(4a-b)$
23 $(3a-b)(2a-3)$
24 $(2m-3n)(1-n)$
25 $(t+r)(t+s)$
26 $(x-1)(x+y)$

27 $2(1-a)(2a+b)$
28 $(x+y)(1-y)$
29 $(2a+3b)(2-3a)$
30 $(a+b)(2a+c)$
31 $2(2x+y)(1-y)$
32 $(x+y)(y+z)$
33 $(x-2)(5-y)$
34 $(a+4)(b-3)$
35 $(x+3)(y-z)$
36 $(p-4)(2-q)$
37 $(a-2)(b-3)$
38 $(a-4)(3-b)$
39 $(m+n)(m+1)$
40 $(a+1)(a-b)$
41 $(2p+1)(p-2q)$
42 $(x+1)(1-y)$
43 $(a+b)(a+1)$
44 $(a-b)(a+1)$
45 $(x+y)(x-1)$
46 $(a-1)(2a+b)$
47 $(x+2y)(5x-1)$
48 $(n-1)(m-1)$
49 $(x-1)(3x+y)$
50 $(2p-1)(p+2q)$
51 $(1-a)(3a+b)$
52 $(1-z)(2x+y)$

Exercise 14c   Page 181

1 $(x+1)(x+2)$
2 $(x+1)(x+5)$
3 $(x+3)(x+4)$
4 $(x+3)(x+5)$
5 $(x+1)(x+20)$
6 $(x+1)(x+7)$
7 $(x+6)(x+2)$
8 $(x+1)(x+12)$
9 $(x+1)(x+15)$
10 $(x+2)(x+10)$

11 $(x+4)(x+4)$
12 $(x+3)(x+12)$
13 $(x+1)(x+18)$
14 $(x+2)(x+20)$
15 $(x+1)(x+8)$
16 $(x+3)(x+3)$
17 $(x+2)(x+18)$
18 $(x+3)(x+6)$
19 $(x+5)(x+6)$
20 $(x+4)(x+10)$

Exercise 14d   Page 182

1 $(x-1)(x-8)$
2 $(x-3)(x-4)$
3 $(x-2)(x-15)$
4 $(x-4)(x-7)$
5 $(x-6)(x-7)$

6 $(x-2)(x-3)$
7 $(x-1)(x-15)$
8 $(x-3)(x-3)$
9 $(x-2)(x-16)$

Exercise 14e   Page 183

1 $(x+2)(x-3)$
2 $(x+5)(x-4)$
3 $(x-4)(x+3)$
4 $(x-4)(x+7)$
5 $(x+5)(x-3)$

6 $(x-6)(x+4)$
7 $(x-3)(x+9)$
8 $(x-11)(x+2)$
9 $(x-7)(x+5)$

Exercise 14f   Page 183

1 $(x+2)(x+7)$
2 $(x-3)(x-7)$
3 $(x+7)(x-2)$
4 $(x+6)(x-5)$
5 $(x+1)(x+8)$
6 $(x-5)(x-5)$

7 $(x+9)(x-1)$
8 $(x-13)(x-2)$
9 $(x+8)(x-7)$
10 $(x+2)(x+30)$
11 $(x+3)(x-9)$
12 $(x+20)(x-4)$

13 $(x+1)(x+13)$
14 $(x-2)(x+14)$
15 $(x+10)(x-8)$
16 $(x-5)(x-6)$
17 $(x-4)(x+12)$
18 $(x+6)(x+12)$

19 $(x+4)(x+13)$
20 $(x+2)(x-14)$
21 $(x+3)(x+8)$
22 $(x+3)(x-14)$
23 $(x-2)(x-16)$
24 $(x+12)(x-5)$

Exercise 14g   Page 184

1 $(x+1)(x+8)$
2 $(x-3)(x-3)$
3 $(x+4)(x+7)$
4 $(4-x)(5+x)$
5 $(x+3)(x+3)$
6 $(x-1)(x-8)$
7 $(x+2)(x+15)$
8 $(9+x)(3-x)$
9 $(x+2)(x+11)$
10 $(x-13)(x+2)$
11 $(x-1)(x-7)$
12 $(x-6)(x+7)$

13 $(x-8)(x+3)$
14 $(x-2)(x-7)$
15 $(x+1)(x+27)$
16 $(x-7)(x+9)$
17 $(x+5)^2$
18 $(x-5)^2$
19 $(x+2)^2$
20 $(x-7)^2$
21 $(x+6)^2$
22 $(x-6)^2$
23 $(x-2)^2$
24 $(x+8)^2$

Exercise 14h   Page 184

1 $(2+x)(1+x)$
2 $(3-x)(2+x)$
3 $(1-x)(4+x)$
4 $(4-x)(2+x)$
5 $(3+x)(2-x)$
6 $(2-x)(1+x)$
7 $(4+x)(2-x)$
8 $(5+x)(1-x)$

9 $(5+x)(2-x)$
10 $(6-x)(2+x)$
11 $(5-x)(1+x)$
12 $(7+x)(2-x)$
13 $(6-x)(1+x)$
14 $(5+x)(4-x)$
15 $(5+x)(3-x)$
16 $(4-x)(3+x)$

Exercise 14i   Page 185

1 $(x+5)(x-5)$
2 $(x+2)(x-2)$
3 $(x+10)(x-10)$
4 $(x+1)(x-1)$
5 $(x+8)(x-8)$
6 $(x+4)(x-4)$
7 $(x+6)(x-6)$
8 $(x+9)(x-9)$
9 $(3+x)(3-x)$

10 $(6+x)(6-x)$
11 $(10+x)(10-x)$
12 $(a+b)(a-b)$
13 $(3y+z)(3y-z)$
14 $(4+x)(4-x)$
15 $(5+x)(5-x)$
16 $(9+x)(9-x)$
17 $(x+y)(x-y)$

Exercise 14j   Page 185

1 $3(x+4)$
2 $5x(5x+2)$
3 $4(3x^2-2)$
4 $7(2x+3)$
5 $2(2x^2+1)$
6 $7(3x-1)$
7 $9x(x-2)$
8 $4(5x+3)$
9 $2(2x-7)$

10 $2(x+3)(x+4)$
11 $3(x-1)(x-8)$
12 $7(x+1)^2$
13 $4(x+3)(x-4)$
14 $5(x+1)(x+7)$
15 $3(x+2)(x+6)$
16 $4(x-3)^2$
17 $5(x+2)(x-3)$
18 $2(x+2)(x-11)$

Exercise 14k   Page 186

1 $(2x+1)(x+1)$
2 $(3x-2)(x-1)$
3 $(4x+3)(x+1)$
4 $(2x-1)(x-3)$
5 $(3x+1)(x+4)$
6 $(3x-2)(x-2)$
7 $(2x+1)(x+4)$
8 $(5x-2)(x-3)$
9 $(2x+3)(x+4)$
10 $(7x-1)(x-4)$

11 $(2x+1)(x-2)$
12 $(3x+4)(x-1)$
13 $(5x+2)(x-3)$
14 $(x+2)(4x-3)$
15 $(3x-2)(x+4)$
16 $(7x+2)(x-3)$
17 $(6x+5)(x-2)$
18 $(5x-4)(x-3)$
19 $(3x+4)(x-5)$
20 $(4x-3)(x+5)$

Exercise 14l   Page 186

1 $(3x + 2)(2x + 1)$
2 $(2x + 3)(3x + 5)$
3 $(3x + 1)(5x + 2)$
4 $(2x + 3)(6x + 5)$
5 $(7x + 2)(5x + 2)$
6 $(3x - 1)(2x - 3)$
7 $(3x - 2)(3x - 4)$
8 $(2x - 1)(8x - 1)$
9 $(5x - 3)(3x - 7)$
10 $(5x - 2)(4x - 3)$
11 $(4x + 1)(2x - 3)$
12 $(5x - 2)(3x + 1)$
13 $(3x + 2)(7x - 4)$
14 $(10x + 3)(8x - 3)$
15 $(3x + 4)(8x - 5)$
16 $(3a - 5)(2a + 3)$
17 $(3t - 2)(2t + 1)$
18 $(3b - 2)^2$
19 $(x - 2y)(5x + 3y)$
20 $(x - 2)(4x - 3)$

29 $4(x + 2y)(x - 2y)$
30 Does not factorise
31 $2(3x + 2)(2x - 5)$
32 $(x - 2)(x + 15)$
33 $(2 - x)(14 + x)$
34 $(a - 7)(a - 9)$
35 $2(3 - 2x)(1 - 2x)$
36 $(1 + 2x)(1 + 4x^2)$
37 $(x + 17)(x - 4)$
38 $(2x - 1)(x^3 + 2)$
39 $3(2x + 1)(x - 2)$
40 $(p + 1)(p^2 + 1)$
41 $(a + b + c)(a + b - c)$
42 $(29x + 1)(4x - 1)$
43 $(a + 16)(a + 7)$
44 $(x^2 + y + 1)(x^2 - y - 1)$
45 $(a - 8)(3a - 7)$
46 $2(x + 7)(x - 11)$
47 $(2x + y - z)(2x - y + z)$
48 $(ab + 18)(ab - 19)$

Exercise 14m   Page 187

1 $(2x + 5)(2x - 5)$
2 $(3x + 2)(3x - 2)$
3 $(6a + 1)(6a - 1)$
4 $(4a + b)(4a - b)$
5 $(3x + 5)(3x - 5)$
6 $(2a + 1)(2a - 1)$
7 $(4a + 3b)(4a - 3b)$
8 $(5s + 3t)(5s - 3t)$
9 $(10x + 7y)(10x - 7y)$
10 $(3y + 4z)(3y - 4z)$
11 $(2x + 7y)(2x - 7y)$
12 $(9x + 10y)(9x - 10y)$
13 $(3a + 2b)(3a - 2b)$
14 $(8p + 9q)(8p - 9q)$
15 $3(a + 3b)(a - 3b)$
16 $2(3t + 5s)(3t - 5s)$
17 $3(3x + y)(3x - y)$
18 $5(3x + 2)(3x - 2)$
19 $5(a + 2)(a - 2)$
20 $5(3 + b)(3 - b)$

21 $\frac{1}{2}(a + 2b)(a - 2b)$

22 $\left(\dfrac{a}{2} + \dfrac{b}{3}\right)\left(\dfrac{a}{2} - \dfrac{b}{3}\right)$ or $\frac{1}{36}(3a + 2b)(3a - 2b)$

23 $\frac{1}{3}(9x + y)(9x - y)$

Exercise 14n   Page 187

1 7.5
2 18.5
3 17.7
4 35.04
5 31.2
6 20.4
7 12.9
8 1000
9 336
10 53.2
11 5.336
12 8
13 140
14 75.8
15 0.526

Exercise 14p   Page 188

1 $5(x + 1)(3x + 2)$
2 $2(x - 2)(2x + 1)$
3 $3(x + 1)(2x + 1)$
4 $3(x - 2)(6x + 5)$
5 $2(x + 5)(4x - 3)$
6 $2(x + 1)(4x + 3)$
7 $5(x - 3)(5x + 2)$
8 $3(x - 1)(3x + 4)$
9 $2(x + 4)(3x + 1)$
10 $5(x + 4)(3x - 2)$
11 $2(3x - 2)(3x - 4)$
12 $3(2x - 1)(8x - 1)$
13 $2(2x + 1)(3x + 2)$
14 $5(4x - 3)(5x - 2)$
15 $4(2x + 1)(3x - 2)$
16 $7(x + 4)(3x - 2)$
17 $(4 + 3x)(1 - 2x)$
18 $(4 - 3x)(3 + 4x)$
19 $(7 - x)(3 + 4x)$
20 $2(2 - x)(6 - x)$
22 $2(4 + x)(2 - 3x)$
22 $(9 - x)(1 + x)$
23 $(12 + x)(1 - x)$
24 $2(2 + 3x)^2$
25 $5(3 - x)^2$

Exercise 14q   Page 189

1 $(x + 5)(x + 8)$
2 $(3x + 1)(2x + 1)$
3 $(x + 6)(x - 6)$
4 Does not factorise
5 $(x - 2)(x - 6)$
6 $(2x - 3)(x + 5)$
7 $(x + 7)(x - 1)$
8 $(5x - 2)(x + 1)$
9 $(x - 3)(x - 8)$
10 $(3x + 2)(x + 3)$
11 $(x + 15)(x - 1)$
12 $(4x - 1)(3x - 1)$
13 $(x + 2)(x + 6)$
14 $(4x + 1)(2x - 1)$
15 $(x + 7)(x - 7)$
16 Does not factorise
17 $(3x + 2)(2x - 5)$
18 $(x + 6)(x + 8)$
19 $(2x + 3y)(2x - 3y)$
20 $(5x - 4)(3x - 2)$
21 $(2x - 3)(3x + 2)$
22 $(x + 13)(x - 2)$
23 $2(3x + 1)(5x - 2)$
24 $(4 + x)(7 - x)$
25 $(2x - 1)(3x + 4)$
26 $5(2x + 1)(3x + 2)$
27 $(x + 2)(x + 9)$
28 $(x - 4)(x - 6)$

Exercise 14r   Page 190

1 a $7a + 21$
   b $3x - 6y$
2 a $x^2 + 14x + 40$
   b $6x^2 - 19x + 15$
3 a $25 + 10x + x^2$
   b $25 - 10x + x^2$
   c $25 - x^2$
4 a $10(a + 2)$
   b $5p(3p - 2)$
5 a $(a + 1)(a^2 + 1)$
   b $(k + l)(2m - n)$
6 a $(x - 3)(x + 9)$
   b $(x - 7)(5x - 7)$
   c $\left(a + \dfrac{b}{2}\right)\left(a - \dfrac{b}{2}\right)$
7 a $(5x + 2)(2x - 3)$
   b $(10a + 9b)(10a - 9b)$

Exercise 14s   Page 190

1 a $5a^2 + 15a$
   b $12x^2 - 8xy$
2 a $y^2 - 9y + 20$
   b $15x^2 - 14xy - 8y^2$
3 a $4p^2 + 12pq + 9q^2$
   b $4p^2 - 12pq + 9q^2$
   c $4p^2 - 9q^2$
4 a $4z^2(2z - 1)$
   b $5y(x - 4z)$
5 a $(m + 1)(2 + 3n)$
   b $(a + 2b)(c - 2d)$
6 a $(x + 3)(x - 9)$
   b $(4x - 1)(x + 7)$
   c $(2m + 9n)(2m - 9n)$
7 a $3(x - 3)(5x - 3)$
   b Does not factorise or $5(3 + 5x - 4x^2)$

Exercise 14t   Page 191

1 a $4a + 28$
   b $6x^2 - 9xy$
2 a $x^2 + 12x + 27$
   b $15x^2 - x - 2$
3 a $25x^2 + 20x + 4$
   b $25x^2 - 4$
   b $25x^2 - 20x + 4$
4 a $6z(2z - 1)$
   b $4y(2x - 3z)$
5 a $(z + 2)(z^2 + 1)$
   b $(3a + b)(c + 2)$
6 a $(x - 6)(x + 4)$
   b $(2a + 5)(2a - 3)$
   c $\left(3m + \dfrac{n}{3}\right)\left(3m - \dfrac{n}{3}\right)$
7 a $(5x - 3)(3x + 2)$
   b $(3 + 5x)(2 - 3x)$
8 a Hint: rewrite $100a + 10b + c$ as $98a + 2a + 2b + 7b + b + c$
   b Hint: rewrite $100a + 10b + c$ as $91a + 9a + 9b + b + c$

Chapter 15

Exercise 15a   Page 193

	a	b	c
1	8	0	0
2	0	5	0
3	0	7	0
4	0	0	3
5	33	0	0
6	-24	0	0
7	70	0	0

STP Caribbean Mathematics 3

Exercise 15b   Page 194

1 0	5 4	9 0
2 0	6 1	10 7
3 0	7 0	11 any value
4 any value	8 2	12 0

13 **a** 0 **b** 0   20 $a = 3$ or $b = 0$
14 **a** 0 **b** 0   21 $a = 9$ or $b = 0$
15 **a** 0 **b** 0   22 $a = 0$ or $b = 4$
16 **a** 0 **b** any value   23 $a = 0$ or $b = 10$
17 $a = 0$ or $b = 1$ 24 $a = 1$ or $b = 0$
18 $a = 0$ or $b = 5$ 25 $a = 7$ or $b = 0$
19 $a = 0$ or $b = 2$

Exercise 15c   Page 195

1 0 or 3	10 1, 2	19 -4 or -9
2 0 or 5	11 5 or 9	20 -1 or -8
3 0 or 3	12 7 or 10	21 p or q
4 0 or -4	13 4 or 7	22 $-a$ or -6
5 0 or -5	14 1 or 6	23 4 or -1
6 0 or 6	15 8 or -11	24 -9 or 8
7 0 or 10	16 3 or -5	25 -6 or -725
8 0 or 7	17 -7 or 2	26 -10 or -11
9 0 or -7	18 -2 or -3	27 a or b

Exercise 15d   Page 196

1 1 or $2\frac{1}{2}$	11 $2\frac{1}{3}$ or 2
2 4 or $\frac{2}{3}$	12 $1\frac{2}{3}$ or $\frac{1}{2}$
3 $\frac{4}{5}$ or $\frac{3}{4}$	13 0 or $\frac{1}{3}$
4 0 or $1\frac{1}{4}$	14 0 or $\frac{3}{7}$
5 0 or $\frac{3}{10}$	15 $-1\frac{1}{2}$ or 3
6 $-\frac{2}{5}$ or 7	16 $-\frac{3}{4}$ or $2\frac{1}{2}$
7 $-\frac{5}{6}$ or $\frac{2}{3}$	17 $-\frac{9}{10}$ or $\frac{4}{5}$
8 $\frac{3}{8}$ or $-2\frac{1}{2}$	18 $\frac{2}{3}$ or $-2\frac{1}{4}$
9 $1\frac{1}{7}$ or $-3\frac{3}{4}$	19 $2\frac{2}{5}$ or $-3\frac{1}{2}$
10 $-\frac{3}{4}$ or $-1\frac{1}{2}$	20 $-1\frac{3}{5}$ or $-\frac{3}{4}$

Exercise 15e   Page 196

1 1 or 2	15 2 or -9	29 -1 or -15
2 1 or 7	16 -1 or 13	30 -3 or -6
3 2 or 3	17 2 or -3	31 ± 1
4 2 or 5	18 -2 or 6	32 ± 3
5 3 or 4	19 4 or -5	33 ± 4
6 1 or 5	20 -3 or 8	34 ± 9
7 1 or 11	21 -1 or -2	35 ± 13
8 2 or 4	22 -1 or -7	36 ± 2
9 2 or 6	23 -3 or -5	37 ± 5
10 1 or 12	24 -2 or -6	38 ± 10
11 1 or -7	25 -2 or -9	39 ± 12
12 4 or -2	26 -1 or -6	40 ± 6
13 3 or -4	27 -2 or -5	
14 5 or -3	28 -1 or -13	

Exercise 15f   Page 198

1 0 or 2	8 0 or -1	15 0 or $\frac{12}{7}$
2 0 or 10	9 0 or $\frac{5}{3}$	16 0 or $-\frac{7}{6}$
3 0 or -8	10 0 or $\frac{7}{5}$	17 0 or $-\frac{7}{12}$
4 0 or $\frac{1}{2}$	11 0 or $-\frac{3}{2}$	18 0 or -4
5 0 or $\frac{5}{4}$	12 0 or $-\frac{5}{8}$	19 0 or $\frac{2}{7}$
6 0 or 5	13 0 or 7	20 0 or $-\frac{3}{14}$
7 0 or -3	14 0 or $-\frac{5}{3}$	

Exercise 15g   Page 198

1 1 (twice)	11 -9 (twice)
2 5 (twice)	12 7 (twice)
3 10 (twice)	13 11 (twice)
4 -4 (twice)	14 -6 (twice)
5 -3 (twice)	15 $\frac{1}{2}$ (twice)
6 3 (twice)	16 -5 (twice)
7 4 (twice)	17 6 (twice)
8 9 (twice)	18 20 (twice)
9 -1 (twice)	19 8 (twice)
10 -10 (twice)	20 $-\frac{2}{3}$ (twice)

Exercise 15h   Page 199

1 $\frac{1}{2}$ and 2	16 $\frac{3}{4}$ and $1\frac{1}{2}$
2 $1\frac{1}{2}$ and 4	17 $-\frac{5}{6}$ and $2\frac{1}{2}$
3 $2\frac{1}{2}$ and 4	18 $-\frac{1}{2}$ and $-1\frac{1}{2}$
4 -1 and $-\frac{2}{3}$	19 $-\frac{2}{3}$ and $-\frac{3}{4}$
5 -7 and $2\frac{1}{2}$	20 $3\frac{1}{2}$ and $-\frac{3}{5}$
6 $\frac{2}{3}$ and 3	21 $\pm\frac{5}{4}$
7 $\frac{1}{3}$ and 2	22 $\pm\frac{9}{10}$
8 $1\frac{1}{2}$ and -4	23 $\pm\frac{5}{2}$
9 $-\frac{2}{3}$ and -3	24 $\pm\frac{4}{3}$
10 $-\frac{2}{3}$ and -5	25 $\pm\frac{12}{5}$
11 $-\frac{1}{2}$ and $\frac{2}{3}$	26 $\pm\frac{2}{3}$
12 $\frac{2}{3}$ and $-1\frac{1}{3}$	27 $\pm\frac{5}{9}$
13 $\frac{1}{3}$ and $\frac{1}{4}$	28 $\pm\frac{2}{3}$
14 $-\frac{1}{3}$ and $2\frac{1}{2}$	29 $\pm\frac{5}{6}$
15 $-\frac{1}{5}$ and $-\frac{3}{4}$	30 $\pm\frac{9}{2}$

Exercise 15i   Page 200

1 -5 and 6	18 1 and 7
2 -2 and 8	19 2 and 4
3 3 and -12	20 3 and 7
4 $\frac{2}{3}$ and -2	21 2 and 6
5 3 and -2	22 4 and 5
6 1 and -7	23 5 and 7
7 $\frac{1}{2}$ and -3	24 3 and 5
8 3 and $-\frac{3}{5}$	25 0 and $\frac{1}{2}$
9 3 or -1	26 2 and 3
10 -4 and 6	27 2 and 6
11 5 and 7	28 -1 and $-\frac{2}{3}$
12 $-\frac{1}{5}$ and $1\frac{1}{2}$	29 $\frac{1}{2}$ and -3
13 -2 and 5	30 0 and 3
14 2 and 4	31 1 and 2
15 $\frac{1}{2}$ and $-\frac{1}{3}$	32 -1 and -2
16 $\frac{1}{3}$ and 4	33 $\frac{1}{3}$ and 2
17 2 and 5	

Exercise 15j   Page 201

1 -4 and 5	8 -5 and 7
2 2 (twice)	9 2 and $-3\frac{1}{2}$
3 $\pm\frac{1}{3}$	10 -3 (twice)
4 0 and $-3\frac{1}{2}$	11 1 and -7
5 -1 and -12	12 $\pm\frac{2}{5}$
6 $\pm\frac{1}{4}$	13 $\pm 2\frac{1}{2}$
7 0 and 6	14 -2 and -9

15 $\frac{1}{2}$ and $-\frac{2}{3}$

16 0 and $2\frac{1}{2}$

17 2 and $-\frac{1}{3}$

18 $-\frac{1}{2}$ and $-1\frac{1}{3}$

19 0 and $1\frac{3}{4}$

20 $\frac{1}{3}$ and $\frac{1}{4}$

21 $\frac{1}{3}$ and $-2\frac{1}{2}$

22 $-\frac{1}{3}$ and 2

23 $-\frac{1}{2}$ and $-1\frac{1}{2}$

24 $\pm\frac{1}{2}$

25 3 and -4

26 3 and -1

27 $\frac{1}{2}$ and $-\frac{1}{3}$

28 1 and 4

29 -3 and 8

30 5 and 7

31 -2 and $\frac{2}{3}$

32 $-\frac{1}{3}$ and 2

33 5 and -10

34 -11 and 8

35 5 and -9

36 -2 and 7

37 7 and -4

38 5 and -11

39 -4 and -5

40 -4 and -5

41 0, 1 and 2

42 0, 3 and -4

43 0, 2 and $2\frac{1}{2}$

44 0, 1 and 1

45 0, $-\frac{1}{2}$ and -4

46 0, 6 and 7

47 0, -2 and 5

48 0, 5 and $-2\frac{1}{3}$

49 0, $\frac{3}{2}$ and $-\frac{3}{2}$

50 0, 2 and 4

Exercise 15k   Page 202

1 -2 or 8

2 -2 or 7

3 -7 or 6

4 $x + (x^2 - 6) = 66$; $x = -9$ or 8; 8 marbles

5 $x + x^2 = 56$; $x = -8$ or 7; Ahmed is 7 and his father is 49

6 $x + (x^2 + 2) = 44$; $x = -7$ or 6; Kathryn is 6 and her mother is 38

7 $x(x + 5) = 84$; $x = 7$ or -12; Peter is 7

8 $x(x - 4) = 140$; $x = 14$ or -10; Ann is 10

9 $x(x + 3) = 28$; $x = 4$ or -7; 4 cm by 7 cm

10 $x(x + 5) = 66$; $x = -11$ or 6; 6 cm by 11 cm

11 $\frac{1}{2}x \times \frac{1}{2}x = 25$; $x = \pm 10$; 5 cm

12 a $A = 20x\,\text{m}^2$, $B = x^2\,\text{m}^2$, $C = 30x\,\text{m}^2$

   b $x^2 + 50x = 104$; $x = 2$ or -52; path is 2 m wide

Exercise 15l   Page 204

1 a -10 **b** 0   **c** 8

2 a 0 or -7 **b** 0 or $\frac{1}{2}$

3 a 3 and 8   **b** 2 and $-\frac{3}{5}$

4 a 7 and -5 **b** 5 and 8

5 a $\frac{1}{2}$ and $\frac{4}{5}$ **b** $\frac{2}{5}$ and $-\frac{1}{3}$ **c** $\pm\frac{2}{3}$

6 a 0 and 2   **b** 0 and $\frac{3}{4}$

7 a $-\frac{1}{3}$ and $\frac{2}{3}$ **b** $\frac{1}{2}$ and $2\frac{1}{3}$

8 a 5 and -9 **b** 5 and -6

Exercise 15m   Page 205

1 a -2 **c** 12

   b 0   **d** 0

2 a 0 and 2   **b** 0 and $-\frac{3}{7}$

3 a 2 and -5 **c** $-1\frac{1}{2}$ and $1\frac{1}{2}$

   b -2 and $1\frac{1}{2}$

4 a -3 and 2   **b** -5 and -6

5 a $\frac{1}{5}$ and $-\frac{3}{4}$ **b** $-\frac{2}{5}$ and $-2\frac{1}{3}$

6 a 0 and $-\frac{2}{3}$ **b** 0 and $-\frac{3}{7}$

7 a 5 and $-\frac{3}{4}$ **b** $-\frac{1}{2}$ and $2\frac{1}{3}$

8 a -4 and 8   **b** -2 and 4

Exercise 15n   Page 205

1 a -11 **b** 0   **c** 0

2 a 0, -7 **b** 0, $\frac{3}{4}$

3 a $-4, 5$ **b** $1\frac{3}{4}, -3$ **c** $\frac{3}{5}, -\frac{3}{5}$

4 a $5, -3$ **b** $-4, -8$

5 a $-\frac{1}{5}, -\frac{3}{4}$ **b** $-\frac{2}{7}, \frac{1}{4}$

6 a $0, -1\frac{1}{3}$ **b** $0, -1\frac{2}{3}$

7 a $\frac{2}{7}, -1$ **b** $-\frac{1}{2}, 1\frac{2}{3}$

8 a $-5, 2$ **b** $-10, 3$

Chapter 16

Exercise 16a   Page 208

1 $\frac{x}{4}$ **5** $\frac{x}{y}$ **9** $\frac{pq}{2}$ **13** $\frac{b}{d}$ **17** $\frac{m}{k}$

2 $\frac{a}{2}$ **6** $\frac{1}{2a}$ **10** $\frac{a}{c}$ **14** $\frac{1}{3x}$ **18** $\frac{s}{4t}$

3 $\frac{p}{q}$ **7** $\frac{a}{2c}$ **11** $\frac{a}{2}$ **15** $\frac{q}{2}$

4 $\frac{a}{b}$ **8** $\frac{2}{q}$ **12** $\frac{z}{2}$ **16** $\frac{2}{3y}$

Exercise 16b   Page 209

1 $\frac{1}{x}$ **7** $p - q$ **13** $\frac{2a}{3(a - b)}$

2 $\frac{t}{s - t}$ **8** $\frac{1}{(4 - a)}$ **14** $\frac{2(x - y)}{3xy}$

3 not possible   **9** not possible   **15** not possible

4 not possible   **10** $\frac{1}{v}$ **16** $u - v$

5 $\frac{x}{2(x - y)}$ **11** $\frac{y}{x + y}$ **17** not possible

6 $\frac{(a + b)}{2ab}$ **12** $\frac{1}{2}$ **18** $\frac{1}{(s - 6)}$

Exercise 16c   Page 210

1 $\frac{2a}{4a - b}$ **12** $\frac{p + q}{5}$ **23** $\frac{3}{x + 3}$

2 $\frac{2q}{p - q}$ **13** $\frac{1}{3}$ **24** $\frac{9}{y + 2}$

3 $\frac{1}{a}$ **14** $\frac{3 + a}{4b}$ **25** $\frac{y}{x - 2}$

4 $\frac{3}{5}$ **15** $\frac{2 - y}{x}$ **26** $\frac{q}{p + 2}$

5 $\frac{2 - x}{3y}$ **16** $\frac{1}{3y}$ **27** $\frac{t}{s - 7}$

6 $\frac{a}{3 - b}$ **17** a **28** $\frac{1}{p + 3}$

7 $\frac{1}{3a}$ **18** $\frac{p}{2}$ **29** $\frac{1}{x + 6}$

8 s **19** $\frac{1}{a - 2}$ **30** $\frac{2}{x - 4}$

9 $\frac{3}{a}$ **20** $\frac{1}{x - 4}$ **31** $\frac{3}{x - 4}$

10 $\frac{2x}{3x - y}$ **21** $\frac{1}{y + 2}$ **32** $\frac{v}{u + 6}$

11 $\frac{3a}{a + b}$ **22** $\frac{2}{a + 3}$ **33** $\frac{y}{x - 2}$

Exercise 16d   Page 211

1 $\dfrac{x+3}{2x-1}$

2 $\dfrac{4}{x+2}$

3 $\dfrac{2x-1}{x-2}$

4 $\dfrac{1}{2-x}$

5 $\dfrac{a+b}{a-b}$

6 $\dfrac{a+b}{2a+b}$

7 $\dfrac{x-y}{3x-2y}$

8 $\dfrac{2-x}{y}$

9 $-a$

10 $\dfrac{y+3}{2y+1}$

11 $\dfrac{x-3y}{x}$

12 $\dfrac{4x+1}{4x}$

13 $\dfrac{2x-3}{x-5}$

14 $\dfrac{-1}{1+a}$

15 $a+b$

16 $\dfrac{-(x+5)}{(x+1)}$

17 $\dfrac{2(2x-1)}{x-3}$

18 $\dfrac{x-2y}{y}$

19 $\dfrac{1-x}{3(x+2)}$

20 $\dfrac{1+y}{x+y}$

Exercise 16e   page 213

1 $\dfrac{ac}{bd}$

2 $\dfrac{ad}{bc}$

3 $\dfrac{5(x-y)}{2x}$

4 $\dfrac{x(x-y)}{10}$

5 $\dfrac{a}{bc}$

6 $\dfrac{ac}{b}$

7 $\dfrac{3(a-b)}{4(a+b)}$

8 $\dfrac{(x-2)(x+3)}{3}$

9 $\dfrac{x-2}{3(x+3)}$

10 $\dfrac{pr}{q}$

11 $\dfrac{6b}{a}$

12 $\dfrac{q}{2p}$

13 $\dfrac{12y}{x}$

14 $\dfrac{2b^2}{5}$

15 $\dfrac{pq}{6}$

16 $\dfrac{x}{2y}$

17 $\dfrac{1}{2b}$

18 $\dfrac{2}{3p}$

19 $\dfrac{a}{4b}$

20 $\dfrac{a^3}{b^3}$

21 $\dfrac{1}{4(b-2)}$

22 $2(x-2)$

23 $2(a+3)$

24 6

25 $x-3$

26 $x-3$

27 $\dfrac{1}{x-2}$

28 $\dfrac{2}{x+4}$

29 $\dfrac{3(x-2)}{5(x+6)}$

30 $\dfrac{2(2x-3)}{9}$

31 $\dfrac{3}{3x+2}$

32 $\dfrac{2x-3}{2}$

33 $\dfrac{2x-1}{6x+1}$

34 a

35 $\dfrac{-c(a+b)}{b}$

36 $(x-4)(x-2)$

Exercise 16f   Page 215

1 pq

2 rst

3 30

4 abc

5 $wxyz$

6 ad

7 uvw

8 168

9 xy

10 $2x^2$

11 $3pq$

12 $2x^2y$

13 abc

14 st

15 $3p^2$

16 $5ab$

17 $3pq^2$

18 $6x$

19 $8x$

20 $18a$

21 60

22 a^2b

23 $30x$

24 $12x$

25 $15y$

26 $12x$

Exercise 16g   Page 216

1 $\dfrac{x+y}{xy}$

2 $\dfrac{3q+2p}{pq}$

3 $\dfrac{2t-s}{st}$

4 $\dfrac{6b+a}{2ab}$

5 $\dfrac{5y-6x}{15xy}$

6 $\dfrac{2b+5a}{2ab}$

7 $\dfrac{2y-3x}{xy}$

8 $\dfrac{4q+6p}{3pq}$

9 $\dfrac{3y-2x}{xy}$

10 $\dfrac{20b+21a}{28ab}$

11 $\dfrac{5}{6x}$

12 $-\dfrac{1}{35x}$

13 $\dfrac{5}{4y}$

14 $\dfrac{1}{8p}$

15 $\dfrac{13}{8a}$

16 $\dfrac{4}{21x}$

17 $\dfrac{6}{35x}$

18 $\dfrac{1}{3y}$

19 $\dfrac{3a+2b}{4ab}$

20 $\dfrac{ab-2a^2}{2b^2}$

21 $\dfrac{3y-4}{xy}$

22 $\dfrac{4-3p}{2p^2}$

23 $\dfrac{9a^2+2b^2}{12ab}$

24 $\dfrac{10q-3p}{4pq}$

25 $\dfrac{2s+ts^2}{2t^2}$

26 $\dfrac{15b+4}{6ab}$

27 $\dfrac{3+2x}{3x^2}$

28 $\dfrac{4y^2-9x^2}{6xy}$

29 $\dfrac{5y+4x}{8xy}$

30 $\dfrac{pq+3p^2}{3q^2}$

31 $\dfrac{10y-3}{14xy}$

32 $\dfrac{18b-3a}{2a^2b}$

33 $\dfrac{3x^2-3y^2}{2xy}$

34 $\dfrac{14q-15p}{18pq}$

35 $\dfrac{5a^2+4ab}{5b^2}$

36 $\dfrac{21+8p}{15pq}$

Exercise 16h   Page 217

1 $\dfrac{9x+3}{20}$

2 $\dfrac{5-x}{12}$

3 $\dfrac{13x+1}{15}$

4 $\dfrac{4x+13}{12}$

5 $\dfrac{1-2x}{35}$

6 $\dfrac{7x+3}{10}$

7 $\dfrac{3x+9}{35}$

8 $\dfrac{5x-3}{42}$

9 $\dfrac{5-22x}{21}$

10 $\dfrac{7x+9}{12}$

11 $\dfrac{22-13x}{6}$

12 $\dfrac{11-7x}{12}$

13 $\dfrac{20-17x}{24}$

14 $\dfrac{22-7x}{20}$

15 $\dfrac{10-5x}{6}$

16 $\dfrac{31x-6}{24}$

17 $\dfrac{11-7x}{10}$

18 $\dfrac{2-11x}{18}$

19 $\dfrac{26x+34}{15}$

20 $\dfrac{17x-1}{12}$

21 $\dfrac{5x-19}{21}$

22 $\dfrac{42x-49}{10}$

23 $\dfrac{27x+3}{14}$

24 $\dfrac{19x-73}{9}$

25 $\dfrac{26x-18}{15}$

26 $\dfrac{-17x+104}{30}$

27 $\dfrac{3a+6}{a(a+3)}$

28 $\dfrac{6x+4}{x(x+2)}$

29 $\dfrac{7x-4}{2x(x-4)}$

30 $\dfrac{2x-3}{4x(2x+1)}$

31 $\dfrac{5a+12}{a(a+4)}$

32 $\dfrac{7x-4}{x(x-1)}$

33 $\dfrac{11x+1}{3x(2x+1)}$

34 $\dfrac{21x-6}{5x(2x+3)}$

Exercise 16i Page 219

1 $\dfrac{2c - ab}{ac}$

2 $\dfrac{qr^2}{p}$

3 $\dfrac{7x - 14}{12}$

4 $\dfrac{a}{a - b}$

5 $\dfrac{1}{12x}$

6 $\dfrac{1}{x + 2}$

7 $\dfrac{-p}{p + q}$

8 $\dfrac{12 - 2x}{3x^2}$

9 $\dfrac{1 - 2x}{x(x + 1)}$

10 $\dfrac{ab}{c}$

11 $\dfrac{8}{15}$

12 $\dfrac{23}{20x}$

13 $\dfrac{3}{10x^2}$

14 $\dfrac{4x + 7}{10}$

15 $\dfrac{(x + 4)(2x - 1)}{50}$

16 $\dfrac{25}{12x}$

17 $\dfrac{25}{24x^2}$

18 $\dfrac{3}{2}$

19 $\dfrac{19x - 1}{3x(x - 1)}$

20 $\dfrac{2}{x(x - 1)}$

21 $\dfrac{-a - 3}{2a(a - 1)}$

22 $\dfrac{3}{a - 1}$

23 $\dfrac{3}{y}$

24 -1

Exercise 16j Page 220

1 8
2 -5
3 6
4 $1\frac{1}{3}$
5 10
6 5
7 $9\frac{3}{5}$
8 $5\frac{1}{4}$
9 -1
10 $8\frac{3}{4}$
11 2
12 -18
13 3
14 -1

15 21
16 $\frac{4}{9}$
17 $-2\frac{1}{2}$
18 -17
19 2
20 4
21 1
22 $-2\frac{1}{19}$
23 $-2, -1$
24 3, 2
25 $-2, -2$
26 $-3, -3$
27 1, -4
28 $-3, -3$

29 1, 1
30 $\frac{2}{3}$, 1
31 2, $-\frac{2}{3}$
32 -2, 1
33 $4\frac{1}{2}$
34 $\frac{2}{5}$
35 2, 1
36 $-2\frac{4}{5}$
37 -40
38 $\frac{2}{5}$
39 0, 4
40 3
41 $\frac{1}{2}$, $-\frac{1}{2}$
42 3

Exercise 16k Page 222

1 a $\dfrac{b}{2}$ b a c $a - b$

2 a $\dfrac{4}{3x}$ b $\dfrac{1}{3x^2}$ c 3

3 a -13 b 3, -1

4 a $\dfrac{5x - 7}{6}$ b $1\frac{7}{10}$

Exercise 16l Page 223

1 a $\dfrac{2x}{y}$ b $\dfrac{x - y}{2x}$ c $x + 3$

2 a $\dfrac{1}{6p}$ b $x - 2$ c $\dfrac{3y}{2x}$

3 a $\frac{8}{9}$ b 7, -2

4 a $\dfrac{x^2 - 2x + 12}{4x}$ b $6\frac{1}{2}$

Exercise 16m Page 223

1 a $\dfrac{v}{uw}$ b $\dfrac{1}{2a - b}$ c $\dfrac{x}{3 - x}$

2 a $18s^2$ b $2(x - 2)$ c $\dfrac{2 - 5x}{x(4x - 1)}$

3 a 4   b 1, 2

4 a $\dfrac{x}{6}$ b 30

5 a $x = 40$ b $x = 21$ c $x = 66$
   x equals the LCM

6 $x = ab$

Chapter 17

Exercise 17a Page 226

1 $(x + 3)^2$
2 $(a + 2)^2$
3 $(p - 5)^2$
4 $(s - 6)^2$
5 $\left(x - \frac{5}{2}\right)^2$
6 $\left(b + \frac{3}{2}\right)^2$
7 $\left(x + \frac{9}{2}\right)^2$
8 $\left(x - \frac{1}{2}\right)^2$
9 $\left(x - \frac{1}{4}\right)^2$

10 $(x + 4)^2$
11 $\left(x + \frac{1}{2}\right)^2$
12 $\left(x + \frac{1}{3}\right)^2$
13 $(p + 9)^2$
14 $\left(a - \frac{2}{5}\right)^2$
15 $\left(t - \frac{3}{4}\right)^2$
16 $(x + b)^2$
17 $(x - c)^2$
18 $\left(x + \dfrac{b}{2a}\right)^2$

19 $(3x + 1)^2$
20 $(2x - 3)^2$
21 $(10x - 3)^2$
22 $(3x - 4)^2$
23 $(2x - 1)^2$
24 $(5x + 2)^2$
25 $(3x - 1)^2$
26 $\left(2x + \frac{1}{2}\right)^2$
27 $\left(\frac{3}{2}x + \frac{2}{3}\right)^2$

Exercise 17b Page 227

1 4
2 16
3 36

4 49
5 $\frac{9}{4}$
6 100

7 $\frac{49}{4}$
8 $\frac{1}{16}$
9 $\frac{9}{16}$

10 $\frac{1}{4}$
11 h^2
12 $\dfrac{b^2}{4a^2}$

Exercise 17c Page 227

1 4   3 25   5 4   7 4   9 $\frac{1}{4}$
2 9   4 9   6 25   8 4

Exercise 17d Page 228

1 2, -4
2 -2, 6
3 -2, 8
4 -16, 4
5 -8, -6
6 -4, 6
7 -9, 5
8 1, 9
9 5, 9

10 -8, 0
11 -8, 2
12 3, 15
13 $-\frac{1}{2}$, $-1\frac{1}{2}$
14 $\frac{1}{2}$, $3\frac{1}{2}$
15 -2, 3
16 $-\frac{3}{2}$, $\frac{5}{2}$
17 $-\frac{7}{3}$, 1
18 -1, $\frac{7}{5}$

19 1, $\frac{5}{3}$
20 $-\frac{12}{7}$, $\frac{8}{7}$
21 $-\frac{7}{2}$, $\frac{5}{2}$
22 -1, $\frac{11}{3}$
23 $-\frac{7}{5}$, $\frac{3}{5}$
24 $\frac{1}{2}$, 1
25 $\frac{1}{3}$, $\frac{7}{9}$
26 $-1\frac{2}{5}$, $\frac{1}{5}$
27 $-\frac{4}{7}$, 2

Exercise 17e Page 229

1 1, -5
2 -1, 7
3 -11, 1
4 -7.61, -0.39
5 0.27, 3.73

6 -8.36, 0.36
7 -1.61, 5.61
8 -8.53, -0.47
9 0.81, 6.19
10 -1.56, 2.56

11 $-9.32, 0.32$
12 $-0.54, -7.46$
13 $-4.10, 1.10$
14 $-0.35, 2.35$
15 $-2.32, 0.32$
16 $-0.85, 2.35$
17 $-4.58, 0.58$

18 $-0.18, 1.85$
19 $-0.52, 1.52$
20 $-1.29, -0.31$
21 $-0.36, 2.11$
22 $-0.17, 1$
23 $-1.41, 0.41$
24 $-0.21, 3.21$

Exercise 17f   Page 232

1 $x = -0.55$ or -5.45
2 $x = -6.37$ or -0.63
3 $x = -3.62$ or -1.38
4 $x = -7.27$ or 0.27
5 $x = -4.65$ or 0.65
6 $x = -7.37$ or -1.63
7 $x = -5.73$ or -2.27
8 $x = -11.32$ or 1.32
9 $x = -6.87$ or 0.87
10 $x = -9.11$ or 0.11
11 $x = -4.19$ or 1.19
23 $x = \frac{1}{2}(7 + \sqrt{61})$ or $\frac{1}{2}(7 - \sqrt{61})$
24 $x = -(4 + \sqrt{11})$ or $\sqrt{11} - 4$
25 $x = 1 - \sqrt{5}$ or $1 + \sqrt{5}$
26 $x = -3.19$ or -0.31
27 $x = -2.78$ or -0.72
28 $x = -1.54$ or -0.26
29 $x = 2.78$ or 0.72
30 $x = 1.64$ or 0.15
31 $x = -1.77$ or -0.57
32 $x = -1.59$ or -0.16
33 $x = 1.54$ or 0.26
34 $x = -2.14$ or 0.47
35 $x = 0.22$ or -2.51

12 $x = -5.32$ or 1.32
13 $x = 0.59$ or 3.41
14 $x = 6.54$ or 0.46
15 $x = 1.27$ or 4.73
16 $x = 5.85$ or -0.85
17 $x = 4.56$ or 0.44
18 $x = 2.62$ or 0.38
19 $x = 9.22$ or -0.22
20 $x = -1.61$ or 5.61
21 $x = -7.27$ or 0.27
22 $x = 2 - \sqrt{7}, 2 + \sqrt{7}$
36 $a = 0.69$ or -0.29
37 $b = -1.99$ or -0.26
38 $S = 2.22$ or -0.22
39 $x = 1.79$ or -1.12
40 $R = -0.30$ or -1.41
41 $n = -5.06$ or 0.06
42 $p = -0.30$ or 0.15
43 $x = -0.95$ or -0.18
44 $A = -0.56$ or -4.44
45 $N = 0.04$ or -6.04

Exercise 17g   Page 234

1 $x = 5.08$ or -1.08
2 $x = 2.32$ or -0.32
3 $x = -2.14$ or 0.47
4 $x = 0.88$ or -0.68
5 $x = 1.39$ or 0.36
6 $x = 4.16$ or -0.16
7 $x = 1.78$ or -0.28
8 $x = -1.55$ or 0.80
9 $x = 2.76$ or 0.24

10 $x = 1.86$ or -0.36
11 $x = 3.27$ or -0.77
12 $x = -1.55$ or 0.22
13 $x = 1.21$ or -0.21
14 $x = -2.59$ or 0.26
15 $x = 0.72$ or 0.28
16 $x = 0.42$ or -0.30
17 $x = 1.89$ or -0.20
18 $x = 0.58$ or -0.14

Exercise 17h   Page 235

1 $x = 0.5$ or -2

2 $x = -1.58$ or -0.42

3 $x = -0.5$ or $-\frac{2}{3}$

4 $x = -2.19$ or 0.69

5 $x = 2.39$ or 0.28

6 $x = 3$ or -0.33

7 $x = 2.19$ or -0.69

8 $x = 0.25$ or -1.5

9 $x = -1.40$ or 0.24

10 $x = \dfrac{-4 \pm \sqrt{30}}{7}$

11 $x = \dfrac{3 \pm \sqrt{29}}{10}$

12 $x = 2$ or $\frac{1}{3}$

13 $x = \dfrac{-6 \pm \sqrt{3}}{11}$

14 $x = \frac{1}{5}$ or $-\frac{3}{4}$

15 $x = 3$ or $\frac{5}{3}$

16 $x = \dfrac{-4 \pm \sqrt{6}}{5}$

17 $x = \dfrac{7 \pm \sqrt{73}}{4}$

18 $x = \frac{1}{2}$ or -5

Exercise 17i   Page 235

1 $x + \dfrac{30}{x} + 11$

2 $1 - 3x + \dfrac{30}{x}$

3 $5x - \dfrac{12}{x} - 17$

4 $3x - 26 + \dfrac{35}{x}$

5 $32 - 5x - \dfrac{48}{x}$

6 $\dfrac{5}{x} - 4x + 8$

7 $\dfrac{4}{x} - 15x + 17$

8 $10x - \dfrac{18}{x} + 3$

9 $3x - 11 + \dfrac{6}{x}$

10 $x = 5$ or $\frac{8}{9}$

11 $x = 2.8$ or -5

12 $x = 10$ or 3

13 $x = 8$ or -20

14 $x = \frac{12}{5}$ or -7

15 $x = -3\frac{1}{2}$ or $-2\frac{2}{9}$

16 $x = -0.8$ or -12

17 $x = -\frac{9}{7}$ or -4

18 a $x^2 - x - 12 = 0$ **b** $x = 4$ or -3
19 a $x^2 - 11x + 2 = 0$ **b** $x = 10.82$ or 0.18
20 a $x^2 - 3x - 5 = 0$ **b** $x = 4.19$ or -1.19
21 a $x^2 - 7x + 2 = 0$ **b** $x = 6.70$ or 0.30
22 a $4x^2 + x - 2 = 0$ **b** $x = -0.84$ or 0.59
23 a $2x^2 + 5x - 5 = 0$ **b** $x = -3.27$ or 0.77
24 $x = 2$ or $-1\frac{4}{7}$ **30** $x = 4$ or $\frac{1}{3}$
25 $x = -1$ or 8 **31** $x = -3.42$ or 2.92
26 $x = 6$ or -3 **32** $x = -4.30$ or -0.70
27 $x = 1.5$ or -2 **33** $x = -4.55$ or 2.80
28 $x = 6$ or $2\frac{2}{3}$ **34** $x = -4.27$ or 3.27
29 $x = 5$ or -1 **35** $x = 0.55$ or 5.45

Exercise 17j   Page 239

1 1.13 and 8.87
2 $\frac{47}{6}$ and $\frac{7}{6}$
3 0.05 and 19.95
4 $6.22\,\text{cm}, 3.22\,\text{cm}$

5 $4\,\text{cm}, 6\,\text{cm}$ and $10\,\text{cm}$
6 $3\,\text{cm}, 5\,\text{cm}$ and $10\,\text{cm}$
7 42 years
8 $6\,\text{cm}$ by $13\,\text{cm}$

9 a $\dfrac{420}{x}$ cents   **b** $\left(\dfrac{420}{x} + 5\right)$ cents

d 14 grapefruit at $30\,\text{c}$ each

10 a $\$\dfrac{240}{x}$ **b** $\$\left(\dfrac{240}{x} - 2\right)$

e 20 books at $\$12$ each

11 $60\,\text{m.p.h.}$ **12** $30\,\text{c}$ **13** $\$6$ or $\$9$
14 $3\,\text{cm} \times 3\,\text{cm}$ and $6\,\text{cm} \times 9\,\text{cm}$ or $\frac{51}{7}\,\text{cm} \times \frac{51}{7}\,\text{cm}$ and $\frac{27}{7}\,\text{cm} \times \frac{18}{7}\,\text{cm}$
15 42 **16** $20\,\text{c}$
17 Len: $3\,\text{m.p.h.}$, 5 hours   Mandy: $4\,\text{m.p.h.}$, $4\frac{1}{2}$ hours
18 George: $6\,\text{km/h}$ Liam: $4\,\text{km/h}$

Exercise 17k   Page 243

1 $-0.6, 1.6$
2 $-2.2, 0.7$

3 $-3.2, 1.2$
4 $-0.4, 2.4$

Chapter 18

Exercise 18a   Page 247

1 No, angles not equal
2 Yes
3 No, sides not equal

4 No, $\begin{cases} \text{sides not equal} \\ \text{angles not equal} \end{cases}$

5 No, $\begin{cases} \text{sides not equal} \\ \text{angles not equal} \end{cases}$

6 No, $\begin{cases} \text{sides not equal} \\ \text{angles not equal} \end{cases}$

7 Yes

8 No, not bounded by straight lines

Exercise 18b   Page 248

1 180°   **2** 360°

3 a $p = 100°$, $r = 135°$, $x = 55°$, $q = 125°$
  b 360°

4 a $w = 120°$, $x = 60°$, $y = 120°$, $z = 60°$
  b 360°

5 a 180°   **b** 540°   **c** 180°   **d** 360°

6 360°

7 a equilateral **b** 60°   **c** 120°
  d 60°   **e** 360°

Exercise 18c   Page 250

1 60°	**6** 90°	**11** $x = 50°$
2 90°	**7** 95°	**12** $x = 30°$
3 50°	**8** 55°	**13** $x = 24°$
4 50°	**9** 30°	**14 a** 5
5 60°	**10** 125°	**b** 8

Exercise 18d   Page 252

1 36°	**4** 60°	**7** 40°
2 45°	**5** 24°	**8** 225°
3 30°	**6** 20°	**9** 18°

Exercise 18e   Page 253

1 720°	**4** 360°	**7** 2880°
2 540°	**5** 900°	**8** 1260°
3 1440°	**6** 1800°	**9** 2340°

Exercise 18f   Page 254

1 a 3240°	**b** 2520°	**c** 1620°
2 80°	**6** 85°	**10** 135°
3 120°	**7** 110°	**11** 144°
4 110°	**8** 108°	**12** 150°
5 105°	**9** 120°	**13** 162°
14 a 18	**b** 24	
15 a 12	**b** 20	
16 a yes, 12	**d** yes, 6	
b yes, 9	**e** no	
c no	**f** yes, 4	
17 a yes, 4	**d** yes, 72	
b yes, 6	**e** yes, 36	
c no	**f** yes, 8	

Exercise 18g   Page 256

1 54°	**8** 135°	**15 a** 36° **b** 36°
2 45°	**9** 100°	**16 a** 128.6° **b** 25.7°
3 150°	**10** 60°	**17** 77.1°
4 72°	**11** 72°	**18 a** 22.5° **b** 22.5°
5 60°	**12** 45°	**19** 22.5°
6 50°	**13** 60°	**20** 45°
7 80°	**14** 36°	

Exercise 18h   Page 260

1 a The interior angles (135°) do not divide exactly into 360°
  b A square
2 a No
  b A regular ten-sided polygon
4 Square, equilateral triangle

Chapter 19

Exercise 19a   Page 264

1 AB, AC, AD, BC, BD, CD. Yes, AC.

Exercise 19b   Page 265

1 Minor arc DC	**6** AB
2 Minor arc BC	**7** Minor arc BE
3 $A\hat{C}B$, $A\hat{D}B$	**8** Minor arc CD
4 $B\hat{A}C$, $B\hat{D}C$	**9** CE
5 DA	**10** DB
11 a $A\hat{C}B$, $A\hat{E}B$	**12 a** $A\hat{B}E$, $A\hat{C}E$, $A\hat{D}E$
b $B\hat{A}C$, $B\hat{E}C$	**b** $C\hat{D}E$, $C\hat{A}E$, $C\hat{B}E$

Exercise 19d   Page 266

1 $h = 38°$	**5** $l = 100°$
2 $i = 39°$, $j = 46°$	**6** $x = 108°$, $y = 26°$
3 $x = 33° = y$	**7** $w = 57°$, $x = 123°$
4 $p = 72°$, $q = 57°$	**8** $c = 114°$

Exercise 19f   Page 268

1 $d = 80°$ **3** $f = 114°$ **5** $g = 98°$ **7** $l = 132°$
2 $e = 64°$ **4** $i = 38°$ **6** $h = 32°$ **8** $m = 102°$

Exercise 19h   Page 270

1 $d = 108°$	**5** $l = 131°$
2 $e = 84°$	**6** $m = 87°$, $n = 112°$
3 $f = 103°$	**7** $g = 121°$, $h = 68°$
4 $k = 115°$	**8** $i = 110°$, $j = 50°$

Exercise 19i   Page 272

5 100°	**7** $p = 54°$, $q = 76°$
6 109°	**8** $r = 126°$, $s = 83°$

Exercise 19j   Page 272

1 $d = 75°$, $e = 65°$, $f = 140°$
2 $p = 60°$, $q = 60°$, $r = 120°$, $s = 60°$
3 $k = 30°$, $l = 30°$, $m = 30°$, $n = 60°$
4 $g = 24°$, $h = 156°$, $i = 74°$
5 $w = 73°$, $x = 34°$, $y = 34°$, $z = 73°$
6 $d = 64°$, $e = 64°$, $f = 116°$, $g = 116°$, $h = 64°$
7 $a = 44°$
8 $c = 60°$, $d = 46°$
9 $g = 116°$
10 $b = 78°$
11 $e = 34°$, $f = 52°$
12 $h = 72°$
13 $l = 154°$, $m = 40°$, $n = 37°$
14 $r = 110°$, $s = 122°$
15 $x = 30°$, $y = 58°$, $z = 88°$
16 $c = 25°$, $d = 25°$, $e = 50°$
17 $h = 116°$, $i = 32°$
18 $l = 126°$, $m = 63°$, $n = 117°$
19 $u = 34°$, $v = 68°$, $w = 56°$, $x = 56°$
20 $k = 62°$, $l = 56°$, $m = 124°$, $n = 16°$

Exercise 19k   Page 275

1 $d = 90° = e$ **2** $f = 90° = g$ **3** $h = 90° = i$

Exercise 19l   Page 275

1 $d = 90°, e = 53°$ **4** $l = 90°, m = 61°$
2 $f = 90°, g = 45°$ **5** $j = 90°, k = 55°$
3 $h = 90°, i = 26°$ **6** $p = 90°, q = 38°$
7 $r = 90°, s = 52°, t = 90°, u = 43°$
8 $d = 90°, e = 45°, f = 90°, g = 18°$
9 $c = 90°, d = 58°, e = 32°$
10 $v = 90°, w = 47°, x = 90°, y = 51°$
11 $j = 90°, k = 33°, l = 33°, m = 57°$
12 $f = 45°, g = 58°, h = 45°, i = 32°$

Exercise 19m   Page 277

2 5.83 cm
3 a 109 cm   **b** 6.3 cm
4 O is the midpoint of AB
   $A\hat{C}B = 90°$ since it is in a semicircle
5 a They all pass through one point
   b It is outside the triangle
6 Yes. It passes through D.
   Because the trapezium is symmetrical about the
   perpendicular bisector of AB
   No

Exercise 19n   Page 278

1 $d = 106°$
2 $e = f = 38°$
3 $d = 34°, e = 68°$
4 $x = 75°, y = 15°, z = 132°$
5 $p = 36° = q, r = 39°$
6 $x = 112°, y = 68°, z = 112°$
7 $g = 54°, h = 120°$
8 $d = 37°, e = 53°, f = 57°, g = 33°$

Review test 2   Page 280

1 B   **3** B   **5** B   **7** B   **9** D
2 C   **4** D   **6** C   **8** C   **10** D
11 Average speed $= 60\frac{2}{3}$ km/h
12 a 13.37 cm
   b i 30°   **ii** 60°   **iii** 75°
13 a i $2x(x-6y)(x+6y)$
   ii $(a-2b)(a+c)$
   iii $(3x-4)(2x+1)$
   b $-0.15, -3.35$
14 a $\begin{pmatrix} -3.4 \\ -0.6 \end{pmatrix}$ **b** $\begin{pmatrix} -3.41 \\ -0.59 \end{pmatrix}$
15 a $\dfrac{8x-11}{(x-1)(x-2)}$ **b** 2, 1

Chapter 20

Exercise 20a   Page 283

1 2 : 3   **9** 2 : 3 : 1   **17** 2.73 : 1
2 1 : 2 : 3   **10** 6 : 11   **18** 0.6 : 1
3 7 : 5   **11** 15 : 4   **19** 2.63 : 1
4 2 : 3   **12** 31 : 4   **20** 1.33 : 1
5 18 : 8 : 9   **13** 5 : 16   **21** 0.75 : 1
6 2 : 3 : 1   **14** 1.5 : 1   **22** 1.43 : 1
7 4 : 9   **15** 2.4 : 1
8 3 : 5 : 4   **16** 0.857 : 1

Exercise 20b   Page 284

1 9 : 2   **5** 9 : 20   **9** 27 : 25
2 2 : 5   **6** 50 : 3   **10** 9 : 10
3 17 : 60   **7** 20 : 19   **11** 25 : 24
4 2 : 125   **8** 36 : 35
12 a 3 : 2   **b** 2 : 3   **c** 3 : 5
13 a 2 : 3   **c** 21 : 23
   b 9 : 5   **d** 6 : 5
14 18 : 25
15 a 1 : 1   **c** 1 : 8   **e** 1 : 3
   b 1 : 2   **d** 1 : 1   **f** 1 : 8
16 a 1 : 9   **b** 1 : 4   **c** 4 : 9

Exercise 20c   Page 286

1 $1\frac{1}{9}$ or 1.11   **9** $1\frac{1}{5}$ or 1.2
2 $\frac{3}{7}$ or 0.429   **10** $7\frac{1}{5}$ or 7.2
3 $7\frac{1}{2}$ or 7.5   **11** $3\frac{1}{3}$ or 3.33
4 $1\frac{3}{7}$ or 1 : 43   **12** $8\frac{4}{7}$ or 8.57
5 24   **13** 12 grandsons; 3 : 7
6 $22\frac{1}{2}$ **14** 152
7 $9\frac{1}{3}$ or 9.33   **15** 101 cm
8 $2\frac{8}{11}$ or 2.73   **16** 264

Exercise 20d   Page 287

1 $20, $25
2 54 m, 42 m
3 0.625 kg, 1.25 kg, 3.125 kg
4 $\frac{1}{2}$ h, $2\frac{1}{2}$ h, 4 h
5 18 boys, 14 girls
6 60°, 50°, 70°
7 9, 12, 9
8 66 hits, 24 misses

Exercise 20e   Page 287

1 9 : 7   **6** 5 : 3
2 30 m, 42 m   **7** 500 : 53
3 $5\frac{1}{4}$ **8** 4 : 3
4 $2\frac{2}{9}$ or 2.22   **9** 3 : 4
5 275 cm   **10** 2 : 3

Exercise 20f   Page 288

1 a $1.80   **b** $7.20
2 a 6 units   **b** $\frac{3}{4}$ unit
3 a 72 km   **b** 1188 km
4 a 35 rows   **b** 42 rows
5 a $1.12\frac{1}{2}$ **b** $5.40

Exercise 20g   Page 288

1 $1.20   **6** $4.20
2 155 km   **7** $8.30
3 $4\frac{1}{3}$ or 4.33 km   **8** 1.5 c
4 65 c   **9** 1.5 m
5 $7.70   **10** 5.5 m²

Exercise 20h   Page 289

1 3.2 litres   **5 a** $30 **b** 35 km
2 3 hours   **6** $96
3 $12\frac{1}{2}$ units   **7** 700
4 3.6 hours   **8** $2.64

9 66 rows
10 2025 cm
11 $168
12 392
13 65.6 km
14 a 2.25×10^7
b 8.1×10^6
c 1.35×10^5

15 15 V
16 247 joules
17 $2.032, $2.03
18 $42.15

Exercise 20i   Page 292

1 $5\frac{1}{2}$ hours
2 12
3 203
4 8 days
5 25 cm
6 20
7 16 cm
8 44
9 49

Exercise 20j   Page 293

1 a TT$15
b 2100 yen
2 $195.70
3 $3\frac{1}{2}$ hours
4 No answer
5 4.46 cm
6 49
7 24
8 34
9 1.44 m
10 6 weeks
11 No answer
12 1.5 amps

Exercise 20k   Page 294

1 3 : 1
2 $3\frac{3}{5}$ or 3.6
3 8 m, 16 m, 32 m
4 114 km (3 s.f.)
5 6 hours 40 mins
6 6 : 2 : 1
7 9
8 $\frac{6}{5}$: 1 or 1.2 : 1

Chapter 21

Exercise 21a   Page 297

1 38.44
2 187.69
3 58 564
4 7 728 400
5 0.5041
6 0.003 481
7 0.000 002 89
8 97 344
9 9.7344
10 0.000 973 44
11 84.64
12 8464
13 27 140 000
14 2714
15 0.2714
16 0.002 714
17 3.142
18 4.461
19 11.14
20 311.1
21 0.2195
22 0.069 43
23 9.798
24 17.92
25 1.619
26 0.2490
27 0.027 93
28 0.7071
29 0.6790
30 2.147
31 21.47
32 0.021 47

Exercise 21b   Page 298

1 10 cm
2 11.7 cm
3 9.43 cm
4 13 cm
5 11.4 cm
6 13.9 cm
7 The square of the third side is equal to the sum of the squares of the other two.

Exercise 21c   Page 299

1 10 cm
2 13 cm
3 20 cm
4 9.85 cm
5 10.8 cm
6 10.6 cm
7 11.7 cm
8 12.6 cm
9 12.1 cm
10 10.4 cm
11 3.61 cm
12 11.4 cm
13 6.40 m
14 11.4 m
15 12.2 cm
16 5.40 cm
17 121 cm
18 3.31 cm
19 9.57 cm
20 44.7 m
21 0.361 cm
22 8.64 cm
23 17.4 m
24 2.61 cm
25 35.0 cm
26 13.0 m
27 12.0 cm

Exercise 21d   Page 302

1 30 cm
2 18.4 cm
3 130 mm
4 7.5 cm
5 26 m
6 32.0 cm
7 $2\frac{1}{2}$
8 12.8 cm

Exercise 21e   Page 303

1 12 cm
2 48 cm
3 24 cm
4 10 m
5 4.90 cm
6 2.65 cm
7 1.73 m
8 4.58 cm
9 7.84 m
10 7.94 cm
11 6.24 cm
12 16.0 cm
13 6.71 cm
14 13.7 m

Exercise 21f   Page 305

1 6.71 cm
2 8.67 cm
3 55 cm
4 5.66 cm
5 11.5 cm
6 9 cm
7 3.46 cm
8 8.15 m
9 88.5 cm
10 7, 24
11 4.9 cm
12 265 cm
13 9.27 cm
14 2.5 cm
15 6.8 m
16 89.6 cm
17 3.51 cm

Exercise 21g   Page 308

1 12.5 cm
2 10 cm
3 15.0 cm, 30 cm
4 4.25 m
5 11.3 cm

Exercise 21h   Page 309

1 2.71 cm
2 4.69 cm
3 10.4 cm
4 16.2 m
5 12.7 cm

Exercise 21i   Page 310

1 2.60 m
2 7.81 cm
3 14.1 cm
4 105 m
5 7.55 m
6 74.3 cm
7 83.1 m
8 1.63 m
9 14.1 cm
10 6.22 km
11 8.94 units
12 38.8 n.m.
13 5.52 m
14 0.598 m
15 21.2 cm
16 a 39.4 cm
17 a 2.4 cm   **b** 4.64 cm
No. $AC^2 = AB^2 + BC^2$
18 c $AC = 7.07$ cm
$AD = 8.66$ cm
$AE = 10$ cm
19 Use 7 cm and 4 cm or 8 cm and 1 cm
$\sqrt{65} = 8.06$
20 8.46 cm
21 21 m and 28 m
22 BNC = 24 cm, AN = 5 cm
23 4 cm

Exercise 21j   Page 313

1 Yes
2 Yes
3 No
4 No
5 Yes
6 No

Exercise 21k   Page 314

1 18.9 cm
2.. 6.52 cm
3 2.02 cm
4 0.0265 cm
5 205 cm
6 4.16 cm
7 0.05 cm
8 130 cm
9 3.58 cm
10 64.5 cm
11 Yes
12 3.13 cm
13 262 cm
14 Yes, $\widehat{M} = 90°$

Chapter 22

Exercise 22a   Page 319

7 $\frac{5}{12}$, 0.4167
8 $\frac{8}{15}$, 0.5333
9 $\frac{3}{4}$, 0.75
10 $\frac{3}{4}$, 0.75
11 $\frac{12}{5}$, 2.4
12 $\frac{35}{12}$, 2.917

Exercise 22b   Page 320

1 1.8807
2 0.2493
3 0.5890
4 0.3019
5 0.0805
6 3.0777
7 4.8716
8 1
9 0.5774
10 1.1184
11 0.0524
12 0.5635
13 10.1°
14 19.6°
15 55.0°
16 23.4°
17 53.7°
18 32.3°
19 42.7°
20 38.7°
21 17.8°
22 69.6°
23 42.7°
24 0.1°

Exercise 22c   Page 321

1 32.0° **4** 35.8° **7** 48.4°
2 63.4° **5** 31.0° **8** 47.7°
3 23.2° **6** 51.3° **9** 34.2°

Exercise 22d   Page 322

1 2.44 cm **8** 22.2 cm **15** 46.6 cm
2 5.40 cm **9** 2.82 cm **16** 10.4 cm
3 2.56 cm **10** 7.54 cm **17** 4.69 cm
4 6.72 cm **11** 3.60 cm **18** 366 cm
5 17.0 cm **12** 11.4 cm **19** 0.976 cm
6 81.8 cm **13** 2.42 cm **20** 69.5 cm
7 5.62 cm **14** 1.76 cm

Exercise 22e   Page 324

1 0.8862 **8** 0.5 **15** 31.6°
2 0.9397 **9** 0.9903 **16** 65.4°
3 0.2470 **10** 0.4664 **17** 41.8°
4 0.1564 **11** 0.2723 **18** 21.8°
5 0.2622 **12** 0.9988 **19** 37.9°
6 0.6088 **13** 15.7° **20** 46.7°
7 0.8625 **14** 26.2° **21** 7.1°

Exercise 22f   Page 325

1 30° **6** 62.7° **11** 4.38 cm **16** 23.2 cm
2 17.5° **7** 44.4° **12** 10.6 cm **17** 6.31 cm
3 48.6° **8** 41.8° **13** 1.46 cm **18** 21.9 m
4 44.4° **9** 23.6° **14** 4.57 cm **19** 3.34 cm
5 14.5° **10** 19.5° **15** 11.7 cm **20** 45.7 cm

Exercise 22g   Page 327

1 0.8480 **7** 0.6143 **13** 51.1°
2 0.7455 **8** 0.6561 **14** 71.6°
3 0.1392 **9** 69.7° **15** 30.1°
4 0.6717 **10** 20.6° **16** 89.2°
5 0.5 **11** 44.0°
6 0.9632 **12** 69.6°

Exercise 22h   Page 327

1 34.9° **8** 66.4° **15** 11.6 cm
2 36.9° **9** 81.4° **16** 38.2 cm
3 45.6° **10** 25.8° **17** 2.90 cm
4 48.2° **11** 34.0° **18** 17.1 cm
5 48.2° **12** 3.50 cm **19** 2.23 cm
6 53.1° **13** 26.9 m **20** 4.12 cm
7 50.2° **14** 1.96 cm **21** 13.5 cm

Exercise 22i   Page 329

1 40.0° **13** 56.9° **25** 6.04 cm
2 33.6° **14** 37.8° **26** 3.50 cm
3 51.3° **15** 39.3° **27** 13.7 cm
4 42.8° **16** 55.6° **28** 3.08 cm
5 35.5° **17** 42.1° **29** 11.3 cm
6 33.7° **18** 66.2° **30** 2.59 cm
7 39.8° **19** 6.69 cm **31** 9.99 cm
8 33.7° **20** 19.3 cm **32** 7.45 cm
9 37.7° **22** 8.03 cm **33** 14.5 cm
10 53.1° **22** 4.86 cm **34** 21.4 cm
11 68.5° **23** 4.48 cm **35** 74.5 cm
12 14.5° **24** 80.5 cm **36** 60.6 cm

Exercise 22j   Page 332

1 4.13 cm **5** 14.9 cm **9** 33.1 cm
2 8.72 cm **6** 17.0 cm **10** 42.6 cm
3 23.3 cm **7** 4.40 cm
4 4.67 cm **8** 14.9 cm

Exercise 22k   Page 333

1 8.99 cm **3** 143 m **5** 61.6° **7** 48.2°
2 47.7 m **4** 39.8° **6** 56.3° **8** 11.3°
9 a 5.30 cm **b** 6.25 cm
10 a 5.20 cm **b** 15.6 cm^2
11 4.66 m
12 a $A\hat{O}B = 72°$, $O\hat{A}B = 54°$
  b 6.88 cm **c** 34.4 cm^2, 172 cm^2

Exercise 22l   Page 335

1 a EA = FB = GC = HD; AB = EF = HG = DC;
    BC = FG = EH = AD; 24 right angles
  b EB = 5 cm, $E\hat{B}A = 36.9°$
  c FC = 12.4 cm, $F\hat{C}B = 14.0°$
2 a AC = 12.6 cm
  b $E\hat{A}C = 90°$, EC = 13 cm, $E\hat{C}A = 13.3°$
3 a FC = 8.25 cm
  b AF = 5.39 cm, $F\hat{A}B = 21.8°$
  c EG = 9.43 cm, 32.0°

Exercise 22m   Page 336

1 a 14.4 cm **b** 15.3 cm **c** 19.1°
2 a 3.61 cm **b** 33.7° **c** 6.71 cm
3 a 10 cm **b** 15.6 cm **c** 39.8°
4 a 14.9 cm **c** 19.1 cm **e** 47.7°
  b 19.1 cm **d** 47.5°
5 24.7 cm
6 a 15 cm **b** 16.6 cm **c** 25.0°
7 a 7.07 cm **b** 7.07 cm **c** 600 cm^2
8 a 33.7° **b** 56.3° **c** 31.4°
9 a 7.07 cm **b** 336 cm^2

Exercise 22n   Page 338

1 a AB = DC = FE, BC = AD, EC = FD,
    14 right angles
  b $E\hat{B}C = 33.7°$, BE = 7.21 cm
  c AC = 11.7 cm, $C\hat{A}B = 31.0°$, Yes
  d AE = 12.3 cm, AE = FB
2 a 3.00 cm **c** 10.9 cm
  b 7.42 cm **d** 15.4°
3 a 27.5 m **d** 49.5 m
  b 48.5 m **e** 11.6°
  c 29.2 m **f** 53.8°
4 a 24.4° **b** 13.9°
5 a 2.62 cm **b** 3.98 cm **c** 5.76 cm

Exercise 22p   Page 339

1 a $A\hat{B}C$, $B\hat{C}D$, $C\hat{D}A$, $D\hat{A}B$, $A\hat{F}B$, $B\hat{F}C$, $C\hat{F}D$, $D\hat{F}A$,
    $B\hat{F}E$, $C\hat{F}E$, $D\hat{F}E$, $A\hat{F}E$. (12). AE = BE = CE = DE
  b AC = 2.83 cm, AF = 1.41 cm
  c EF = 5.83 cm, $E\hat{C}F = 76.4°$
2 a AC = 5.66 cm, AF = 2.83 cm
  b AE = 5.74 cm, $E\hat{A}F = 60.5°$
  c EG = 5.39 cm, $E\hat{G}F = 68.2°$
3 a $E\hat{B}A = 36.9°$, $E\hat{D}A = 45°$
  b 5 cm **c** 5.83 cm
4 a PR = 8.54 cm **c** 54.5°
  b PY = 4.27 cm **d** 7.37 cm

Exercise 22q   Page 340

1 **a** 7.28 m
   b 31.2°
   c 23.3 m, 17.3°
2 **a** AC = CD$'$ = AD$'$ = 5.66 cm.
      Equilateral triangle
   b Rectangle; AC$'$ = A$'$C = BD$'$ = DB$'$ = 6.93 cm
3 **a** BD = 8.49 m, BE = 4.24 m
   b EF = 4.24 m. Height = 8.49 m
   c 45°
4 **a** 7.07 cm   **c** 4.85 cm
5 **a** BD = 8.94 cm
   b DB̂A = 26.6°
   c 11.3 cm
   d DC = BD = 8.94 cm
   e DĈA = DB̂A = 26.6°

Chapter 23

Exercise 23a   Page 343

10 062°   12 328°   14 249°   16 154°
11 098°   13 262°   15 254°   17 050°
22   25

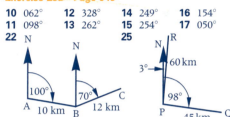

23   26

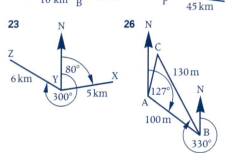

24   27

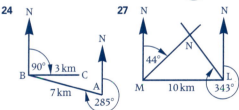

Exercise 23b   Page 346

1 240°   3 342°   5 172°
2 112°   4 032°   6 305°

Exercise 23c   Page 346

1 98°   4 128°   7 45°
2 91°   5 both 60° 8 131°
3 32°   6 129°, 11°, 40°

Exercise 23d   Page 347

1 029°
2 90°, 29.7 km

3 6.40 km
4 **a** 5.81 km   **b** 144°
5 **a** 46°, 44°, 90°   **b** 14.4 km
6 **a** 7.51 km   **b** 332°
7 19.5 km, 31.7 km
8 **a** 54.5°   **b** 186.5°   **c** 007°
9 **a**

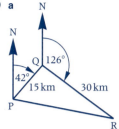

   b **i** 34.92 km **ii** 100.7° **iii** 280.7°
10 **a**

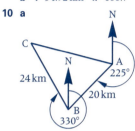

   b 26.97 km (2 d.p.) on a bearing of 284.3°
11 138.9 km on a bearing of 345.9°
12 **a** OR

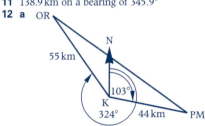

   b 92.81 km on a bearing of 306.07°

Chapter 24

Exercise 24a   Page 351

1 **a** 33, no mode, 24
   b 3.8 m, no mode, 3.75 m
2 **a** 10, 10, 10   **b** $\frac{1}{3}$
3 12
4 **a** 420
   b **i** increase, 24 is larger than 15.
      ii 15.4
5 **a** 5.3 cm, 5.2 cm, 5.2 cm   **b** 4.9 cm
   c 5.14 cm   **d** $\frac{6}{10}$
6 53
7 9
8 **a** 393 (nearest person), 402, no mode
9 3

Exercise 24b   Page 354

1 **a** 2   **b** 2 children, 2 children
2 **a** 1.37   **b** 2, 2
3 **a** 23   **b** 480   **c** 0.575   **d** $\frac{5}{8}$
4 **a** 30   **b** 2, 2   **c** $17.25   **d** 57.5 c

Exercise 24c   Page 356

1 a 74   **b** 24

2

1–10	5
11–20	13
21–30	10

3 a

1–5	24
6–10	13
11–15	2

b 39

4 a Frequencies: 7, 13, 9, 1   **b** 20

5 a 5–9   **b** 20

6 a 108   **b** 210

  c i 14   **ii** $\frac{7}{105}$

7 b Frequencies 7, 12, 14, 6

  c 39

Exercise 24d   Page 359

1 4.2   **2** 7.1 cm   **3** $50\frac{1}{2}$ c

4 Number of defective

screws per box	0–2	3–5	6–8	9–11
Frequency	10	7	2	1

, 3.1

5 160 cm   **6** 55.4   **7** 90.2 c

Chapter 25

Exercise 25a   Page 365

1 a infinite   **b, c** and **d** finite

2 a i 21   **ii** 11

  b i 4   **ii** 11, 13, 17, 19; no

    iii $\varnothing$ or { } null or empty set

3 $A = \{11, 13, 17, 19, 23, 29\}$, $B = \{12, 18, 24\}$,

  $C = \{18\}$

4 $\{1, 2, 3, \ldots, 10\}$

5 a $\{2, 4, 6, 8, 10, 12, 14, 16, 18, 20\}$

  b infinite

6 $\{-1, 0, 1, 2\}$

7 $\{(-1, -3), (0, 0), (1, 3)\}$

8 $n(A) = 4$

9 $P = \{3, 6, 9, 12\}$, $Q = \{2, 4, 6, 8, 10, 12, 14\}$,

  $R = \{5, 10\}$

10 $A = \{-6, -4, -2, 0, 2, 4\}$,

  $B = \{-6, -5, -4, -3, -2, -1 \}$,

  $C = \{2, 3, 5 \}$

Exercise 25b   Page 366

1 $A' = \{10, 20\}$

2 $B' = \{5, 6, 11\}$

3 $V' = \{\text{consonants}\}$

4 $P' = \{\text{vowels}\}$

5 $A' = \{\text{Sunday, Tuesday, Thursday, Saturday}\}$

6 $X' = \{\text{adults}\}$

7 $M' = \{\text{foreign motor cars}\}$

8 $S' = \{\text{female tennis players}\}$

9 $C' = \{\text{Caribbean towns not in Jamaica}\}$

10 $D' = \{\text{quadrilaterals which are not squares}\}$

11 $E' = \{\text{adults 80 years old or younger}\}$

12 $F' = \{\text{female doctors}\}$

13 $U = \{\text{homes}\}$

14 $U = \{\text{letters of the alphabet}\}$

15 $U = \{a, b, c, d, e, f, g, h, i, j\}$

16 a   **b**

Exercise 25c   Page 367

1

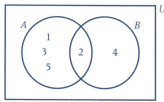

a $A' = \{4\}$

b $B' = \{1, 3, 5\}$

c $A \cup B = \{1, 2, 3, 4, 5\}$

d $(A \cup B)' = \{ \}$ or $\varnothing$

e $A' \cap B' = \{ \}$ or $\varnothing$

2

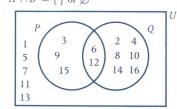

a $P' = \{1, 2, 4, 5, 7, 8, 10, 11, 13, 14, 16\}$

b $Q' = \{1, 3, 5, 7, 9, 11, 13, 15\}$

c $P \cup Q = \{2, 3, 4, 6, 8, 9, 10, 12, 14, 15, 16\}$

d $(P \cup Q)' = \{1, 5, 7, 11, 13\}$

e $P' \cap B' = \{1, 5, 7, 11, 13\}$

3

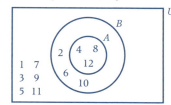

a $A' = \{1, 2, 3, 5, 6, 7, 9, 10, 11\}$

b $B' = \{1, 3, 5, 7, 9, 11\}$

c $A \cup B = \{2, 4, 6, 8, 10, 12\}$

d $(A \cup B)' = \{1, 3, 5, 7, 9, 11\}$

e $A' \cap B' = \{1, 3, 5, 7, 9, 11\}$

4

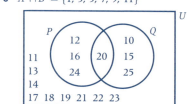

a $P' = \{10, 11, 13, 14, 15, 17, 18, 19, 21, 22,$ $23, 25\}$

b $Q' = \{11, 12, 13, 14, 16, 17, 18, 19, 21, 22,$ $23, 24\}$

c $P \cup Q = \{10, 12, 15, 16, 20, 24, 25\}$

d $(P \cup Q)' = \{11, 13, 14, 17, 18, 19, 21, 22, 23\}$

e $P' \cap Q' = \{11, 13, 14, 17, 18, 19, 21, 22, 23\}$

5

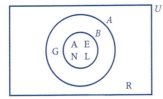

a $A' = \{R\}$
b $B' = \{G, R\}$
c $A \cap B = \{A, E, L, N\}$
d $A \cup B = \{A, E, G, L, N\}$
e $(A \cap B)' = \{G, R\}$
f $A' \cap B' = \{R\}$

6

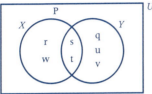

a $X' = \{p, q, u, v\}$
b $Y' = \{p, r, w\}$
c $X' \cap Y' = \{p\}$
d $X \cup Y' = \{p, r, w\}$
e $(X \cup Y)' = \{p\}$
   $X' \cap Y'$ and $(X \cup Y)'$

7

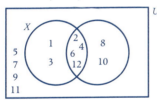

a $X' = \{5, 7, 8, 9, 10, 11\}$
b $Y' = \{1, 3, 5, 7, 9, 11\}$
c $X' \cap Y' = \{5, 7, 9, 11\}$
d $X' \cup Y' = \{1, 3, 5, 7, 8, 9, 10, 11\}$
e $X \cup Y = \{1, 2, 3, 4, 6, 8, 10, 12\}$
f $(X \cup Y)' = \{5, 7, 9, 11\}$
   $X' \cap Y'$ and $(X \cup Y)'$

8 a b

c d

e f

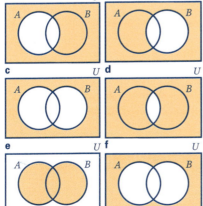

9

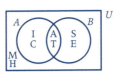

a $A' = \{E, H, M, S\}$
b $B' = \{C, H, I, M\}$
c $A \cup B = \{A, C, E, I, S, T\}$
d $(A \cup B)' = \{H, M\}$
e $A' \cup B' = \{C, E, H, I, M, S\}$
f $A' \cap B' = \{H, M\}$

10 a $P' = \{$pupils in my class without compasses$\}$
b $Q' = \{$pupils in my class without protractors$\}$
c $P' \cap Q' = \{$pupils in my class with neither compasses nor protractors$\}$
d $(P \cup Q)' = \{$pupils in my class with neither compasses nor protractors$\}$
e $P \cup Q = \{$pupils in my class with either compasses and/or a protractor$\}$

11 a $A' = \{1, 2, 4, 5, 7, 8, 10, 11\}$
b $B' = \{1, 3, 5, 7, 9, 11\}$
c $A' \cap B'\{1, 5, 7, 11\}$
d $(A \cup B)' = \{1, 5, 7, 11\}$

12 a $P' = \{-6, -4, -2, 0, 2, 4, 6\}$
b $Q' = \{-6, -4, -1, 0, 1, 4, 6\}$
c $P' \cap Q' = \{-6, -4, 0, 4, 6\}$
d $P' \cup Q' = \{-6, -4, -2, -1, 0, 1, 2, 4, 6\}$
e $(P \cup Q)' \{-6, -4, 0, 4, 6\}$

13 a $A' = \{10, 11, 13, 14, 15, 17, 18, 19, 21, 22, 23\}$
b $B' = \{10, 11, 13, 14, 16, 17, 19, 20, 22, 23\}$
c $A' \cap B' = \{10, 11, 13, 14, 17, 19, 22, 23\}$
d $(A \cup B)' = \{10, 11, 13, 14, 17, 19, 22, 23\}$

Exercise 25d   page 370

	$n(A)$	$n(B)$	$n(A \cup B)$	$n(A \cap B)$
1	5	7	9	3
2	4	5	7	2
3	3	6	7	2
4	6	4	8	2

	$n(X)$	$n(Y)$	$n(X \cup Y)$	$n(X \cap Y)$
5	9	7	12	4
6	5	2	7	0
7	13	10	16	7
8	12	12	20	5

	9	10	11	12	13	14	15	16
$n(A)$	3	3	6	11	8	9	8	4
$n(B)$	5	1	5	13	5	5	2	5
$n(A')$	9	5	4	15	6	6	4	8
$n(B')$	7	7	5	13	9	10	10	7
$n(A \cup B)$	7	4	8	18	10	12	8	9
$n(A \cap B)$	1	0	3	6	13	2	2	0
$n(A' \cup B')$	11	8	7	20	11	13	10	12
$n(A \cap B)'$	11	8	7	20	11	13	10	12

Exercise 25e   page 372

1 a 3   **b** 4   **c** 12   **d** 13
2 a 27   **b** 14   **c** 8   **d** 19
3 a 11   **b** 13   **c** 19

4	a	41	b	20	c	29

5 23

6	a	19	b	9	c	23

7 3

8	a	8	b	11	c	23
9	a	32	b	20	c	17
10	a	28	b	20		
11	a	15	b	37	c	23
12	a	15	b	21	c	7
13	a	13	b	18		

Exercise 25f   page 376

1 a $x + 8$
  b $2x + 3 + x + 5 + x - 5 = 43$, i.e. $4x = 40$
  c 20   d 10

2 a
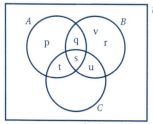

  c p, q, s, t, u d 2

3
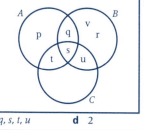

4 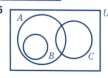   **6**

5

7 a $A \cup B = \{1, 2, 3, 4, 5, 6\}$,
    $B \cup C = \{1, 2, 4, 5, 6, 8, 10\}$
    $(A \cup B) \cup C = \{1, 2, 3, 4, 5, 6, 8, 10\}$
    $A \cup (B \cup C) = \{1, 2, 3, 4, 5, 6, 8, 10\}$
  b $A \cap B = \{1, 2\}$ $B \cap C = \{2, 6\}$
    $(A \cap B) \cap C = \{2\}$ $A \cap (B \cap C) = \{2\}$

8
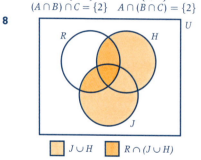

□ $J \cup H$ □ $R \cap (J \cup H)$

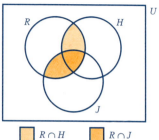

□ $R \cap H$ □ $R \cap J$

9
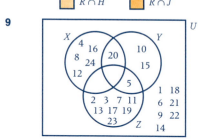

  a $\{4, 5, 8, 10, 12, 15, 16, 20, 24\}$
  b $\{2, 3, 5, 7, 11, 13, 17, 19, 23\}$
  c $\{1, 6, 9, 14, 18, 21, 22\}$
  d $\{5, 10, 15, 20\}$
10 a $\{E, G, M, R, T\}$ b 11
11 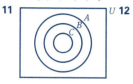   12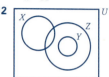

  a Isosceles right-
    angled triangles
  b Equilateral triangles
  c Equilateral triangles
    and right-angled
    triangles

13
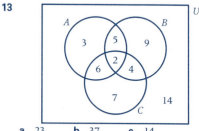

  a 23   b 37   c 14
14
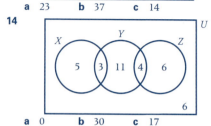

  a 0   b 30   c 17
15 a 5   b 16   c 11   d 0
16 a 5   b 26   c 17   d 7
17 a, b, f

Chapter 26

Exercise 26a   page 382

1 a 7   b 11   c $1\frac{1}{6}$
2 a $\frac{5}{9}$ b 3   c $\frac{8}{11}$
3 a -21 b 3   c $\frac{3}{16}$
4 a $1\frac{1}{5}$ b $-2\frac{2}{9}$ c $-\frac{3}{2}$
5 a 10   b 16   c 22
6 a -18 b 2303   c -144
7 a 2   b $-1\frac{1}{3}$ c $-\frac{1}{3}$
8 a -8 b -3 c 12
9 a $15\frac{1}{6}$ b $1\frac{1}{4}$ c $\frac{1}{a}+\frac{1}{b^2-c}$
10 a 16   b 28   c $2r(p+2q)$

Exercise 26b   page 384

1 a e.g. $3-2\neq 2-3$
   b no, e.g. $3+(2-1)=4$ but $3+2-3+1=3$
2 a no, e.g. $2\div 3\neq 3\div 2$
   b yes   c yes
3 a no, e.g. $4\div(2\times 1)\neq 4\div 2\times 4\div 1$
   b no, e.g. $4\times(2\div 1)\neq 4\times 2\div 4\times 1$
4 a i $\begin{pmatrix}3 & 9 \\ 6 & 8\end{pmatrix}$ ii $\begin{pmatrix}3 & 9 \\ 6 & 8\end{pmatrix}$; yes
   b i $\begin{pmatrix}17 & 34 \\ 12 & 24\end{pmatrix}$ ii $\begin{pmatrix}14 & 18 \\ 21 & 27\end{pmatrix}$; no
5 a A   b A   c yes
6 d It is commutative, associative and distributive across addition
7 a no   b no
8 a no   b no

Exercise 26c   page 386

1 a 1   b $-8\notin\mathbb{N}$
   c i yes   ii yes   iii no
2 a 1   b $\frac{1}{3}$ c yes
3 a $\begin{pmatrix}0 & 0 \\ 0 & 0\end{pmatrix}$ b $\begin{pmatrix}1 & 0 \\ 0 & 1\end{pmatrix}$ c $\frac{1}{13}\begin{pmatrix}-2 & 5 \\ 3 & -1\end{pmatrix}$
4 $\mathbf{A}+(-\mathbf{B})=0$ but $|\mathbf{B}|=0$ so $\mathbf{B}^{-1}$ does not exist
5 For $a=3$ and $b=0$, $a*b=3(3-0)=9$ and 9 is not in the set
6 $5\sim 5=0$ and $0\notin A$

Exercise 26d   page 387

1 $\{(1,1),(2,8),(3,27),(4,64)\}$
2 $\{(2,4),(-2,4),(3,9),(-3,9)\}$
3 a $\{(2,1),(3,1),(3,2)\}$
   b $\{2,3\},\{1,2\}$

Exercise 26e   page 388

1 $(-1,5),(0,0),(1,5),(2,10)$; yes
2 $(1,4),(1,6),(4,6)$; no, 1 occurs twice as the first number
3 $(0,1),(2,5),(4,17)$; yes
4 $(3,15),(7,35),(9,45)$
5 $(0,-1),(1,0),(2,7)$
6 $(-1,3),(0,0),(2,0)$
7 $(2,6),(4,16),(6,26)$
8 a no   b yes   c yes
9 a yes   b no   c yes
10 a -1 b 10   c -8
11 a 2   b 0   c -2
12 a 0   b 1   c 6

Exercise 26f   page 390

1 a   b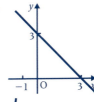

2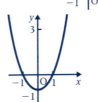

Chapter 27

Exercise 27b   page 395

1 The square root of any number between 0 and 1.
2 Not when $x=4$.
3 Diagonals of a rhombus cut at right angles and a rhombus is a parallelogram.
4 e.g. $120°$, $40°$ and $40°$
5 square root step; i.e. $2-\frac{5}{2}=\pm(3-\frac{5}{2})$
6 cannot divide by $x-y$ since $x-y=0$

Exercise 27d   page 401

1 $OB=5\,cm$, $CB=2\,cm$
2 $x=30$
3 $x=50$, $y=40$
4 $x=20$, $y=70$
5 a $40°$ b $50°$
6 $AB=12\,cm$, $O\hat{B}A=22.6°$
7 $x=30$, $y=60$, $z=60$
8 $30°$ 9 $5\,cm$ 11 $9.80\,cm$

Exercise 27e   page 404

1 $p=q=65$
2 $e=f=67$, $g=i=23$, $h=134$
3 a $50°$
   b kite (2 pairs of adjacent sides equal) and cyclic quadrilateral (opposite angles supplementary)
4 a $96°$ b $48°$
5 a $36°$ b $36°$
6 a $6\,cm$ b $73.7°$ to 3 s.f.
7 $8\,cm$
8 yes
9 $5.77\,cm$ to 3 s.f.
10 $30°$, $13.9\,cm$ to 3 s.f.
11 $13.7\,cm$, $13.7\,cm$, $19.3\,cm$

Exercise 27f   page 407

4 $d=73$, $e=26$, $f=81$
5 $p=q=60$, $r=s=60$
6 $k=64$, $l=64$, $m=64$, $n=52$
7 $u=67$, $v=67$
8 $p=57$, $q=57$
9 $w=90$. $x=27$, $y=117$
10 $d=47$, $e=47$, $f=47$, $g=86$
11 $e=54$, $f=54$, $g=54$, $h=72$
12 $k=74$, $l=53$, $m=53$
13 $r=90$, $s=55$, $t=35$
14 $s=30$, $t=60$, $u=60$, $v=10$

15 $x = 28, y = 62, z = 62$
16 a $40°$ **b** $40°$ **c** $100°$
   d $80°$ **e** $65°$ **f** $45°$
17 a $45°$ **b** $55°$ **c** $80°$ **d** $80°$
18 a $25°$ **b** $50°$ **c** $65°$
19 a $52°$ **b** $42°$ **c** $10°$
20 a i $x°$ **ii** $(180 - x)°$ **iii** $90°$
   b i parallel   **ii** diameter of circle
21 a i $48°$ **ii** $75°$ **iii** $30°$
   b isosceles
22 a $36°$ **b** $108°$ **c** $72°$ **d** $18°$
23 a $38°$ **b** $20°$ **c** $46°$
24 a $20°$ **b i** $20°$ **ii** $140°$ **iii** $60°$

12 a 9.5 cm   **b** 13 cm   **c** $101°$
13 b 21.9   **c** 11–20
14 a John \$4000, James \$5000, Mary \$7000
   b no because (3, 1) and (3, 2) are both members of the relation, so a member of the domain is paired with more than one member of the range.
15 a $\left(\dfrac{b - c}{c} \right)\left(\dfrac{ac - b + c}{c} \right)$ **b** $\dfrac{b(a - b) - c}{c}$
16 a $80°$ **b** $73°$

Review test 3   Page 411

1 C   **3** D   **5** C   **7** C   **9** C
2 A   **4** A   **6** A   **8** A   **10** C
11

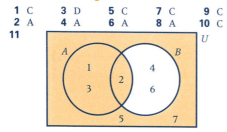

Multiple Choice Questions   page 413

1 C	**12** C	**23** D	**34** A	**45** D
2 A	**13** D	**24** B	**35** D	**46** C
3 D	**14** C	**25** D	**36** C	**47** C
4 D	**15** A	**26** B	**37** B	**48** A
5 B	**16** B	**27** B	**38** A	**49** C
6 D	**17** B	**28** A	**39** C	**50** B
7 B	**18** D	**29** B	**40** D	**51** D
8 D	**19** C	**30** C	**41** B	**52** C
9 A	**20** B	**31** D	**42** A	**53** A
10 D	**21** B	**32** D	**43** B	**54** A
11 B	**22** B	**33** C	**44** C	**55** A

INDEX